Manual of Soil Laboratory Testing

Manual of Soil Laboratory Testing

Volume 3: Effective Stress Tests

Second Edition

K. H. Head, *MA (Cantab), C.Eng, FICE, FGS*

JOHN WILEY & SONS

Chichester · New York · Weinheim · Brisbane · Singapore · Toronto

© 1998 K. H. Head

Published by   John Wiley & Sons Ltd,
Baffins Lane, Chichester,
West Sussex PO19 1UD, England

National   01243 779777
International (+44) 1243 779777
e-mail (for orders and customer service enquiries): cs-books@wiley.co.uk
Visit our Home Page on http://www.wiley.co.uk
or http://www.wiley.com

First edition published 1992 by Pentech Press Limited

Other Wiley Editorial Offices

John Wiley & Sons, Inc., 605 Third Avenue,
New York, NY 10158-0012, USA

WILEY-VCH Verlag GmbH, Pappelallee 3,
D-69469 Weinheim, Germany

Jacaranda Wiley Ltd, 33 Park Road, Milton,
Queensland 4064, Australia

John Wiley & Sons (Canada) Ltd, 22 Worcester Road,
Rexdale, Ontario M9W 1L1, Canada

John Wiley & Sons (Asia) Pte Ltd, 2 Clementi Loop #02-01,
Jin Xing Distripark, Singapore 129809

British Library Cataloguing in Publication Data

A catalogue record for this book is available from the British Library

ISBN 0 471 97795 0

Typeset in 10/12pt Times by Dobbie Typesetting Limited, Tavistock, Devon

Printed and bound by Antony Rowe Ltd, Eastbourne

This book is printed on acid-free paper responsibly manufactured from sustainable forestation, for which at least two trees are planted for each one used for paper production

Preface

The first edition of this volume completed the series on routine soil laboratory testing. This revised edition, like the second editions of Volumes 1 and 2 of this series already published, takes into account the requirements of BS 1377: 1990 and subsequent amendments, and relates to Parts 1, 6 and 8 of that Standard.

This volume covers relatively complex tests, in most of which the measurement of pore water pressure is an essential feature. These measurements enable the 'effective' stresses, which are fundamental to the understanding of soil behaviour, to be determined. The scope of this volume is therefore considerably wider than that of Volume 2, and it is intended not only for technicians and supervisors responsible for executing laboratory tests but also as a work of reference for geotechnical engineers and designers, and for undergraduate and postgraduate students. I hope that teachers will also find it to be of value.

The first edition included descriptions of some procedures and concepts which are more advanced than those commonly needed in commercial laboratories. The scope of the tests covered in this second edition is confined to those procedures which are included in Parts 6 and 8 of BS 1377: 1990, together with a few other related procedures. After some deliberation I decided that procedures more advanced than these, including those referred to in the first edition, should not be included in this volume because they require a different approach. The intention is to cover the more advanced procedures in a separate publication, to be written by several contributing authors, each being a recognised authority on the relevant topic. This will ensure that details of current practice, as well as the latest developments, will be adequately covered.

The format differs in some ways from that of the previous volumes. The general arrangement, which can be considered in four parts, is explained in the Introduction. The sections on theory are not intended to take the place of a textbook on theoretical soil mechanics, several of which are quoted from among the numerous works that are available.

In presenting the widely-used 'basic' triaxial tests I have tried to show that some generally accepted practices (especially with regard to specimen saturation) may not necessarily be the most appropriate in all circumstances. In complex tests of this kind it is not practicable to cover every eventuality by written procedures; often a decision based on the engineer's judgement is necessary. From the 'basic' tests many other triaxial test procedures have been developed, and some of those that seem likely to be most often needed in practice are presented here. I have introduced the use of stress paths as a means of presenting triaxial test data and results, which I suggest is more explicit than the conventional Mohr–Coulomb presentation.

Consolidation testing of clays and silts in the Rowe hydraulic cell has now become routine practice in many laboratories. The guidance I have given is combined with an attempt to bring together the relevant theory and empirical methods of analysis.

Modern soil testing, in common with many other processes, is already making use of electronic instrumentation and microprocessors. However, in this book I have described the use of manual observation and recording of data because I believe it is the only way to learn and understand the principles and procedures. Reference to the use of automatic data-logging facilities, and automatic control for certain tests, is included, and these capabilities are now becoming more widely used.

I am very grateful to all who have contributed in various ways to this book, including those who have sent me their personal comments and suggestions on the first edition. I am particularly indebted to Mr John Masters for reviewing my original drafts for this edition, and for contributing his valuable comments; and to Mr John Turner for the assistance and suggestions he provided for Chapters 16 and 17.

I hope that the completion of the revision of this series will prove to be a useful guide to the practical interpretation of BS 1377: 1990, and that it will contribute to the achievement of consistently high standards in soil testing. I hope that it will be well used, and I shall welcome comments or suggestions from those who use it.

K. H. Head
Cobham, Surrey
September 1997

Acknowledgements

The author acknowledges the contributions made by numerous individuals during the preparation of this second edition. In particular, valuable comments and support have been provided by Professor J. H. Atkinson, Dr B. G. Clark, Mr J. D. Harris, Dr D. W. Hight, Mr J. R. Masters and Mr J. M. Turner.

The author is very grateful to ELE International Limited for supplying many of the photographs used in this volume. Photographs, diagrams and tables from other sources are acknowledged individually. Contributors were: British Standards Institution; Imperial College, London; University of Manchester; University of Newcastle-on-Tyne; Auto Valve Systems Ltd; Legris Ltd; LTG Laboratories; Pneumatic Engineering and Distribution Ltd; Sage Engineering Ltd.

Contents

CONTENTS

Summary of test procedures described in Volume 3

Test designation	Section
Chapter 18	
Triaxial compression:	
Consolidated-undrained (CU)	18.6.1, 2, 3, 5; 18.7.1, 2, 3, 5
Consolidated-drained (CD)	18.6.1, 2, 4, 5; 18.7.1, 2, 4, 5
Chapter 19	
Triaxial compression:	
Multistage	19.2
Unconsolidated-undrained (UU)	19.3.2
Consolidated quick-undrained (C-QU)	19.3.3
Consolidated constant volume (CCV)	19.3.4
Consolidated-undrained (CU) on compacted clay	19.3.5
Chapter 20	
Triaxial consolidation:	
Isotropic, vertical drainage	20.2.1
Isotropic, horizontal drainage	20.2.2
Anisotropic	20.3
Triaxial permeability:	
With two back pressure systems	20.4.1
With one back pressure system	20.4.2
With two burettes	20.4.3
Very small rates of flow	20.4.4
Chapter 21	
Use of stress paths	21.3
Pore pressure coefficient $\bar{B}$	21.4
Chapter 22	
Consolidation in the Rowe hydraulic cell:	
Vertical (one way) drainage	22.6.2
Vertical (two way) drainage	22.6.3
Radial drainage to periphery	22.6.4
Radial drainage to central well	22.6.5
Permeability in the Rowe cell:	
Vertical	22.7.2
Horizontal	22.7.3

Introduction to Volume 3

SCOPE

The significance of the shear strength of soils was introduced in Volume 2. Determination of shear strength in the triaxial apparatus was described in Chapter 13 but with reference only to the total stresses applied to a test sample. This volume extends the scope of shear strength testing of soils to cover the concept of 'effective stress', which is of fundamental importance in soil mechanics. Effective stress cannot be measured directly, but is determined indirectly by measuring the pressure of the fluid within the pore spaces of the soil (the 'pore pressure') as well as the total applied stress.

Measurement of pore pressure is an essential feature of many of the tests described in this volume. These comprise various kinds of triaxial compression tests for measuring soil strength characteristics; consolidation tests (one-dimensional and three-dimensional) for determining soil compressibility and rate of consolidation; and measurement of permeability.

This revised volume deals only with the range of tests specified in Parts 6 and 8 of BS 1377 : 1990, together with a few other related procedures which do not require the use of additional specialised equipment. Tests of a more advanced nature, including those that are now becoming part of commercial testing practice, are beyond the scope of this volume. Procedures of this kind that were included in the first edition have been removed, and will be updated elsewhere by others who are recognised authorities in their respective fields.

TRIAXIAL TESTS

A considerable proportion of this volume is concerned with the measurement of soil strength. The 'strength' of a soil, unlike that of other materials such as concrete or steel, is not a unique property but varies within wide limits depending on the conditions imposed, whether in-situ or in a laboratory test.

The triaxial test offers the most satisfactory way of measuring the shear strength of soil for many engineering purposes. The triaxial test principle is versatile, and procedures can be related to numerous types of practical problems. Its outstanding advantages are the facilities for the control of the magnitude (though not the orientation) of the principal stresses; the control of drainage; and the measurement of pore water pressure. Information derived from these tests can provide data for the understanding of basic soil behaviour as well as soil properties for use in design.

The triaxial apparatus can also be used for the measurement of other soil properties such as consolidation characteristics and permeability.

CONSOLIDATION TESTS

Compressibility and time-dependent consolidation properties of clay soils were described in Volume 2, Chapter 14, with reference to the standard oedometer consolidation test. Data obtained from consolidation tests are of greater fundamental value when pore pressures are measured, so that effective stresses can be determined. This facility is one feature of the advanced type of consolidation test apparatus which is now widely used, and which also provides control of drainage and enables large specimens to be tested.

SYNOPSIS OF CHAPTERS

This volume can be thought of as comprising four parts, as follows.

The first part is Chapter 15, which covers the principle of effective stress and other theoretical concepts that are necessary for a proper understanding of the tests described later. It includes a summary of some practical applications of the various tests.

The second part consists of Chapters 16 and 17, covering test equipment and its calibration and use. Chapter 16 describes the equipment used for all types of triaxial test dealt with in this volume, including preparation of apparatus for test and pre-test checks. Conventional measuring equipment and electronic instrumentation are covered, and an outline of recommended good laboratory practice is given. Chapter 17 deals with the calibration of measuring instruments and other equipment, including hydraulic consolidation cells. Grouped together in this chapter are details of corrections which have to be made to observed data in order to allow for various experimental factors, relating to both triaxial apparatus and consolidation cells.

The third part, Chapters 18 to 21, gives details of triaxial test procedures. Chapter 18 deals with the two types of test most widely used for the determination of effective shear strength, the consolidated-undrained (CU) and the consolidated-drained (CD) compression tests. These are regarded here as 'basic' procedures, from which numerous modifications have been developed to suit particular requirements. Some of those that seem likely to be of most use in commercial testing are described in Chapter 19. Triaxial test procedures for determining consolidation and permeability characteristics are given in Chapter 20. Presentation of shear strength test data in the form of stress paths is outlined in Chapter 21.

The fourth part (Chapter 22) is devoted to the Rowe hydraulic cell, which provides for control of effective stresses and measurement of pore pressure. Consolidation test procedures using various loading and drainage conditions (vertical and radial drainage) are described, as well as tests for measuring vertical and horizontal permeabilities.

REFERENCES TO OTHER SOURCES

Extensive references are made to the current British Standard, BS 1377 : 1990 (with amendments) — Part 6 and Part 8, for the test procedures, and Part 1 for calibration and other requirements. References are also made to US Standards, ASTM Designations D 4767 and D 5084, in Volume 04.08 : Soils and Rock; Dimension Stone; Geosynthetics, of the *Annual Book of ASTM Standards*.

The recognised standard work on triaxial testing to which reference is made is the book by Bishop and Henkel, *The Measurement of Soil Properties in the Triaxial Test* (1962 edition, reprinted 1972), which regrettably is no longer available. Some of the items of equipment which they developed and described are still in use today, and detailed descriptions are not repeated here. Some of the triaxial test procedures described in this volume are based on their methods, suitably adapted to conform to the author's presentation.

Textbooks on the theory and applications of soil mechanics are suggested in Section 15.1.1. Throughout the book references are also made to other works, and to numerous papers in *Géotechnique*, in the *Proceedings of the American Society of Civil Engineers* (Geotechnical Division), in International Conference Proceedings and in other journals.

Chapter 15

Effective stress testing — principles, theory and applications

15.1 INTRODUCTION

15.1.1 Scope

The concept of effective stress is the most significant principle in the branch of engineering science known as soil mechanics, and is the basis on which modern soil mechanics has been built up. The principle of effective stress in soils is set out in this chapter; its historical development is outlined and its practical importance is explained. Most of the chapter deals with theoretical aspects of effective stress relevant to laboratory tests for the measurement of the shear strength of soils in various types of triaxial test.

In commercial practice most effective stress tests are carried out on samples of soil that are fully saturated, whether in the natural state or as a result of a laboratory-controlled process. The emphasis in this chapter is on soil behaviour in the saturated condition. Partial saturation is mentioned briefly only to illustrate its effect on pore pressure coefficients. Tests on partially saturated soils are beyond the scope of this volume.

The aim of this chapter is to explain in simple terms some of the physical theories relating to the mechanical behaviour of soils when they are subjected to shearing or compression. The theoretical concepts presented are those needed for an understanding of soil behaviour in the laboratory tests described in subsequent chapters. For a more complete presentation of theoretical soil mechanics reference should be made to textbooks such as those by C. R. Scott (1980), Lambe and Whitman (1979) and Atkinson and Bransby (1978).

15.1.2 Presentation

After the brief historical outline given below, definitions of the terms used in this volume are summarised in Section 15.2. The principle of effective stress is stated in Section 15.3, followed by explanations of the pore pressure coefficients which need to be understood in laboratory tests. In Section 15.4 the shear strength of soil is discussed and the criteria used for defining 'failure' are reviewed, including reference to the 'critical state' concept. The idealised behaviour of typical soil types in triaxial tests is described in Section 15.5, together with the derivation of data necessary for properly conducting the tests, which are tabulated for reference. The use of back pressures in triaxial and consolidation testing, and the need for obtaining full saturation of test samples, are discussed in Section 15.6.

1

This chapter concludes with a brief outline summary, in Section 15.7, of some of the applications of laboratory tests to practical engineering problems. This is presented only by way of illustration, because selection of test procedures appropriate to in-situ conditions is beyond the scope of the laboratory technician and should be controlled by an experienced geotechnical engineer.

15.1.3   Historical Note

The principle of effective stress, though basically very simple, is of fundamental significance in soil mechanics. Many of the eminent engineers of the past (including Coulomb, Collin and Bell who were referred to in Volume 2) had failed to grasp this principle, which is possibly why the development of soil mechanics as a branch of engineering science was delayed until as late as the 1920s.

However, at least two scientists of the nineteenth century seem to have realised some of its implications. Sir Charles Lyell (1871) appreciated that the pressure of deep water does not consolidate the sand and mud of the sea bed. In 1886 Osborne Reynolds demonstrated the dilatancy of dense sand by compressing a rubber bag full of saturated sand. He observed a large negative pressure in the pore water and the resulting high strength, most of which was lost on releasing the negative pressure to atmosphere.

The significance of pore pressure in soils, and the principle of effective stress, were first fully appreciated by Karl Terzaghi about 1920. He realised that the total stress normal to any plane was shared between the grain structure and the pressure of the water filling all or most of the void spaces between the grains. The effective stress was defined as the difference between the total stress and the pressure of the water in the pore spaces, which he called the 'neutral stress' and is now generally known as the pore water pressure. Terzaghi found that the mechanical properties of soils (notably strength and compressibility) are directly controlled by the effective stress, that is to say they are related entirely to the solid phase of the soil. He first used the effective stress equation $\sigma' = \sigma - u$ in 1923, and published the principle in 1924 which marked the beginning of a true understanding of the shearing behaviour of soils.

The effective stress principle was clearly stated by Terzaghi in English at the First International Conference on Soil Mechanics held at Cambridge, Mass., USA (Terzaghi, 1936). The first part of the statement defines effective stress; the second part defines its importance.

(1) 'The stresses in any point of a section through a mass of soil can be computed from the total principal stresses σ_1, σ_2 and σ_3 which act at this point. If the voids of the soil are filled with water under a stress u the total principal stresses consist of two parts. One part u acts in the water and in the solid in every direction with equal intensity. It is called the neutral stress (or the pore pressure). The balance $\sigma_1' = \sigma_1 - u$, $\sigma_2' = \sigma_2 - u$, $\sigma_3' = \sigma_3 - u$, represents an excess over the neutral stress u and it has its seat exclusively in the solid phase of the soil. This fraction of the total principal stress will be called the effective principal stress.'

(2) 'All measurable effects of a change of stress, such as compression, distortion and a change of shearing resistance, are exclusively due to changes in the effective stresses.'

From 1921 to 1925 Terzaghi carried out and published results of 'consolidated-undrained' shearbox tests on clays which were normally consolidated, from which he obtained values of the effective 'angle of internal friction', denoted by ϕ', ranging from 14°

to 42° (Terzaghi, 1925 a and b). The first triaxial tests on clay, both drained and undrained, with measurement of pore pressure were carried out in Vienna under Terzaghi's direction by Rendulic from 1933 (Rendulic, 1937). The results of these tests gave direct evidence of the validity of the principle of effective stress in saturated soils. Tests in the USA by Taylor (1944) provided similar evidence.

A detailed historical survey of Terzaghi's work is given by Skempton in *From Theory to Practice in Soil Mechanics* (1960). Terzaghi presented his concepts in London in the James Forrest Lecture to the Institution of Civil Engineers (Terzaghi, 1939), and in his address to the Fourth International Conference (1957).

The first triaxial tests to be carried out in Britain with measurement of pore water pressure were probably those on saturated sand at Imperial College reported by Bishop and Eldin (1950). These were followed by tests with pore pressure measurement on samples of undisturbed clay from the site of a dam, the results from which were the first of their kind in Britain to play a major part in a foundation design (Skempton and Bishop, 1955). Many developments in analysis and design using the effective stress principle, and in the design of testing equipment, were made at Imperial College during the next few years. Following the publication of *The Measurement of Soil Properties in the Triaxial Test* (Bishop and Henkel, 1957), effective stress testing began to be used as standard procedure in commercial laboratories. The second edition (1962) of that book is still widely used, although regrettably it is no longer available. The conference on Pore Pressure and Suction in Soils (London, 1960) provided a further stimulus for the recognition of these procedures. Today most soil testing laboratories of significance throughout the world, including site laboratories on many important earthwork projects, are equipped to carry out effective stress triaxial tests.

15.2 DEFINITIONS

Most of the terms used in this volume, except the basic terms defined in Volume 1, are defined below. Some of these definitions supplement those found in Volume 2. The symbols generally used are also shown.

SOIL SKELETON   The assemblage of solid particles which can transmit stresses through the points of intergranular contact.

POROSITY (n)   The volume of voids between solid particles expressed as a fraction of the total volume of a mass of soil.

VOIDS RATIO (e)   The ratio of the volume of voids (water and air) to the volume of solid particles in a mass of soil.

DEGREE OF SATURATION (S)   The volume of water contained in the void space between soil particles expressed as a percentage of the total volume of voids.

FULLY SATURATED SOIL   A soil in which the void spaces are entirely filled with water (i.e. $S = 1$, or 100%).

PARTIALLY SATURATED SOIL   A soil in which the void spaces contain both free air and water.

PARTIAL SATURATION   The state of saturation in a partially saturated soil.

TOTAL STRESS (σ)   The actual stress in a soil mass due to the application of an applied pressure or force.

NEUTRAL STRESS (u)   The term originally used by Terzaghi for the pore water pressure.

PORE PRESSURE (u)   The pressure of the fluid contained in the void spaces.

PORE WATER PRESSURE (u_w)   The hydrostatic pressure of the water contained in the void spaces; usually referred to as pore pressure; abbreviated p.w.p.

PORE AIR PRESSURE (u_a)   The pressure of the air in the voids of a partially saturated soil; generally greater than the pore water pressure.

EFFECTIVE STRESS (σ')   The difference between the total stress and the pore pressure:

$$\sigma' = \sigma - u$$

INTER-GRANULAR STRESS   The stress carried by the soil skeleton, in terms of the total area of cross-section. Approximates closely to the effective stress.

INTER-PARTICLE STRESS   The stress transmitted by solid particles at their points of contact.

EXCESS PORE PRESSURE ($u - u_0$)   The increase in pore pressure above the static value u_0 due to the sudden application of an external load or pressure. Also referred to as the excess hydrostatic pressure.

DISSIPATION   Decay of excess pore pressure due to drainage of pore fluid.

PORE PRESSURE DISSIPATION (U)   The ratio of excess pore pressure lost after a certain time due to drainage ($u_i - u$) to the initial excess pore pressure ($u_i - u_0$) at any instant during consolidation. It is usually expressed as a percentage, and is also referred to as the degree of consolidation.

$$U = \frac{u_i - u}{u_i - u_0} \, (\times 100\%)$$

CONFINING PRESSURE (σ_3 or σ_h)   The all-round pressure applied to a test specimen in the cell chamber of the triaxial apparatus.

VERTICAL (AXIAL) STRESS (σ_1 or σ_v)   The stress applied to a test specimen in the vertical (axial) direction.

PRINCIPAL STRESSES   The normal stresses which act on principal planes, i.e. the three mutually perpendicular planes on which the shear stresses are zero.

MAJOR PRINCIPAL STRESS (σ_1 or σ_1')   The largest of the three principal stresses. In triaxial compression, $\sigma_1 = \sigma_v$ and $\sigma_1' = \sigma_v'$.

MINOR PRINCIPAL STRESS (σ_3 or σ_3')   The smallest of the three principal stresses. In triaxial compression, $\sigma_3 = \sigma_h$ and $\sigma_3' = \sigma_h'$.

INTERMEDIATE PRINCIPAL STRESS (σ_2 or σ_2')   The principal stress which is intermediate between σ_1 and σ_3.

DEVIATOR STRESS ($\sigma_1 - \sigma_3$)   The stress due to the axial load applied to a triaxial test specimen in excess of the confining pressure; positive for compression.

SHEAR STRENGTH   The maximum resistance of a soil to shear stresses on a potential rupture surface within the soil mass.

FAILURE   Failure in shear occurs in a soil when its shear resistance is exceeded, resulting in relative sliding of soil particles along a surface of rupture. Failure can also be defined as a state at which stresses, or strains, reach a critical or limiting value.

BACK PRESSURE (u_b)   The pressure of water applied to the pore fluid in a test specimen.

NORMALLY CONSOLIDATED   A normally consolidated soil has never been subjected to a greater pressure than the present overburden pressure.

OVERCONSOLIDATED CLAY   A clay which in the past has been consolidated under a pressure greater than the present overburden pressure, usually by overlying deposits which have since been eroded away.

PRECONSOLIDATION PRESSURE   The maximum effective pressure to which an overconsolidated clay was subjected.

OVERCONSOLIDATION RATIO (OCR)   The ratio of the preconsolidation pressure to the effective overburden pressure.

ISOTROPIC CONSOLIDATION   Consolidation under the influence of an all-round hydrostatic pressure, in which the vertical effective stress (σ'_v) and the horizontal effective stresses (σ'_h) are equal.

ANISOTROPIC CONSOLIDATION   Consolidation under stresses in which the vertical effective stress is different from the horizontal effective stresses.

K_0 CONDITION   Vertical compression or consolidation in which there is no lateral displacement.

COEFFICIENT OF EARTH PRESSURE AT REST (K_0)   The ratio of the horizontal to the vertical effective stress in a mass of soil in equilibrium under its own weight.

$$K_0 = \frac{\sigma'_h}{\sigma'_v}$$

15.3 EFFECTIVE STRESS THEORY

15.3.1 Principle of Effective Stress

SOIL CONSTITUENTS

In Volume 1, Section 1.1.6, soil was defined as an assemblage of discrete particles, together with variable amounts of water and air. The solid particles are in contact with one another, forming an uncemented skeletal structure, and the spaces between them form a system of interconnecting voids or 'pores' (Volume 1, Section 3.2). In a saturated soil the voids are completely filled with water; in a dry soil they contain only air. If both air and water are present the soil is said to be 'partially saturated', and the condition is that of 'partial saturation'. The interaction between the soil structure and the pore fluid, whether water or a combination of water and air, is responsible for the behaviour of a soil mass, especially its time-dependent properties. Saturated soil is considered below and in Section 15.3.2. Reference is made to partially saturated soil in Section 15.3.3.

SIGN CONVENTION FOR STRESS AND STRAIN

Stress is defined as intensity of loading, i.e. force per unit area (Volume 2, Section 12.2). Normal stresses act normal (perpendicular) to a given plane of section, and shear stresses act tangentially to the plane. The sign convention adopted for normal stresses is that compressive forces and stresses are positive, because soils cannot usually sustain tensile stresses. It follows that compressive displacement and strains are also positive, and therefore changes that show decreases in length, area or volume are measured as positive changes.

   In the pore fluid between soil particles, pressure is positive and suction is negative.

STRESSES IN DRY SOIL

In a mass of dry soil, forces are transmitted between particles at their points of contact. The resulting local inter-particle stresses are very much higher than the stresses applied to the soil mass. However, these inter-particle stresses are of interest only in special studies of the interactions between particles. For most practical purposes the soil is treated as if it were a continuum, and the intergranular stress on a plane of section is defined as the mean stress on the total area of section. This assumption is satisfactory as long as the dimensions of particles are considerably smaller than the test specimen.

TRANSMISSION OF NORMAL STRESSES IN SATURATED SOIL

In a saturated soil that is in equilibrium under an applied load the total stress (σ) normal to a given plane is carried partly by the solid particles at their points of contact and partly by the pressure of the water in the void space, referred to as the pore water pressure (u_w). This is illustrated by the cross-section of a triaxial test specimen shown in Fig. 15.1 (a). The cell confining pressure σ applies a total stress σ normal to the specimen boundary. The pressure u_w of the water in the specimen voids (which is less than σ) is a hydrostatic pressure and acts equally in all directions. It exerts a pressure of u_w normal to the specimen boundary. The difference between the two stresses, $\sigma - u_w$, is transmitted across the boundary (Fig. 15.1 (b)) and is carried by the soil skeleton.

The difference between the total stress and the pore water pressure is known as the 'effective stress' and is denoted by σ' (or sometimes by $\bar{\sigma}$). The equation defining effective stress is

$$\sigma' = \sigma - u_w \qquad (15.1)$$

This equation, the most fundamental in soil mechanics, was first stated by Terzaghi in 1924 (see Section 15.1.3). Although the principle is very simple, it is of the utmost importance and must be properly understood. The significance of effective stress is emphasised by Terzaghi's second statement that all measurable effects of a change of stress are due exclusively to changes in the effective stresses. When dealing with the engineering

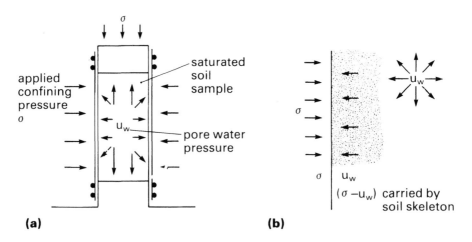

(a)   **(b)**

Fig. 15.1   Total stress and pore water pressure in a triaxial test specimen

behaviour of soil the effective stresses, and changes in effective stresses, must always be considered.

Effective stress is sometimes defined as being equal to the 'intergranular' stress, i.e. the stress transmitted between solid particles. While this may be useful as an aid to visualising the physical significance of effective stress, it is only an approximation. However, within the range of stresses generally applicable to soils the approximation is very close.

COMPRESSIBILITY

The change in volume of a soil subjected only to normal stresses is governed by the change in effective stress, not total stress. The relationship between volume change (expressed as volumetric strain, $\Delta V/V$) and change in normal effective stress is given by the equation

$$\frac{\Delta V}{V} = -C_s \Delta \sigma' \tag{15.2}$$

where ΔV is the change in volume, V the initial volume, C_s the volume compressibility of the soil skeleton, and $\Delta \sigma'$ the change in effective stress. The minus sign appears because mathematically an increase in stress produces a decrease in volume.

A change in volume is always accompanied by a change in effective stress, even when the applied total stress remains constant if the pore pressure changes due to drainage. This point is illustrated by the Terzaghi model of the consolidation process described in Volume 2, Section 14.3.2. In the spring and piston analogy shown in Fig. 14.4, the load applied to the piston represents the total stress, which sets up a pressure in the water in the cylinder (equivalent to the excess pore pressure). Opening the valve represents drainage, which causes a gradual increase of the load carried by the spring (the effective stress) with a corresponding downward movement (change in volume).

Relative volume compressibilities of the soil structure, solid grains and water are indicated in Table 15.1. These approximate values show that a soil of 'low' compressibility is about 100 times more compressible than water, and at least 1600 times more so than the constituent soil particles. For most practical purposes the compressibilities of water and the solid particles can be neglected.

Table 15.1. APPROXIMATE VOLUME COMPRESSIBILITIES

Material	Volume compressibility	
	m^2/MN	relative to water
Soil particles	$1.5–3.0 \times 10^{-5}$	0.03–0.06
Water	0.0005	1
Soils: low compressibility	0.05	100
high compressibility	1.5	3000

TRANSMISSION OF SHEAR STRESS IN SATURATED SOIL

A fluid cannot provide resistance to shear. Shear stresses in soil are transmitted entirely by friction forces at the points of intergranular contact, i.e. by the soil skeleton itself. Therefore resistance to the shear stress along a given plane depends on the effective stress normal to that plane, not on the total normal stress. The Coulomb equation (Volume 2, Section 12.3.6, Equation (12.7)) which gives the maximum resistance to shear (τ_f) on a plane of failure was modified by Terzaghi to the form

$$\tau_f' = c' + (\sigma - u_w) \tan \phi'$$

i.e.
$$\tau_f' = c' + \sigma' \tan \phi' \tag{15.3}$$

in which τ_f' is the shear stress on the plane (in terms of effective stress), σ' the effective stress normal to that plane, u_w the pore water pressure, c' the apparent cohesion, and ϕ' the angle of shear resistance, the two last items being effective stress parameters for the soil.

Shear stresses imposed on a soil can cause changes in the effective stress when there is no change in volume. This occurs for example when drainage of pore water is prevented during an undrained compression test on a saturated specimen. But the shear strains resulting from the shear stresses cause deformation of the soil, and the specimen undergoes a change of shape.

Changes in effective stress are always accompanied by deformation of the soil structure. Deformation may consist of volumetric strain (under drained conditions, referred to above under Compressibility) or shear strain, or a combination of both.

Results of shear strength tests carried out in the triaxial apparatus must be interpreted in terms of effective stress, for which the Mohr–Coulomb failure theory is represented by Equation (15.3). Effective stresses cannot be measured directly but are obtained from values of total stress and pore water pressure as measured in the tests to be described. Data obtained from these tests enable pore pressure changes to be related to changes in the applied stresses by using the experimental coefficients described in the next section.

15.3.2   Pore Pressure Coefficients in Saturated Soil

TOTAL STRESS AND PORE PRESSURE CHANGES

The relationship between pore pressure changes and changes in total stress in a soil can be expressed in terms of the pore pressure coefficients A and B which were defined by Skempton (1954). Their meaning and function are explained below by considering the stress system of a triaxial compression test on a saturated soil.

An element of soil in a triaxial test specimen is initially under zero all-round pressure and the internal pore water pressure is zero (Fig. 15.2 (a)). In a triaxial test stresses are applied to the specimen in two stages.

(1) The all-round pressure, i.e. the cell pressure, is increased by $\Delta\sigma_3$ (Fig. 15.2 (b)) (an isotropic stress increase) which causes the pore pressure to rise by Δu_c.

(2) An additional axial stress increment (deviator stress) of $(\Delta\sigma_1 - \Delta\sigma_3)$ is applied (Fig. 15.2 (c)), which causes a further pore pressure increase of Δu_d. The total resultant change in pore pressure (Fig. 15.2 (d)) is equal to

$$\Delta u = \Delta u_c + \Delta u_d$$

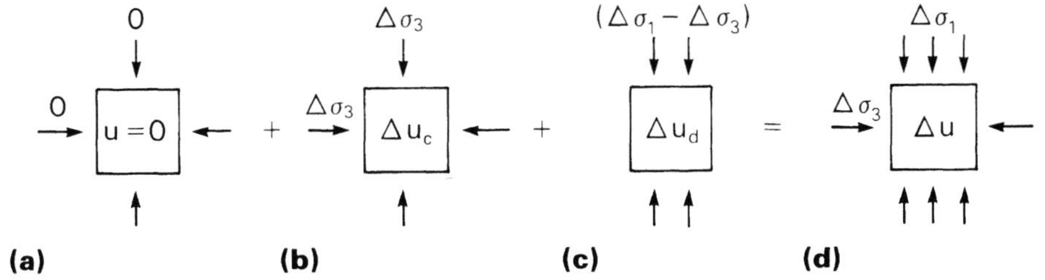

$$\text{Horizontal stress}: \Delta\sigma_3 + 0 = \Delta\sigma_3$$
$$\text{Vertical stress}: \Delta\sigma_3 + (\Delta\sigma_1 - \Delta\sigma_3) = \Delta\sigma_1$$
$$\text{Pore pressure}: \Delta u_c + \Delta u_d = \Delta u$$

Fig. 15.2 Stresses applied to an element of soil in a triaxial test specimen: (a) initially, (b) cell pressure increment, (c) increment of deviator stress, (d) resultant change

The component of pore pressure change Δu_c due to the isotropic stress change is related to that change by the coefficient B, defined by the equation

$$\Delta u_c = B . \Delta\sigma_3 \qquad (15.4)$$

The component of pore pressure change Δu_d due to the increment of deviator stress depends on the coefficient B and also on coefficient A, as explained later.

The coefficients B and A are discussed under separate headings below. These coefficients apply to both increasing and decreasing stresses, but their values for a stress change of a certain magnitude might depend on the sign of that change.

COEFFICIENT B IN SATURATED SOIL

In a saturated soil from which no drainage is permitted, the compression of the soil structure under stress is the same as the compression of the water in the voids, assuming that the compressibility of the soil grains can be neglected (see Table 15.1). In an element of soil of volume V, the volume of the void spaces, and therefore of the water (V_w), is equal to nV, where n is the porosity (Volume 1, Section 3.3.2).

An increase in the isotropic stress of $\Delta\sigma_3$ causes an increase in pore pressure of Δu_c, and the change in effective stress from Equation (15.1) is equal to

$$\Delta\sigma_3' = \Delta\sigma_3 - \Delta u_c$$

The compressibility of the soil skeleton as a whole is denoted by C_s, and from Equation (15.2) the resulting unit volume change of $\Delta V/V$ is equal to $-C_c . \Delta\sigma_3'$, i.e. it depends on the change of *effective* stress. Therefore

$$\Delta V = -C_s V \Delta\sigma_3'$$
$$= -C_s V(\Delta\sigma_3 - u_c)$$

Denoting the compressibility of water by C_w, the unit volume change of the water in the voids, $\Delta V_w/V_w$, due to the increase in water pressure Δu_c is equal to $-C_w . \Delta u_c$, i.e.

$$\Delta V_w = -C_w V_w \Delta u_c$$

and since $V_w = nV$,

$$\Delta V_w = -C_w n V \Delta u_c$$

If the solid particles are incompressible, ΔV and ΔV_w are equal, i.e.

$$-C_s V(\Delta\sigma_3 - \Delta u_c) = -C_w n V \Delta u_c$$

Re-arranging,

$$\Delta u_c = \frac{1}{1 + \dfrac{nC_w}{C_s}} \Delta\sigma_3 \tag{15.5}$$

Comparing with Equation (15.4), the coefficient B is defined by the relationship

$$B = \frac{1}{1 + \dfrac{nC_w}{C_s}} \tag{15.6}$$

Referring to the values given in Table 15.1, the ratio C_w/C_s is very small and therefore the value of B for a saturated soil is very close to 1; for practical purposes $B = 1$ for 100% saturation. This is the same as saying that the increment of total stress $\Delta\sigma_3$ is carried entirely by the pore pressure increment Δu_c if no drainage is permitted.

COEFFICIENT A

The shear stresses induced in a soil specimen by the application of a deviator stress tend to cause the specimen to change its volume. Loose sands and soft clays tend to collapse, and dense sands and stiff clays tend to expand (dilate). In a saturated soil this causes water to be expelled from or sucked into the soil if flow of water is allowed. If it is prevented, the attempt to change volume results in a change in pore pressure. In loose or collapsing soils this leads to a positive change (increase) in pore pressure $(+\Delta u_d)$; in dense or dilatant soils it leads to a negative change (decrease) $(-\Delta u_d)$. This behaviour is accounted for in a simple manner by means of the pore pressure coefficient A, which is applied as follows.

If a saturated soil behaved as a perfectly elastic material, the pore pressure change Δu_d according to elastic theory would be related to the deviator stress change $(\Delta\sigma_1 - \Delta\sigma_3)$ by the equation

$$\Delta u_d = \tfrac{1}{3}(\Delta\sigma_1 - \Delta\sigma_3)$$

For soils generally, for any value of B the equation would be

$$\Delta u_d = \tfrac{1}{3}B(\Delta\sigma_1 - \Delta\sigma_3)$$

However, soils do not usually behave in an elastic manner, and this equation has to be re-written

$$\Delta u_d = A . B(\Delta\sigma_1 - \Delta\sigma_3) \tag{15.7}$$

where A is a coefficient, to be derived experimentally, which relates pore pressure changes to changes of deviator stress. The value of A depends on the type of soil and the amount of shear strain to which it is subjected, and generally lies in the range -0.5 to $+1.0$.

GENERAL PORE PRESSURE EQUATION

The combined effect of the stress changes illustrated in Fig. 15.2 is found by adding together the two components represented by Equations (15.4) and (15.7):

$$\Delta u = \Delta u_c + \Delta u_d = B.\Delta\sigma_3 + A.B(\Delta\sigma_1 - \Delta\sigma_3)$$

i.e.
$$\Delta u = B[\Delta\sigma_3 + A(\Delta\sigma_1 - \Delta\sigma_3)] \tag{15.8}$$

For the special case of a fully saturated soil, i.e. $B = 1$, Equation (15.8) becomes

$$\Delta u = \Delta\sigma_3 + A(\Delta\sigma_1 - \Delta\sigma_3) \tag{15.9}$$

In a triaxial test in which σ_3 remains constant ($\Delta\sigma_3 = 0$) and when $B = 1$, Equation (15.9) becomes

$$\Delta u = A(\Delta\sigma_1 - \Delta\sigma_3)$$

and the coefficient A at any instant is defined by the equation

$$A = \frac{\Delta u}{\Delta\sigma_1 - \Delta\sigma_3} \tag{15.10}$$

where $(\Delta\sigma_1 - \Delta\sigma_3)$ is the change of deviator stress (equal to the applied deviator stress $((\sigma_1 - \sigma_3)$ in a normal compression test), and Δu is the corresponding pore pressure change (from the start of loading in a compression test).

The above relationships relate to conditions in which the major principal stress remains vertical and the minor principal stress remains horizontal. The application of these parameters to cases involving rotation or interchange of the principal stresses was discussed by Law and Holtz (1978).

COEFFICIENT A_f

At failure, often the most significant condition in a triaxial test, the value of A is denoted by A_f, i.e.

$$A_f = \frac{\Delta u_f}{(\Delta\sigma_1 - \Delta\sigma_3)_f} \tag{15.11}$$

where Δu_f is the pore pressure change from start of test to failure, i.e. $\Delta u_f = u_f - u_0$, and $(\sigma_1 - \sigma_3)_f$ is the deviator stress at failure ('peak' value).

The value of A_f depends not only on the type of soil but also on its previous stress history. Some typical values from compression tests are summarised in Table 15.2. It is related to the overconsolidation ratio (OCR) i.e. the ratio of the maximum stress to which the soil has been subjected in the past to the present in-situ pressure. A typical relationship is shown in Fig. 15.3, in which an OCR of 1 represents normally consolidated clay. Above a certain OCR the value of A_f becomes negative, which means that at the point of failure the tendency to dilate is strong enough to reduce the pore pressure to a level below that at the start of the compression test.

Table 15.2. TYPICAL VALUES OF PORE PRESSURE COEFFICIENT A_f

Type of soil	Volume change due to shear	A_f
Highly sensitive clay	large contraction	$+0.75$ to $+1.5$
Normally consolidated clay	contraction	$+0.5$ to $+1$
Compacted sandy clay	slight contraction	$+0.25$ to $+0.75$
Lightly overconsolidated clay	none	0 to $+0.5$
Compacted clay gravel	expansion	-0.25 to $+0.25$
Heavily overconsolidated clay	expansion	-0.5 to 0

(After Skempton, 1954)

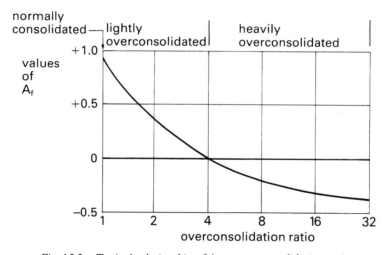

Fig. 15.3   *Typical relationship of* A_f *to overconsolidation ratio*

15.3.3   Pore Pressure Coefficients in Partially Saturated Soil

EFFECT OF PARTIAL SATURATION

A soil that is partially saturated consists of a three phase system, containing water and gas (air or water vapour) between the solid particles. The analysis of partial saturation is much more complex than for full saturation for two main reasons:

(a) gas is highly compressible;

(b) the pressures in the two fluids are not equal.

Item (a) obviously has a very important effect on the compressibility of soil. Item (b) is a result of the effect of surface tension of water.

The word 'air', as applied to the voids between solid particles in soil, is used in this book to signify a gas, usually air or water vapour or a mixture of both.

At an air–water interface the meniscus effect results in the pore air pressure (u_a) being always greater than the pore water pressure (u_w). The significance of the two pore fluids in terms of engineering behaviour is related to the extent to which the voids are saturated, and on this basis partially saturated soils can be divided into two categories.

(1) Soils with a relatively high degree of saturation, in which the water in the voids is continuous and air exists as isolated pockets or bubbles.

(2) Soils with a low degree of saturation, in which the air is continuous and water is present as thin layers around and between the solid particles.

The critical degree of saturation between these two conditions is generally about 20% for sands, 40%–50% for silts and 85% or more for clays (Jennings and Burland, 1962). Below the critical value, collapse of the grain structure of a loaded cohesive soil can occur on wetting (Burland, 1961).

Measurement of pore air pressure is beyond the scope of this book, but some understanding of the nature of a two-fluid system is useful for understanding the process of saturation.

COEFFICIENT B IN PARTIALLY SATURATED SOIL

For a partially saturated soil containing air and water in the pore spaces, the composite pore fluid as a whole is very much more compressible than water alone, and the value of B is less than unity. An isotropic total stress increment $\Delta\sigma_3$ is then carried partly by the pore pressure increment and partly by an increment of effective stress σ_3' in the soil skeleton.

The value of B depends to some extent on the compressibility of the soil skeleton as well as on the degree of saturation, both of which determine the overall compressibility of the composite pore fluid (denoted here by C_f). Using the same reasoning as in Section 15.3.2, and replacing C_w by C_f, Equation (15.6) defining the coefficient B becomes

$$B = \frac{1}{1 + nC_f/C_s} \tag{15.12}$$

In saturated soils ($S = 100\%$), as explained above, C_f is negligible compared with C_s and hence $B = 1$. At the other extreme, in dry soils C_f/C_s becomes very large indeed because air is vastly more compressible than the soil structure, and therefore B can be taken as zero when $S = 0$.

For partially saturated soils B lies between 0 and 1. The relationship between B and the degree of saturation S is not linear, but is typically of the form shown in Fig. 15.4. Compacted soil at 'optimum' moisture content and dry density generally has a B value in the range 0.1 to 0.5. Procedures for bringing a soil to a fully saturated condition are discussed in Section 15.6.

OVERALL COEFFICIENT $\bar{B}$

When considering pore pressures in earth dams or embankments the relationship of pore pressure to total major principal stress, σ_1, is significant. It is then convenient to re-write Equation (15.8) in the form

$$\Delta u = B[\Delta\sigma_3 - (1 - A)(\Delta\sigma_1 - \Delta\sigma_3)]$$

hence

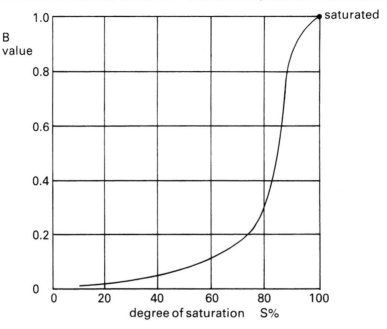

Fig. 15.4   *Relationship between pore pressure coefficient* B *and degree of saturation*

$$\frac{\Delta u}{\Delta \sigma_1} = B\left[1 - (1 - A)\left(1 - \frac{\Delta \sigma_3}{\Delta \sigma_1}\right)\right] \tag{15.13}$$

The right-hand side of Equation (15.13) is denoted by the symbol $\bar{B}$. This is called the 'overall' pore pressure coefficient, and depends on the principal stress ratio.

If the ratio of the major and minor principal effective stresses is denoted by K,

$$K = \frac{\Delta \sigma_3'}{\Delta \sigma_1'} = \frac{\Delta \sigma_3 - \Delta u}{\Delta \sigma_1 - \Delta u}$$

Hence

$$\frac{\Delta \sigma_3}{\Delta \sigma_1} = K + \frac{\Delta u}{\Delta \sigma_1}(1 - K)$$

Substituting in the right-hand side of Equation (15.13) and rearranging modifies that equation to

$$\frac{\Delta u}{\Delta \sigma_1} = B \frac{1 - (1 - A)(1 - K)}{1 - B(1 - A)(1 - K)} = \bar{B} \tag{15.14}$$

A laboratory test for the determination of $\bar{B}$ can be carried out by maintaining the ratio of principal effective stresses σ_3'/σ_1' at a constant value, K. The selected value of K depends on the required factor of safety against failure, and is intermediate between K_0, the coefficient of earth pressure at rest (Section 15.3.4) and K_f, the ratio at failure. For the specified ratio,

$$\Delta u = \bar{B}\Delta \sigma_1 \tag{15.15}$$

The test is described in Section 21.4.

For many compacted soils placed on the wet side of the optimum moisture content, typical values of $\bar{B}$ range from about 0.6 to 0.8. The value tends to increase as the stresses increase.

COEFFICIENT $\bar{A}$

For a partially saturated soil the product $A . B$ in Equation (15.7) is replaced by the coefficient $\bar{A}$, giving the equation

$$\Delta u = B . \Delta\sigma_3 + \bar{A}(\Delta\sigma_1 - \Delta\sigma_3) \tag{15.16}$$

In a triaxial test on partially saturated soil, $\bar{A}$ is defined in the same way as A for saturated soil, and the value at failure is denoted by $\bar{A}_f$. If A is to be calculated from $\bar{A}$, the value of B appropriate to the pressure range during the compression test should be used.

15.3.4 Stresses in a Soil Element

SOIL IN THE GROUND

An element of soil of unit cube in its natural state in the ground at depth z below the surface is indicated in Fig. 15.5 (a). The soil is saturated and in equilibrium, and the static water table is at a depth h below the surface. Above the water table the mass density of the soil is the bulk density, ρ. Below the water table the total mass per unit volume is the saturated density, ρ_{sat}.

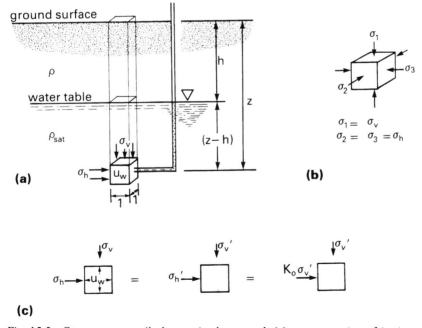

Fig. 15.5 *Stresses on a soil element in the ground: (a) representation of in-situ condition, (b) principal stresses, (c) effective stresses*

The major total principal stress on the soil element σ_1 acts vertically and is denoted by σ_v. The intermediate and minor principal stresses, σ_2 and σ_3, are assumed to be equal and act horizontally, and are denoted by σ_h (Fig. 15.5 (b)).

The total vertical stress σ_v acting on the element is equal to the weight (force) of the column of soil supported by its unit area (Fig. 15.5 (a));

i.e.
$$\sigma_v = h\rho g + (z - h)\rho_{sat} g \qquad (15.17)$$

where g is the acceleration due to gravity. If the soil element lies below layers of soil of differing densities, the contributions of all layers to the total vertical stress are summed when calculating σ_v.

The pore water pressure u_w within the soil element is equal to the pressure of the head of water $(z - h)$ above it, i.e.

$$u_w = (z - h)\rho_w g \qquad (15.18)$$

The effective vertical stress σ'_v is

$$\sigma'_v = \sigma_v - u_w$$

i.e.
$$\sigma'_v = h\rho g + (z - h)\rho_{sat} g - (z - h)\rho_w g$$
$$= [h\rho + (z - h)(\rho_{sat} - \rho_w)]g \qquad (15.19)$$

The horizontal effective stress σ'_h is related to the vertical stress by the coefficient of earth pressure at rest, K_0, in the equation

$$\sigma'_h = K_0 \sigma'_v$$

(Fig. 15.5 (c)). The pore water pressure u_w acts equally in all directions, and the total horizontal stress is therefore

$$\sigma_h = \sigma'_h + u_w$$
$$= K_0 \sigma'_v + u_w \qquad (15.20)$$

The mean total principal stress, p, is equal to $\frac{1}{3}(\sigma_1 + \sigma_2 + \sigma_3)$, i.e.

$$p = \tfrac{1}{3}(\sigma_v + 2\sigma_h) \qquad (15.21)$$

The mean effective principal stress p' is given by

$$p' = p - u_w$$

or

$$p' = \tfrac{1}{3}(\sigma'_1 + 2\sigma'_3)$$
$$= \tfrac{1}{3}\sigma'_v(1 + 2K_0) \qquad (15.22)$$

PRACTICAL APPLICATION

In practice the following units are used:

Depths z, h	metres.
Mass densities ρ, ρ_{sat}, ρ_w	Mg/m^3.
Pore water pressure u_w	kPa.
Density of water $\rho_w = 1$	Mg/m^3.

Acceleration due to gravity $g = 9.81$ m/s^2 (can often be taken as (10) m/s^2.)
Stresses σ_v, σ_h etc.   kPa.

The stresses calculated from the above equations are in kPa without further correction, i.e.

$$\sigma_v = (10)[h\rho + (z - h)\rho_{sat}] \quad \text{kPa}$$
$$u_w = (10)(z - h) \quad \text{kPa}$$
$$\sigma'_v = (10)[h + (z - h)(\rho_{sat} - 1)] \quad \text{kPa}$$
$$\sigma'_h = K_0\sigma'_v \quad \text{kPa}$$
$$\sigma_h = K_0\sigma'_v + u_w \quad \text{kPa}$$

EFFECT OF SAMPLING

When the soil element is removed from the ground as part of a sample the confining stresses are relieved and virtually reduced to zero. The change in the applied stress induces a change in the pore pressure which can be calculated theoretically from Skempton's equation

$$\Delta u = B[\Delta\sigma_3 + A(\Delta\sigma_1 - \Delta\sigma_3)] \tag{15.8}$$

Since the soil is saturated, $B = 1$, and the value of A for a normally consolidated clay may be up to about 0.5 for the small strains envisaged.
   Substituting the above stress values,

$$\Delta\sigma_3 = 0 - \sigma_h = -\sigma_h$$
$$\Delta\sigma_1 = 0 - \sigma_v = -\sigma_v$$

Hence

$$\Delta u = -\sigma_h + A(-\sigma_v + \sigma_h)$$
$$= -A\sigma_v - (1 - A)\sigma_h$$

The pore pressure u_i in the sample is then theoretically equal to $(u_w + \Delta u)$, i.e.

$$u_i = u_w - A\sigma_v - (1 - A)\sigma_h$$

in which σ_v, u_w, σ_h are derived from Equations (15.17), (15.18) and (15.20). The resulting pore pressure is negative and provides a 'suction' imparting an internal effective stress which holds the soil structure together. The negative pore pressure in a sample, even if it can be measured accurately, does not necessarily indicate the value of the mean effective stress in-situ.
   In practice a sample undergoes some deformation in being removed from the ground and some additional change in the effective stress is inevitable. Due to the relief of stress, dissolved gas can escape from solution and form bubbles — the presence of organic matter usually aggravates this tendency. If gas is released the sample is no longer fully saturated and there is a change of volume. For the reasons given in Section 15.6.2 it is desirable to counteract these reactions before carrying out an effective stress test in which pore pressures are to be measured.

Pore pressure changes in the test specimen due to reapplication of stress may not be equal in magnitude to those which occurred when stresses were relieved. Consequently, reimposition of the mean in-situ total stress $p = \frac{1}{3}(\sigma_v + 2\sigma_h)$ as an isotropic stress in the triaxial cell will not necessarily bring the specimen back to the in-situ condition of effective stress. Suitable adjustment of the back pressure may be necessary to achieve this.

15.4 SHEAR STRENGTH THEORY

15.4.1 Shear Strength of Soil

MEANING OF SHEAR STRENGTH

The shear strength of a soil is measured in terms of a limiting resistance to deformation offered by a soil mass or test specimen when subjected to loading or unloading. The limiting shearing resistance, corresponding to the condition generally referred to as 'failure', can be defined in several different ways, which are reviewed in Section 15.4.2. Criteria of failure commonly used as a basis for stability calculations in foundation design are not necessarily the same as a criterion that is relevant to fundamental properties of soils in general.

Shear strength is not a unique property of a soil but depends on many factors. The shear strength of a test specimen is measured in the laboratory by subjecting it to certain defined conditions and carrying out a particular kind of test. Failure can occur in the soil as a whole, or within limited narrow zones referred to as failure planes. Some of the factors on which the strength of soil as measured in a laboratory test depend are as follows.

(a) Mineralogy of grains.

(b) Particle shape, size distribution and configuration.

(c) Voids ratio and water content.

(d) Previous stress history.

(e) Existing stresses in-situ.

(f) Stress changes imposed during sampling.

(g) Initial state of the sample and test specimen.

(h) Stresses applied prior to test.

(j) Method of test.

(k) Rate at which loading is applied.

(l) Whether or not drainage is allowed during the test.

(m) Resulting pore water pressure.

(n) Criterion adopted for determining the shear strength.

Items (a) to (e) are related to natural conditions which cannot be controlled but can be assessed from field observations, measurements and geological evidence. Items (f) and (g) depend on the quality of sampling and the care taken in handling and specimen preparation, but (g) can be controlled with remoulded or compacted specimens (e.g. density and moisture content). Testing methods, items (h) to (l), can vary considerably and determine (m). Certain triaxial test procedures are recognised as normal practice and the most common routine tests are described in Chapter 18.

Regarding item (n), several criteria for determining the point of failure are given in Section 15.4.2. The traditional criterion most often used is the maximum axial stress which a sample can sustain, referred to as the 'peak' deviator stress. This condition is not necessarily related to basic soil properties.

STRESS–STRAIN RELATIONSHIPS

Relationships between stress and strain in soils for applying to in-situ behaviour cannot be obtained from routine compression tests. Determination of realistic stress–strain relationships requires special equipment and procedures, which are beyond the scope of this volume.

SHEAR STRENGTH PARAMETERS

For the specified failure criterion it is necessary to be able to relate the shear stress τ on a potential failure surface to the stress normal to that surface, denoted by σ_n (total stress) or σ'_n (effective stress). This is made possible by using the Mohr–Coulomb relationship with the relevant shear strength parameters for the soil.

In terms of total stress the relationship between τ_f and σ_n when failure occurs on the plane, as suggested by Coulomb, was the equation

$$\tau_f = c + \sigma_n \tan \phi \qquad (12.7)$$

(Volume 2, Section 12.3.6). However, Equation (12.7) is used only for saturated clays for which $\phi_u = 0$ and $c = c_u$, the undrained shear strength in terms of total stresses.

To take account of effective stresses on a plane of failure the Coulomb equation as modified by Terzaghi (Section 15.3.1) is used:

$$\tau'_f = c' + (\sigma_n - u_w) \tan \phi'$$

i.e.

$$\tau'_f = c' + \sigma' \tan \phi' \qquad (15.3)$$

This equation is in accordance with the fundamental principle that changes of shearing resistance are due only to changes in the effective stresses.

Equation (15.3) is of fundamental importance in the analysis of shear strength in soils. It can be represented by a Mohr–Coulomb envelope using effective stresses. The 'angle of shearing resistance', ϕ', relating to effective stresses (which are analogous to intergranular stresses) is a measure of internal friction between the grains, which is present in all soils. The shear strength parameters c', ϕ', can be obtained from a set of triaxial compression tests by plotting the Mohr circles of effective stress representing the selected failure condition and drawing the envelope to them.

The shear stress on the failure plane at failure, τ'_f (the shear strength of the soil) for a particular test, is derived as shown in Fig. 15.6. The value of τ'_f is given by the ordinate of the point P at which the Mohr circle of failure (in terms of effective stresses) touches the strength envelope, denoted by PQ. Since $PQ = PC \cos \phi'$

$$\tau'_f = \tfrac{1}{2}(\sigma'_1 - \sigma'_3)_f \cos \phi' \qquad (15.23)$$

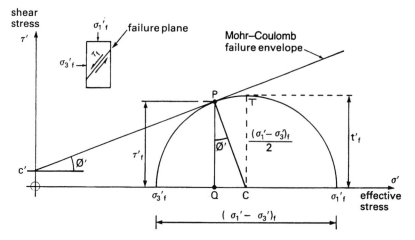

Fig. 15.6   *Derivation of soil shear strength*

For simplicity the shear strength is often assumed to be equal to half the compressive strength, i.e. the radius TC, denoted by t'_f, where

$$t'_f = \tfrac{1}{2}(\sigma'_1 - \sigma'_3)_f \tag{15.24}$$

TOTAL STRESS AND EFFECTIVE STRESS RELATIONSHIPS

Shear stress is denoted above by τ in terms of total stress, and τ' in terms of effective stress. These are equal, since the pore water cannot transmit shear, and the symbol τ is generally used for both purposes.

DRAINED AND UNDRAINED CONDITIONS

In terms of effective stress the shearing behaviour is fundamentally similar for all soil types. The differences that are observed between sands and clays, for instance, are essentially only of magnitude, not of kind, and are due to their widely differing drainage characteristics quantified by permeability. Volume changes in response to changes of stress require movement of water into or out of the soil void spaces. The time scale required for such movement can vary by a factor of about 10^6, depending on the soil permeability. In comparison the time scale during which loadings cause shear deformation in-situ can vary from a few seconds (for transient earthquake shocks) to several decades or even centuries (e.g. beneath a large foundation on a thick clay stratum), a variation factor of 10^9 or more. Within an intermediate time scale, such as for typical short-term construction loading conditions, the drained state can usually be assumed for sands and the undrained state for clays. But under the transient loading of an earthquake shock there is not enough time for even a sand to dissipate pore pressures by drainage, and the undrained condition applies. At the other extreme, as a clay gradually consolidates during a period of many years under a foundation loading or in an embankment it behaves as a drained material.

In laboratory tests on clays either undrained or drained conditions can apply, depending on the test duration. Drained conditions usually apply in tests on sands, but the undrained state is also possible with very rapid loading in special test procedures.

15.4.2 Criteria for Determining Shear Strength

Five different criteria of 'failure', from which the shear strength of a soil is determined, are discussed below. They are illustrated in Fig. 15.7, which is a composite diagram not relating to a specific soil type.

(1) Peak deviator stress.

(2) Maximum principal stress ratio.

(3) Limiting strain.

(4) Critical state.

(5) Residual state.

The first three of these criteria are discussed below. Criteria (4) and (5) are related to fundamental soil properties, and are described in Sections 15.4.3 and 15.4.4 respectively.

(1) MAXIMUM DEVIATOR STRESS

The maximum or 'peak' deviator stress criterion is the one that is traditionally associated with 'failure' in the testing of soil specimens. It is the condition of maximum principal stress difference (Fig. 15.7(1)). If the vertical and horizontal total principal stresses are denoted by σ_1 and σ_3 respectively, the peak deviator stress is written $(\sigma_1 - \sigma_3)_f$, and the corresponding strain is denoted by ε_f. In an undrained test the pore pressure at that strain is denoted by u_f, from which the principal effective stresses at peak can be calculated. In a drained test the pore pressure remains constant at the initial value, and only σ_1' varies while σ_3' remains constant.

(2) MAXIMUM PRINCIPAL STRESS RATIO

If the principal effective stresses σ_1', σ_3' are calculated for each set of readings taken during an undrained test, values of the principal stress ratio σ_1'/σ_3' can be calculated and plotted against strain as shown in Fig. 15.7. The ratio is equal to 1 at the start of the test because at that point $\sigma_1 = \sigma_3$ and therefore $\sigma_1' = \sigma_3'$. The maximum value of the ratio does not necessarily occur at the same strain as the peak deviator stress.

The maximum stress ratio criterion (2) is preferable to the peak stress criterion in some ways because it can provide a better correlation of shear strength with other parameters, or between different types of test. It is particularly useful for clays in which the deviator stress continues to increase at large strains. It can also be used as a criterion in multistage undrained triaxial tests.

The stress ratio is not used for drained tests, except perhaps to quote the value at 'peak' deviator stress, because the effective stress changes are equal to the total stresses changes, and the stress ratio curve is the same shape as the deviator stress plot.

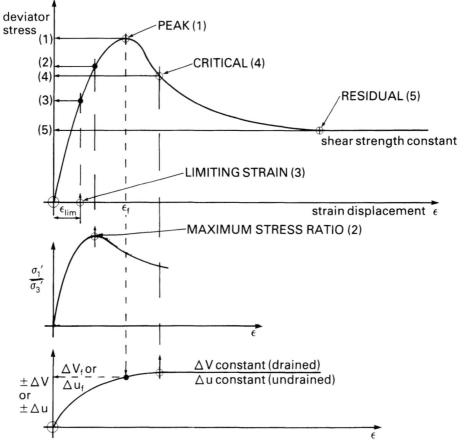

Fig. 15.7   *Idealised failure criteria for soils*

(3) LIMITING STRAIN

For soils in which a very large deformation is needed to mobilise the maximum shear resistance, a limiting strain condition might be more appropriate than a maximum shear stress. 'Failure' is then defined by a strain such as ε_{lim} at (3) in Fig. 15.7, and the corresponding Mohr circles can be drawn for deriving an envelope.

This criterion can be useful in multistage drained triaxial tests.

15.4.3   Critical State

STRESS, STRAIN AND VOLUME CHANGE

The relationship between shear strength and density (or voids ratio) in sands was outlined in Volume 2, Section 12.3.7. These principles were extended by Roscoe, Schofield and Wroth (1958) and others at Cambridge University, to provide relationships between shear strength, principal stresses and voids ratio for soils when sheared under drained or undrained conditions. This has led to a unification of the observed shear strength and deformation characteristics of soils into a coherent whole within the concept of the 'critical state', which provides a fundamental approach to the understanding of idealised soil behaviour. It depends on basic properties of the soil constituents, and applies to both 'cohesive' and 'non-cohesive' soils which were treated separately in Volume 2, Chapter 13, in terms of total stresses only. A detailed account of this principle is beyond the scope of this book, and the reader is referred to textbooks that cover the subject, such as Schofield and Wroth (1968) and Atkinson and Bransby (1978). However, the following simplified explanation is offered to provide a basic understanding from the standpoint of laboratory testing.

LOOSE AND DENSE SANDS

Curves relating shear stress, volume change and voids ratio to displacement in the shearbox test, similar to those shown in Volume 2, Fig. 12.20, are redrawn in Fig. 15.8 (b), (e) and (d), for loose sand (L) and dense sand (D), sheared under a given normal stress σ'_n.

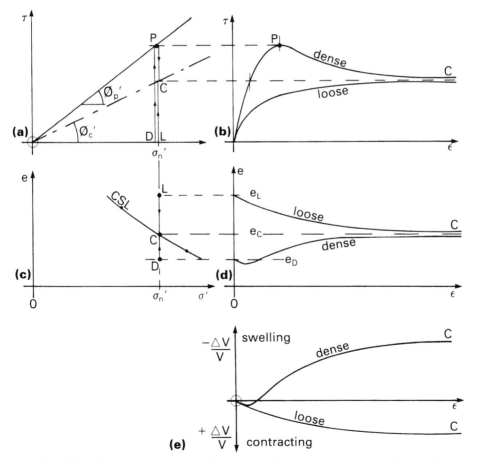

Fig. 15.8 *Shear characteristics of dense and loose sands: (a) Coulomb plot, (b)
shear stress against displacement, (c) voids ratio changes during shear, (d) voids
ratio change against displacement, (e) volume change against displacement*

In the Coulomb plot relating shear resistance to normal stress (Fig. 15.8 (a)), the sharp rise to the peak strength at P for dense sand is represented by DP, giving a peak angle of shear resistance ϕ'_p. The shear strength then falls to C and the angle reduces to ϕ_c. In contrast, the angle of shear resistance of the loose sand rises slowly to a maximum value ϕ'_c after a very large displacement without first attaining a peak value. For both specimens the condition at C is marked by a flattening of the volume change or voids ratio change curves (diagrams (e) and (d)), indicating that shearing is then taking place at constant volume. Both specimens have reached the same density, and therefore the same voids ratio (the 'critical voids ratio', e_c). The condition at C is known as the 'critical state' for that applied normal stress.

The shear strength at the critical state is a fundamental property of a particular soil and depends only on the effective stress, not on the initial density. In contrast, the 'peak' strength is dependent on the initial density (i.e. voids ratio). The angle of shear resistance at peak is made up of two components: the frictional constant value ϕ'_c and a variable dilatancy component related to initial voids ratio. The latter is positive for sands that are initially more dense than the critical density (e_0 less than e_c), and negative for sands that are less dense (e_0 greater than e_c).

CLAYS

The relationships between shear strength, principal stresses and deformations in normally consolidated clays under drained conditions are similar to the relationships referred to above for loose sands, and the relationships for overconsolidated clays are similar to those of dense sands. For clays generally the critical state is the condition in which the clay continues to deform at constant volume under constant effective stress. The critical state concept represents idealised behaviour of remoulded clays, but it is assumed to apply also to undisturbed clays in triaxial compression tests.

Further discussion of the critical state concept is beyond the scope of this book.

15.4.4 Residual State

The concept of the residual state is theoretically applicable to all soils but it is of practical significance only in clay soils because of the nature of clay mineral particles. It has been widely recognised, if not always fully understood, for at least 30 years (Skempton, 1964).

If the shear displacement of a clay is continued under constant normal stress beyond the critical state condition, the shear resistance continues to decrease until a constant value is reached, as shown at (5) in Fig. 15.7. This represents the 'residual' condition, and the constant value of shear resistance is known as the 'residual strength'. It requires a very large shear displacement to achieve — typically 100 to 500 mm, but perhaps over 1 m — unless the clay has been sheared previously, either naturally or artificially. The gradual reduction of strength beyond the critical value, from (4) to (5) in Fig. 15.7 is due to reorientation of the plate-like clay particles adjacent to the surface of shear, until they lie parallel to the plane of shear giving the characteristic 'polished' appearance. The residual condition therefore occurs only within a narrow shear zone. The residual angle of shear resistance is denoted by ϕ'_r, and c'_r is often taken as zero.

Residual strength is applicable to in-situ conditions where there have already been large displacements on shear surfaces, such as those caused by landslides or tectonic earth movements. The measurement of residual strength in the shearbox apparatus was described in Volume 2, Section 12.7.5.

15.5   THEORY RELEVANT TO TRIAXIAL TESTS

15.5.1   General Principles

FEATURES OF THE TRIAXIAL TEST

The basic features of the triaxial test apparatus were described in Volume 2, Section 13.6. The principle of the 'quick-undrained' triaxial compression test described in that chapter is indicated in Fig. 15.9 (a). The vertical and horizontal stresses σ_1 and σ_3 acting on the sample are shown in Fig. 15.9 (b). Three sets of readings are observed during this test (in addition to time):

    cell confining pressure (held constant)
    axial load (deviator stress)
    axial deformation (strain).

It is assumed that the reader is already familiar with the principles and operation of this test.

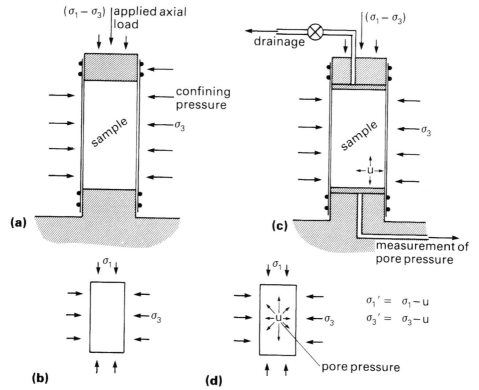

Fig. 15.9   Principles of triaxial compression tests: (a) application of stresses, (b) representation of principal stresses, (c) usual arrangement for effective stress tests, (d) representation of total and effective stresses

For carrying out effective stress triaxial tests, two additional features are needed in the cell:

provision for measurement of pore water pressure
provision for drainage.

Details of the equipment are given in Chapter 16, but the main features are outlined below to provide a basis for describing test principles.

Pore water pressure is generally measured at the base of the specimen. It is assumed that drainage takes place from the top of the specimen, which is the procedure normally referred to in this book, as shown in Fig. 15.9 (c). Alternatively drainage can also be taken from the base, but then the pore pressure within the specimen cannot be measured.

The drainage line may be connected either to a burette open to atmosphere, or to a controlled source of pressure (independent of the cell pressure system) to provide a 'back pressure', usually incorporating a volume-change indicator to measure the movement of water out of or into the specimen.

The immediate connections to a specimen set up in the triaxial cell are shown diagrammatically in Fig. 15.10, to which reference is made in describing the test principles. The following valve designations are used to enable the principles to be easily visualised, and these designations are used consistently throughout this volume.

Valve 'a' is connected to the housing block in which the pore pressure transducer is mounted (Section 16.6.3).

Valve 'a_1' isolates the pore pressure housing block from the flushing system.

Valve 'b' isolates the drainage line connected to the top end of the specimen from the drainage or back pressure system when the 'undrained' condition is required.

Valve 'c' is connected to the cell chamber pressurising system.

Valve 'd' is a spare connection to the specimen base.

Valve 'e' is the cell chamber air bleed.

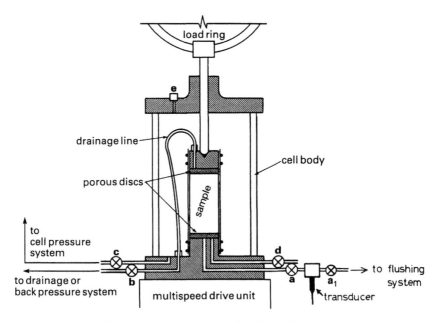

Fig. 15.10   *Connections to triaxial cell for effective stress tests*

The load ring or force measuring device for measuring the externally applied axial force, and the dial gauge or transducer for measuring axial strain, are similar to those referred to in Chapter 13. The axial stress caused by the application of the external axial force is referred to as the deviator stress.

The stresses on the sample are represented diagrammatically in Fig. 15.9 (d).

CATEGORIES OF TEST

Triaxial compression tests for the determination of the shear strength properties of soils can be divided into three main categories:

(1) Quick-undrained (QU) tests, in which no attempt is made to measure pore pressure and only total stresses are considered.

(2) Undrained tests, in which no drainage is permitted during the application of the deviator stress. There is no dissipation of excess pore pressure, which varies during the test and is measured.

(3) Drained tests, in which drainage is permitted and measured, and no excess pore pressure develops during the application of the deviator stress.

Tests of the first category were described in Volume 2, Chapter 13. This volume deals with tests of categories (2) and (3).

Undrained tests with pore pressure measurement can be carried out in three ways:

(a) By allowing no drainage during the application of confining pressure either prior to or during axial compression. This is referred to as an 'Unconsolidated-Undrained' (UU) test.

(b) By allowing drainage to consolidate the specimen under the confining pressure, and then applying compression quickly without allowing further drainage. This is referred to as a 'Consolidated Quick-Undrained' (C-QU) test.

(c) By allowing drainage to consolidate the specimen after applying the confining pressure, and then preventing further drainage during compression, which is applied slowly enough to enable pore pressures to equalise so that they can be measured with reasonable accuracy. This is the 'Consolidated-Undrained' (CU) test.

Drained tests with measurement of volume change are carried out by consolidating the specimen under the confining pressure, and allowing further drainage during the compression stage. This is the 'Consolidated-Drained' (CD) test. Long time periods are needed with soils of low permeability to allow for consolidation and dissipation of excess pore pressures developed during compression.

The drainage conditions relating to these tests are summarised in Table 15.3.

In this volume the CU and CD tests on saturated soils are treated as 'basic' effective stress, and are covered in detail in Chapter 18. They are the procedures specified in BS 1377 : Part 8 : 1990.

15.5.2 Tests on Saturated Soil

UNDRAINED TESTS ON SATURATED SOIL (UU TEST)

This is similar to the QU test except that it is run slowly enough to allow for the measurement of pore pressure. The procedure is given in Section 19.3.2.

Table 15.3. DRAINAGE CONDITIONS DURING TRIAXIAL COMPRESSION TESTS

Type of test	Application of confining pressure	Application of deviator stress	Remarks	Parameters obtained
Quick-undrained (QU)	No drainage	No drainage	Rate of strain usually to reach failure in about 10 minutes	Total stress c_u
Unconsolidated-undrained (UU)	No drainage	No drainage	Rate of strain slow enough to allow pore pressure equalisation and measurement	Effective stress c', ϕ'
Consolidated quick-undrained (C-QU)	Full drainage, usually consolidated under in-situ effective stress	No drainage	Rate of strain as for QU test. Usually three different cell pressures for compression	Total stress c_u
Consolidated-undrained with pore pressure measurement (CU)	Full drainage, three specimens usually consolidated under different effective cell pressures	No drainage	Rate of strain slow enough to allow pore pressure equalisation and measurement	Effective stress c', ϕ'
Consolidated-drained with volume change measurement (CD)	Full drainage, as for CU test	Drainage allowed	Rate of strain must be slow enough to prevent pore pressure build-up	Effective stress c_d, ϕ_d

The following description of the principles relates to a specimen of saturated clay. Throughout the test, drainage from the specimen is prevented; in the arrangement shown in Fig. 15.10, valves 'b' and 'd' remain closed. The pore water pressure at the base of the specimen is measured by the pressure transducer, connected to the cell base at valve 'a' and described in Chapter 16, which permits no movement of water from the specimen. Since the water and the soil grains are virtually incompressible, and the specimen contains no air, there can be no volume change in the specimen under applied stresses.

When a cell confining pressure is applied to the specimen the additional stress is carried entirely by the water in the voids, and the pore pressure is seen to rise by an amount practically equal to the cell pressure increment in accordance with Equation (15.4) with $B = 1$. Increasing the total stress does not alter the effective stress, and in an ensuing compression test the measured shear strength does not depend on the total stress. Several specimens tested under different confining pressures all give practically the same peak deviator stress at failure. Mohr circles of total stress (such as circles (1), (2), (3) in Fig. 15.11) give a horizontal envelope ($\phi_u = 0$) characteristic of saturated clays, the cohesion intercept c_u being the undrained shear strength.

Measurement of pore pressure during the compression test enables the changes in pore pressure due to the increasing deviator stress to be measured and plotted, as represented in

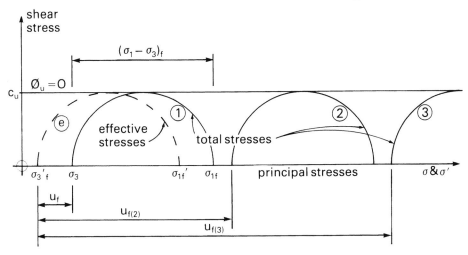

Fig. 15.11 *Mohr circles of total and effective stress for undrained tests on saturated soil*

Fig. 15.12 (a) and (b). At any point the pore pressure coefficient A can be calculated from the change of pore pressure from the initial pore pressure u_0, and the deviator stress, using Equation (15.10). Values of A are plotted against strain as shown in Fig. 15.12 (c). At failure the deviator stress attains the peak value $(\sigma_1 - \sigma_3)_f$, and the measured pore pressure is denoted by u_f. Hence from Equation (15.11),

$$A_f = \frac{u_f - u_0}{(\sigma_1 - \sigma_3)_f}$$

The Mohr circle of effective stress at failure is shown by (e) in Fig. 15.11, and is found to be virtually the same circle for all values of total stress. Hence it is not possible to derive an effective stress envelope from this type of test. To obtain several effective stress circles consolidated-undrained (CU) or consolidated-drained (CD) tests are necessary.

It should be noted that the deviator stress at any instant is the same whether expressed in terms of total stress or effective stress, because the pore pressure cancels out:

$$(\sigma_1' - \sigma_3') = (\sigma_1 - u) - (\sigma_3 - u) = (\sigma_1 - \sigma_3)$$

The undrained shear strength c_u of normally consolidated clays (for which $\phi_u = 0$) is found to increase uniformly with decreasing water content, which decreases uniformly with increasing depth below the ground surface if one-dimensional consolidation applies. The undrained shear strength therefore increases with increasing depth, and it is found that for a particular clay the ratio c_u/σ_v' is constant, where σ_v' is the vertical effective stress at the depth considered. Different clays give different values for the ratio, and Skempton (1957) showed that the ratio is related to the plasticity index (PI) of the clay by the equation

$$\frac{c_u}{\sigma_v'} = 0.11 + 0.0037 \,(\text{PI}) \tag{15.25}$$

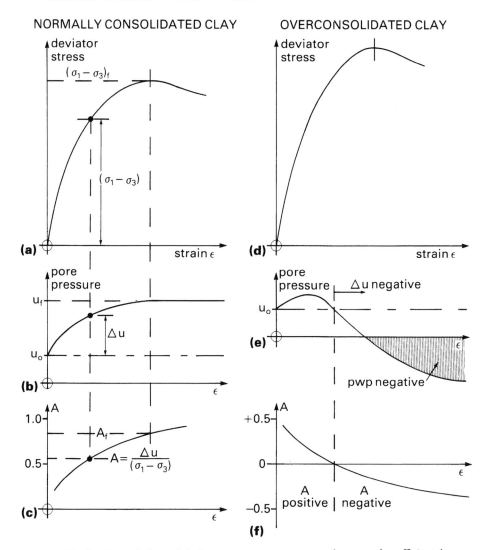

Fig. 15.12   *Typical plots of deviator stress, pore pressure change and coefficient* A *against strain from undrained triaxial compression tests on saturated soils: (a), (b), (c); normally consolidated clays; (d), (e), (f); overconsolidated clays*

This relationship applied only to saturated normally consolidated clays. A more complex theoretical relationship has been derived for overconsolidated clays (Atkinson and Bransby, 1978).

CONSOLIDATED-UNDRAINED TEST ON SATURATED SOIL

This is referred to as the CU test, and is one of the basic procedures described in Chapter 18. The specimen, here assumed to be of normally consolidated saturated clay, is consolidated to the desired effective stress before applying the deviator stress and loading to failure.

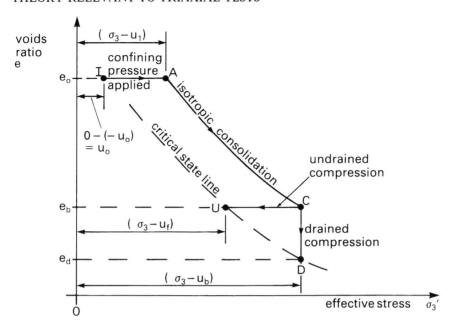

Fig. 15.13   *Voids ratio changes plotted against effective confining pressure during isotropic consolidation and compression (undrained and drained)*

To allow for consolidation the drainage line from valve 'b' (Fig. 15.10) is connected either to an open burette, or to a back pressure system set at a pressure u_b which is less than the pore pressure in the specimen, u_1, after raising the cell pressure. If drainage is to atmosphere, $u_b = 0$.

Changes in effective stress are illustrated in Fig. 15.13, in which voids ratio e is plotted against the effective confining pressure, σ'_3. Initially the specimen has a voids ratio e_0 and is under zero stress, but typically it may have a small negative pore pressure $(-u_0)$ (point I). Application of a confining pressure σ_3 causes an increase in pore pressure to u_1 with no change in voids ratio, and the effective stress is then $(\sigma_3 - u_1)$ (point A).

Consolidation is started by opening valve 'b' (Fig. 15.10). Water drains out of the specimen into the back pressure system until the pore pressure equalises with the back pressure. This requires some time because of the low permeability of clay. The specimen has by then been consolidated to a higher effective stress, equal to $(\sigma_3 - u_b)$ and the voids ratio e_b is less than e_0 (point C in Fig. 15.13). The specimen is therefore more dense and more stiff than originally.

When consolidation is complete, valve 'b' is closed to prevent further drainage and the confining pressure remains unchanged. A compression test is then carried out slowly enough to allow at least 95% equalisation of pore pressure. During compression the volume, and therefore the voids ratio, remain constant, but the pore pressure increases and the effective stress decreases. At failure the pore pressure is denoted by u_f and the effective confining pressure is $(\sigma_3 - u_f)$ (point U in Fig. 15.13). When continued compression causes no further change in pore pressure, the point U represents the critical state for that effective stress, and lies on the 'critical state line'.

The deviator stress at failure is higher than that for the unconsolidated sample, as shown by the Mohr circle of total stress for peak deviator stress marked (5) in Fig.

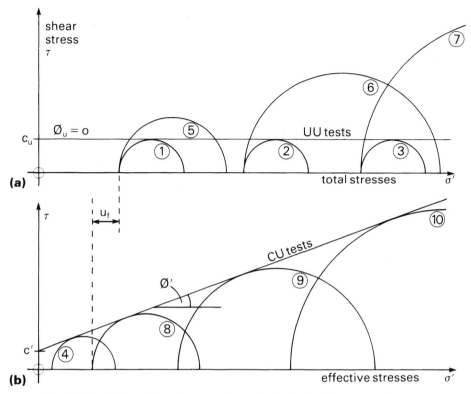

Fig. 15.14   Mohr circles diagrams for undrained triaxial compression tests on saturated soil in terms of (a) total stresses, (b) effective stresses

15.14 (a), compared with circle (1) taken from Fig. 15.11. The Mohr circle of effective stress at failure is plotted as circle (8) in Fig. 15.14 (b). Additional specimens consolidated under higher cell pressures, giving higher initial effective stresses, provide additional effective stress Mohr circles as represented by circles (9) and (10). Values of A and A_f can be obtained for each specimen, as described for the UU test.

A straight line envelope drawn to these circles gives a slope ϕ' and an apparent cohesion intercept c'. These are the effective stress parameters which can be used in Equation (15.3), and which have extensive applications to practical problems. The effective stress circle (e) in Fig. 15.11 is shown for comparison as circle (4), and represents the condition of zero consolidation.

DRAINED TEST ON SATURATED SOIL

This is referred to as the CD test, and is the second of the two basic procedures described in Chapter 18.

The specimen, assumed to be of normally consolidated saturated clay, is consolidated by the cell pressure exactly as described above for the CU test (line AC in Fig. 15.13). But then during the compression stage, further drainage is allowed to take place by leaving valve 'b' open (Fig. 15.10). The applied rate of strain must be slow enough to prevent the

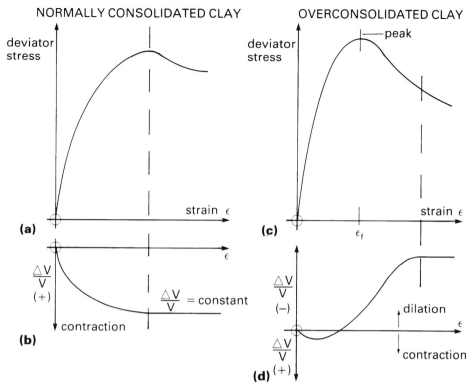

Fig. 15.15 *Typical plots of deviator stress and volume change against strain from drained triaxial compression tests on saturated soils: (a) and (b); normally consolidated clays, (c) and (d); overconsolidated clays*

development of a significant excess pore pressure, that is to ensure that the pore pressure in the specimen remains virtually equal to the value immediately after consolidation. Changes in effective stresses are then equal to changes in total stresses. The voids ratio decreases as water drains out of the specimen, and the volume of that water is equal to the change in volume of the specimen.

At failure (point D in Fig. 15.13) the voids ratio has decreased to e_d but the effective confining pressure remains unchanged at $(\sigma_3 - u_b)$. If compression continues at constant volume the point D lies on the 'critical state line'.

The general form of stress–strain and volume change curves during drained compression of a normally consolidated clay are indicated in Fig. 15.15 (a) and (b), and of an overconsolidated clay in Fig. 15.15 (c) and (d). In each case the volume change curves reflect the shape of the pore pressure change curves seen in the undrained test (Fig. 15.12).

A set of three identical specimens consolidated to three different effective stresses gives a set of Mohr circles of effective stress at failure as shown in Fig. 15.16. The envelope to these circles is inclined at an angle ϕ_d, and gives an apparent cohesion intercept of c_d. These are the drained strength parameters; and can be used in the effective stress Equation (15.3). The value of ϕ_d for the critical state is identical to the corresponding value of ϕ' obtained from CU tests. For most practical purposes it can be accepted that the parameters (c_d, ϕ_d) are the same as (c', ϕ'), and the latter symbols are the ones generally used.

The shearing behaviour of overconsolidated clays is discussed in Section 15.5.4.

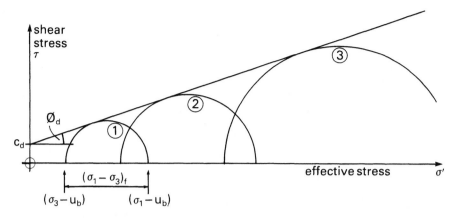

Fig. 15.16   *Mohr circle diagram for drained triaxial compression tests in terms of effective stresses*

15.5.3   Partially Saturated Soil

Application of a cell pressure increment σ_3 to a partially saturated sample, for instance of compacted clay, causes the pore pressure to increase by an amount less than σ_3, the value of B being less than 1. Consequently there is an increase in effective stress equal to $(1 - B)\sigma_3$, and an increase of strength with increasing values of σ_3, up to the point at which the specimen becomes saturated. The value of B can be calculated from the observed pore pressure changes, using Equation (15.4).

Tests on partially saturated soils are beyond the scope of this volume.

15.5.4   Overconsolidated Clay

OVERCONSOLIDATION

An overconsolidated clay is one that has been consolidated in the past under an effective stress p_c' greater than the present effective overburden stress p_0'. Part of a normal consolidation curve during deposition and consolidation to p_c' is represented in Fig. 15.17 by the curve AEC. Subsequent reduction of the vertical effective stress, for example by erosion of some of the overburden, causes the soil at a certain depth to swell back under the lower stress p_0' to the overconsolidated condition represented by the point B. The overconsolidation ratio (OCR) at B is equal to p_c'/p_0'. The overconsolidated clay denoted by B has a lower voids ratio, a higher density, and is stiffer, than the normally consolidated clay under the same effective stress represented by A.

DRAINED SHEAR STRENGTH

In the laboratory an effective stress p_1', greater than p_0' but possibly appreciably less than p_c', is applied to the specimen, which as a result is consolidated to the point D in Fig. 15.17. This is also represented in Fig. 15.18 (d), where effective stresses on the σ' axis represent stresses normal to the plane of shear failure. The stress–strain curve from a consolidated-drained test is represented in Fig. 15.18 (a) (using $t = \frac{1}{2}(\sigma_1 - \sigma_3)$ for the ordinate) by curve D. The corresponding volume change due to shear is indicated in diagram (c).

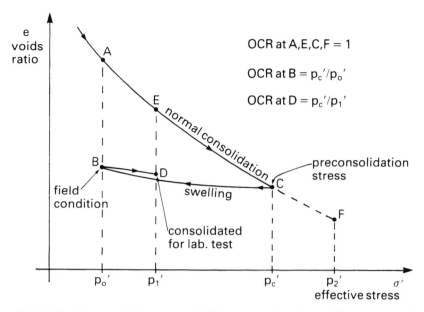

Fig. 15.17   *Overconsolidation ratio (OCR) in terms of voids ratio/effective stress plot*

Curve E in diagram (a) represents the strength of the clay when normally consolidated at the same effective stress p_1'. The excess of the peak strength in curve D over the maximum strength of curve E indicates the 'dilatancy component' of shear strength referred to in Section 15.4.3.

The form of the effective strength envelope is represented in diagram (b). For effective pressures less than the preconsolidation pressure p_c' the envelope is slightly curved, giving a cohesion intercept c'. For pressures exceeding p_c' (e.g. p_2', point F) the envelope is the same as for the normally consolidated clay, inclined at an angle ϕ' and passing through the origin if produced backwards. The point C corresponds to the stress p_c'.

When a soil is taken from the ground the stresses are removed and even a normally consolidated clay has a slight degree of overconsolidation. As a result, tests carried out under effective confining pressures less than the mean in-situ effective stress often indicate a cohesion intercept.

UNDRAINED SHEAR STRENGTH

Stress–strain characteristics from an undrained test on an overconsolidated clay are indicated in Fig. 15.12 (d), and diagrams (e) and (f) indicate changes in pore pressure and A value. When the pore pressure returns to its initial value u_0 as a result of dilatancy, the pore pressure change, and the value of the coefficient A, are zero. If the initial pore pressure u_0 is not high enough, continuing dilatancy might cause the actual pore pressure to fall to zero and then become negative, as shown in Fig. 15.12 (e). This is very undesirable, and is avoided by using a back pressure as explained in Section 15.6.

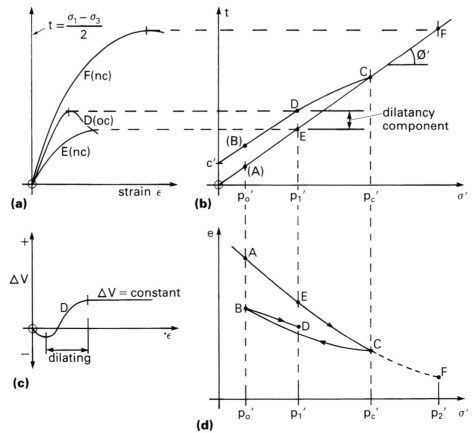

Fig. 15.18   *Relationships during a drained shear test on overconsolidated clay:* (a) *shear stress on failure surface against strain,* (b) *shear stress against normal effective stress,* (c) *volume change against strain,* (d) *voids ratio against effective normal stress*

15.5.5   Triaxial Consolidation

PRINCIPLES

Consolidation of a saturated test specimen is carried out in a triaxial cell before starting a consolidated-drained or a consolidated-undrained compression test. Usually the specimen is first subjected to a saturation stage to ensure that it is indeed fully saturated, with a B value close enough to unity for the type of soil. Saturation is discussed and described in Section 15.6.

During consolidation in the triaxial cell, equal pressure is applied to all faces of the specimen, and the consolidation is therefore described as 'isotropic'. Drainage of excess pore water usually takes place from the upper end of the specimen, as indicated in Fig. 15.19 (a). The rate of consolidation of low-permeability soils can be increased by fitting filter-paper side drains to allow drainage from the radial boundary, as shown in Fig. 15.19 (b). Usually the impervious layer is omitted so that drainage takes place from the top end as well (Fig. 15.19 (c)). Measurement of pore water pressure is usually made at the base, which is therefore a non-drainage surface.

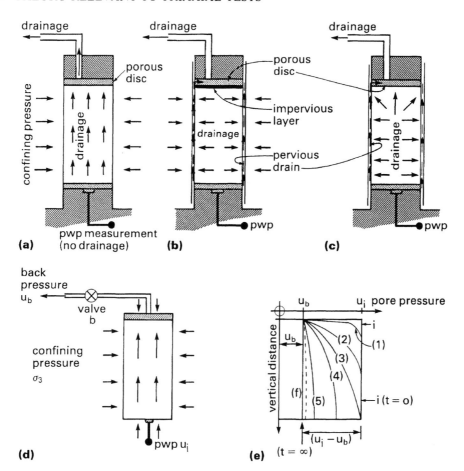

Fig. 15.19 *Consolidation of triaxial test sample: (a) vertical drainage from upper end, (b) drainage from radial boundary, (c) drainage from end and radial boundary, (d) representation of usual test conditions, (e) pore pressure isochromes during consolidation with vertical drainage*

Triaxial consolidation tests can also be carried out specifically for the determination of consolidation properties, using either isotropic or anisotropic stress conditions. Isotropic consolidation tests, in which the radial and axial stresses are equal, are described in Section 20.2. An anisotropic test for the condition in which the axial stress exceeds the radial stress is outlined in Section 20.3. Other types of anisotropic consolidation test are beyond the scope of this volume.

The following theoretical principles relate to isotropic consolidation and the derivation of factors needed for a triaxial compression test. The principles of consolidation, and the delayed time effects in soils of low permeability, are explained in Volume 2, Section 14.3.

PORE PRESSURE DISTRIBUTION

A cylindrical test specimen of clay provided with top end drainage is represented in Fig. 15.19 (d). Immediately before consolidation the internal pore water pressure, under a confining pressure σ_3, is u_i, and is uniform throughout the height of the specimen as indicated by the line 'i' in diagram (e). Beyond the drainage valve 'b' (see also Fig. 15.10), which is closed, is a pressure system maintained at a constant pressure u_b less than u_i.

Consolidation of the specimen is started by opening valve 'b' and almost immediately the excess pressure $(u_i - u_b)$ in the top porous disc falls to the value u_b, as does the pressure at the top surface of the specimen. The pore pressure distribution through a vertical section of the specimen is then as indicated by curve (1) in Fig. 15.19 (e). Because of the low permeability of clay, water can drain out only slowly. The excess pore pressure within the specimen therefore decreases slowly as water drains out, the longest delay to pore pressure change being at the base. Pore pressure distributions (isochrones) at successive intervals are represented by curves (2), (3), (4), (5) in Fig. 15.19 (e). The theoretical final condition of a uniform pore pressure equal to u_b (curve f) may never be fully achieved, but a small residual excess pore pressure at the base of up to 5% of the pressure dissipated (i.e. $0.05(u_i - u_b)$) is usually acceptable at the end of consolidation. This represents 95% dissipation of the excess pore pressure.

During consolidation the distribution of pore pressure within the specimen is not uniform, nor linear with depth, but is often assumed to be parabolic. If the pore pressure measured at the base at any time after the start of consolidation is denoted by u, the mean pore pressure $\bar{u}$ within the specimen is approximately equal to

$$\bar{u} = \tfrac{2}{3}u + \tfrac{1}{3}u_b \tag{15.26}$$

For many practical purposes, if the excess of u over u_b is not large, a linear distribution is assumed for which

$$\bar{u} = \tfrac{1}{2}(u + u_b) \tag{15.27}$$

The percentage pore pressure dissipation, $U\%$, at any time is, for convenience, related to the base pore pressure and is given by the equation

$$U = \frac{u_i - u}{u_i - u_b} \times 100\% \tag{15.28}$$

CALCULATION OF c_{vi}

Consolidation is isotropic and the coefficient of consolidation is denoted by c_{vi}. It can be determined from the consolidation stage data by using a graphical procedure which differs from the square-root time curve fitting method described in Volume 2, Section 14.3.7. The volume of water draining out of the specimen during consolidation (equal to the specimen volume change for a saturated specimen) is recorded and plotted against square-root time (minutes) as indicated in Fig. 15.20. The initial part of this plot, to about 50% consolidation, is approximately linear for all drainage conditions. The straight line portion is extended to intersect the horizontal line representing the end of consolidation at the point X. The end of consolidation should represent at least 95% dissipation of the excess pore pressure. At X the value of $\sqrt{t_{100}}$ is read off, which multiplied by itself gives t_{100}, i.e. the time representing theoretical 100% consolidation.

The value of c_{vi} can then be calculated from the equation

$$c_{vi} = \frac{\pi D^2}{\lambda t_{100}} \tag{15.29}$$

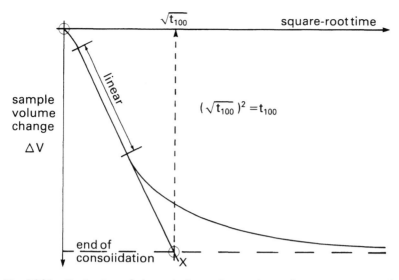

Fig. 15.20 *Derivation of theoretical* t_{100} *from volume change square-root time consolidation curve for a triaxial specimen*

where D is the diameter of specimen, L the length of specimen, and λ a constant depending on drainage boundary conditions. In SI units, D is measured in mm, c_{vi} is expressed in m^2/year, and t_{100} is usually measured in minutes. Therefore

$$c_{vi} = \frac{\pi \left(\dfrac{D}{1000} \right)^2}{\lambda(t_{100}/60 \times 24 \times 365.2)} \ m^2/\text{year}$$

$$= \frac{1.652D^2}{\lambda t_{100}} \ m^2/\text{year} \tag{15.30}$$

Values of λ are given in Table 15.4 for five different drainage boundary conditions, both

Table 15.4. FACTORS FOR CALCULATING c_{vi} AND TIME TO FAILURE

Drainage conditions during consolidation	Values of η	Values of λ $L/D=2$	Values of λ $L/D=r$	Values of F (for $r=2$) drained test	Values of F (for $r=2$) undrained* test
from one end	0.75	1	$r^2/4$	8.5	0.53
from both ends	3.0	4	r^2	8.5	2.1
from radial boundary only	32.0	64	64	12.7	1.43
from radial boundary and one end	36	80	$3.2(1+2r)^2$	14.2	1.8
from radial boundary and two ends	40.4	100	$4(1+2r)^2$	15.8	2.3

*Applies only to plastic deformation of non-sensitive soils. Values correspond to those in Table 1 of BS 1377:Part 8:1990.

in terms of any L/D ratio (denoted by r), and for specimens of the usual proportions ($r = 2$). For drainage from the radial boundary only, λ is independent of r.

Equation (15.30) forms the basis of the data summarised in Table 18.5, Section 18.7.2, for calculating c_{vi}. The value of c_{vi} derived in this way should be used only for the estimation of time to failure in the triaxial test, as described in Section 15.5.6.

CALCULATION OF COMPRESSIBILITY

The coefficient of volume compressibility for isotropic consolidation, denoted here by m_{vi}, is the change in volume per unit volume, $\Delta V/V$, for a unit change of effective stress; i.e.

$$m_{vi} = -\frac{\Delta V}{V} \bigg/ \Delta\sigma'$$

In SI units, with $\Delta\sigma'$ in kPa,

$$m_{vi} = -\frac{\Delta V}{V} \times \frac{1000}{\Delta\sigma'} \; \text{m}^2/\text{MN} \tag{15.31}$$

The coefficient m_{vi} is similar to, but not the same as, the coefficient of volume compressibility, m_v, derived from oedometer consolidation tests (Volume 2, Section 14.3.10). A typical approximate relationship is given by the equation

$$m_{vi} = 1.5 m_v$$

CALCULATION OF PERMEABILITY

A value of the coefficient of vertical permeability, k_v, of the soil can be calculated from the equation

$$k_v = 0.31 c_{vi} m_{vi} \times 10^{-9} \; \text{m/s} \tag{15.32}$$

which is similar to Equation (14.29) in Section 14.3.11 of Volume 2. A calculated value has little meaning when side drains are used.

15.5.6 Time to Failure in Compression Tests

DISSIPATION OF EXCESS PORE PRESSURE DURING COMPRESSION

In a triaxial compression test on a saturated soil, an excess pore pressure is induced by the increasing deviator stress in accordance with Equation (15.7). A fully drained test on clay must be run slowly enough to allow this excess pore pressure to dissipate, by the provision of drainage, before 'failure' is reached. In an undrained test the excess pore pressure is allowed to build up and enough time must be allowed to enable this pressure to equalise throughout the specimen if representative measurements are to be made. These time effects are discussed below.

The drained compressive strength, with 100% dissipation of excess pore pressure, of a specimen of normally consolidated clay is denoted by s_d. The undrained strength, with zero dissipation of excess pore pressure, is denoted by s_u. For similar specimens starting at the same initial pore pressure and effective confining pressure, s_u will be less than s_d. A similar test allowing partial dissipation of pore pressure will give a measured strength between s_d and s_u. If the percentage pore pressure dissipation is $U\%$, the deviator stress at failure, $(\sigma_1 - \sigma_3)_f$, is given by

$$(\sigma_1 - \sigma_3)_f = s_u + \frac{U}{100}(s_d - s_u) \tag{15.33}$$

The theory of consolidation was applied to the problem of the dissipation of excess pore pressures in triaxial compression by Gilbert and Henkel (1954). They showed that the average degree of dissipation at failure, $\bar{U}_f\%$, can be expressed in the form

$$\frac{\bar{U}_f}{100} = 1 - \frac{L^2}{4\eta c_{vi} t_f} \tag{15.34}$$

where L is the length of specimen ($= 2h$ in Bishop and Henkel), c_{vi} the coefficient of isotropic consolidation, t_f the time to failure, and η a factor depending upon drainage conditions at the specimen boundaries.

Values of η are quoted by Bishop and Henkel, and are summarised in Table 15.4 for five drainage conditions. For end drainage the specimen proportions are immaterial, but for radial drainage the value of η is based on the length being twice the diameter.

TIME TO FAILURE IN DRAINED TESTS

A theoretical degree of dissipation of 95% of the excess pore pressure is generally acceptable for deriving the drained strength parameters. Putting $U_f = 95\%$ in Equation (15.34), and rearranging, the time required to failure in a drained test is equal to

$$t_f = \frac{L^2}{0.2\eta c_{vi}} \tag{15.35}$$

By combining Equations (15.35) and (15.29), the time required for failure, t_f, can be calculated directly from t_{100} without first having to determine c_{vi}.

$$t_f = \left[\frac{5r^2}{\pi}\frac{\lambda}{\eta}\right] t_{100} \tag{15.36}$$

For the usual type of specimen in which $r = 2$, this becomes

$$t_f = \left(\frac{20}{\pi}\frac{\lambda}{\eta}\right) t_{100} \tag{15.37}$$

The multiplier $(20\lambda/\pi\eta)$ is denoted by the factor F, values of which are included in Table 15.4. For other values of r the factor can be calculated from Equation (15.36), taking the relevant values of λ and η from Table 15.4.

TIME TO FAILURE IN UNDRAINED TESTS

Guidance on the time required to failure in undrained tests, based on 95% pore pressure equalisation within the specimen, was given by Blight (1964). The relationship between t_f and c_{vi} depends on whether or not side drains are fitted.

For tests without side drains, Blight's equation is $t_f = 1.6H^2/c_{vi}$ and putting $H = L/2$:

$$t_f = \frac{0.4L^2}{c_{vi}} \tag{15.38}$$

Substituting for c_{vi} from Equation (15.29)

$$t_f = \frac{0.4L^2}{\pi D^2} \cdot \lambda t_{100} \tag{15.39}$$

and putting $L/D = r$ as above,

$$t_f = 0.127 r^2 \lambda t_{100}$$

From Table 15.4, $\lambda = r^2/4$ for drainage from one end; therefore

$$t_f = 0.0318 r^4 t_{100} \tag{15.40}$$

For the usual specimen in which $r = 2$,

$$t_f = 0.127 \times \tfrac{16}{4} t_{100} = 0.508 t_{100}$$

For tests with side drains and drainage from the radial boundary only in the consolidation stage, the corresponding relationships are as follows. From Blight's equation

$$t_f = \frac{0.0175 L^2}{c_v}$$

i.e.

$$t_f = \frac{0.0175\lambda}{\pi} \cdot \frac{L^2}{D^2} t_{100} \tag{15.41}$$

Putting $L/D = r$, and $\lambda = 64$ (from Table 15.4)

$$t_f = \frac{0.0175 r^2 \times 64}{\pi} t_{100}$$

i.e.

$$t_f = 0.3565 r^2 t_{100} \tag{15.42}$$

If $r = 2$, this gives $t_f = 1.43 t_{100}$.

For the conditions of drainage from the radial boundary with drainage additionally from one end or from both ends, the above factor is increased in proportion to the value of λ, giving the values of F shown in the last two lines of Table 15.4.

All values of the factor (F) relating t_f to t_{100} given in Table 15.4 are for specimens in which the height is equal to twice the diameter ($r = 2$). For other values of r, the factor can be calculated from Equations (15.40) or (15.42).

15.5.7   Estimation of Compressive Force

The theoretical peak deviator stress which a specimen can sustain, and hence the axial force required to cause failure, can be estimated for a particular value of effective confining pressure if a peak ϕ' value is assumed. An estimation of this kind provides a guide to the selection of the capacity of load ring to use for the test (see Section 18.3.6).

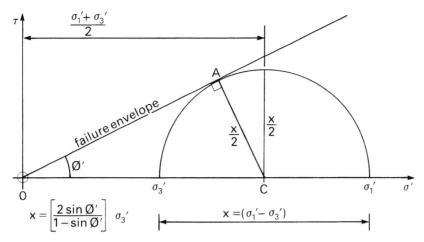

Fig. 15.21   *Estimation of peak deviator stress from shear strength parameter ϕ'*

In the Mohr–Coulomb diagram, Fig. 15.21, the circle represents the 'peak' failure condition under an effective confining pressure σ_3'. It is required to express the value of the deviator stress $(\sigma_1 - \sigma_3)$, denoted by x, in terms of σ_3' and ϕ'.

Since OA is a tangent to the circle centre C, and $OC = (\sigma_1' + \sigma_3')/2$ then

$$\sin\phi' = \frac{AC}{OC} = \frac{x/2}{(\sigma_1' + \sigma_3')/2} = \frac{x}{\sigma_1' + \sigma_3'}$$

But since $x = \sigma_1' - \sigma_3'$,

$$\sigma_1' = x + \sigma_3'$$

Therefore

$$\sin\phi' = \frac{x}{x + 2\sigma_3'}$$

Hence

$$x = \frac{2\sin\phi'}{1 - \sin\phi'}\,\sigma_3' \tag{15.43}$$

The axial force P required to produce this stress is equal to $A \cdot x$, where A is the cross-sectional area of the specimen at failure. This is greater than the initial area A_0, due to the effect of barrelling. Allowing for an area increase of 20% (which corresponds to a strain of about 17%) gives $A = 1.2A_0$, and the axial force P is equal to $1.2A_0x$, i.e.

$$P = 1.2A_0 \frac{2\sin\phi'}{1 - \sin\phi'}\,\sigma_3'$$

If A is in mm^2, σ_3' is in kPa, and P is in newtons, this relationship becomes

$$P = \frac{2.4\sin\phi'}{1 - \sin\phi'}\frac{A_0}{1000}\,\sigma_3' \quad N \tag{15.44}$$

This equation is the basis from which the graphs (Fig. 18.7, Section 18.3.6) are derived.

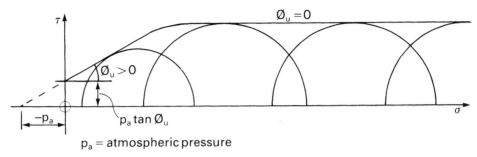

p_a = atmospheric pressure

Fig. 15.22   Mohr circles from undrained triaxial tests on a saturated sand

15.5.8   Influence of Type of Test

SATURATED CLAYS AND SANDS

Comparing the results of the tests described above with those obtained from the total stress 'quick-undrained' test described in Volume 2, Chapter 13, it can be seen that the measured strength of a soil is very much dependent upon the type of test used. For instance a saturated clay, in terms of total stress only, gives an apparent cohesion c_u equal to half the undrained compressive strength, and a ϕ_u of zero. But when related to effective stresses an appreciable value of ϕ' is obtained (though not as high as a typical ϕ' for sand), with a cohesion intercept c' considerably smaller than c_u. Pore pressures, and therefore the effective stresses, are unknown in the 'quick' test.

The converse effect can be seen in a saturated sand if tested under undrained conditions. For confining pressures above a certain value, the measured compressive strength is independent of confining pressure, i.e. $\phi_u = 0$ just as for a saturated clay. But if the applied pressure is below that value, pore pressure decrease during compression is limited by the liberation of water vapour as the specimen tries to dilate. An envelope with $\phi_u > 0$ is then obtained within this range of confining pressure, as indicated in Fig. 15.22 (Bishop and Eldin, 1950; Penman, 1953). The shear strength would theoretically be zero under a negative pressure equal to $(-p_a)$, where p_a is atmospheric pressure.

In many respects stress–strain, pore pressure and volume change characteristics of loose sands are similar to those of normally consolidated clays. Dense sands behave in ways similar to overconsolidated clays.

COMPARISON OF MEASURED STRENGTHS

For identical specimens of a normally consolidated soil the relationship between results obtained from undrained, consolidated-undrained and consolidated-drained tests, all carried out at the same confining pressure σ_3, is illustrated by the Mohr circles shown in Fig. 15.23. The effective stress circles at failure must all be tangential to the Mohr–Coulomb envelope defined by the soil parameters c', ϕ'.

The total stress circle at failure for the unconsolidated-undrained test is denoted by UU. From setting up the specimen to the point of failure the pore pressure change Δu_f is made up of two components, the first (Δu_c) due to the application of the confining pressure and the second (Δu_d) due to the deviator stress. The effective stress circle at failure, UU′, is displaced from the total stress circle to the left by a distance equal to

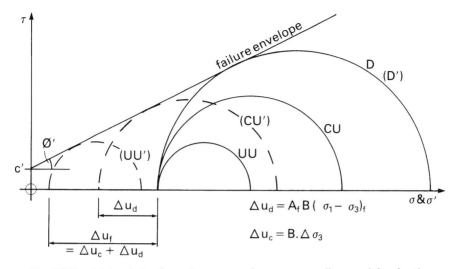

Fig. 15.23 *Mohr circles from three types of test on normally-consolidated soil*

$$\Delta u_f = \Delta u_c + \Delta u_d$$

where

$$\Delta u_c = B.\Delta\sigma_3 \qquad (15.4)$$

and

$$\Delta u_d = A_f.B(\sigma_1 - \sigma_3)_f \qquad (15.7)$$

The consolidated-undrained test circle of total stress is denoted by CU. The only pore pressure change is that due to the deviator stress (Δu_d) because the excess pore pressure caused by the application of the confining pressure is dissipated during the isotropic consolidation stage. The effective stress circle CU′ is displaced by only Δu_d and is therefore larger than UU′.

In the drained test, circle D, there is no change in pore pressure after consolidation, so the effective stress circle coincides with circle D, which is therefore the largest of the three circles.

UNDRAINED AND DRAINED PARAMETERS

For the failure criterion of peak deviator stress the effective stress parameters c', ϕ' from CU tests, and c_d, ϕ_d from CD tests, are theoretically not identical because of the different nature of the two types of test.

In a drained test the specimen volume changes and if it dilates at failure additional work must be applied against the confining pressure. This appears as the dilatancy components of shear resistance mentioned in Section 15.4.3, and gives an increase in the measured strength. The opposite applies to a soil that contracts at failure. However, the differences between c', ϕ' and c_d, ϕ_d are significant only in highly dilatant soils such as heavily over-consolidated clays, and for many applications they can be treated as being equal.

15.6 SATURATION AND USE OF BACK PRESSURE

15.6.1 Principles

REASON FOR SATURATION

In triaxial compression tests on saturated soils, standard equipment and procedures are readily available for the measurement of pore water pressure. But when dealing with partially saturated soils there is the added complication of the pore air pressure which differs from the pore water pressure. Measurement of pore air pressure presents difficulties and is beyond the scope of this volume. In the majority of effective stress triaxial tests carried out in practice these difficulties are avoided by saturating the specimen as the first stage of a test.

There are exceptions in which the achievement of full saturation by the normal procedure is not necessary, or even desirable. However, when the primary purpose of the test is to measure the shear strength at failure in soils that are not initially fully saturated, saturation is normally carried out as a first step. Even with soils that are nominally fully saturated initially, the degree of saturation is usually checked as a routine procedure, and further saturation is applied if necessary. Detailed procedures are described in Chapter 18.

PRINCIPLE OF SATURATION

Saturation is effected by raising the pore pressure to a level high enough for the water to absorb into solution all the air originally in the void spaces. At the same time the confining pressure is raised in order to maintain a small positive effective stress in the specimen. Several ways of achieving this are described in Chapter 18, Section 18.6.1. Ideally the two pressures are raised simultaneously and continuously, maintaining a constant difference between them.

The most usual method in practice is to apply a 'back pressure' to the pore fluid incrementally, alternating with increments of confining pressure. The back pressure is always a little less than the confining pressure to ensure that the effective stress remains positive. The value of the coefficient B can be checked each time the confining pressure is raised.

MAINTAINING SATURATION

Once a specimen has been saturated, the elevated pore water pressure should if possible be maintained at that level. Reduction of pore pressure below about 150 kPa could lead to the dissolved air coming out of solution again in the form of bubbles. For a drained test the drainage line should be connected to a back pressure system in which a pressure of at least 300 kPa is maintained throughout the test.

DIFFUSION OF AIR INTO WATER

Saturation by the application of back pressure not only dissolves air contained in the specimen, but also eliminates any air bubbles in the drainage line and pore pressure connections which could not be flushed out. Lee and Black (1972) investigated the diffusion of bubbles into the pore water under an applied back pressure. The time required for solution of bubbles in drainage tubes depends on their initial length and the tube

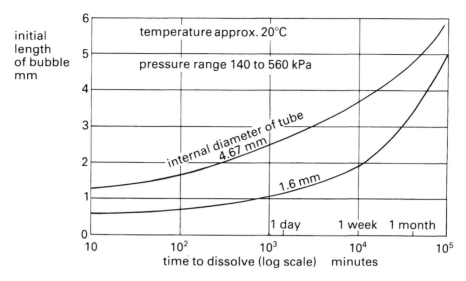

Fig. 15.24 *Rates of diffusion of air into water under pressure in small-bore tubes*

diameter. Large bubbles are reduced to smaller bubbles before they are absorbed. Bubbles in a small bore tube take much longer to dissolve than bubbles of similar volume in a large bore tube because there is less surface area of air in direct contact with water. The time effect for absorption of bubbles in tubes of two diameters, as reported by Lee and Black, is summarised by the graphical relationship shown in Fig. 15.24. The rate of diffusion is little affected by normal temperature variations, or by water pressure over the range 140–560 kPa.

Bubbles must be eliminated from drainage lines in small-bore tubes because they can obstruct the flow of water.

15.6.2 Advantages of Applying a Back Pressure

The advantages of using an elevated back pressure to obtain full saturation are summarised as follows.

(1) Air in the void spaces within the specimen is forced into solution under the applied pressure when full saturation is reached. There is then no separate air phase in the voids which might otherwise give erroneous pore pressure measurements.

(2) Any air trapped between the membrane and the specimen is also dissolved.

(3) In a specimen that dilates during shear, water can be freely sucked in during a drained test without the movement being impeded by an airlock due to air bubbles.

(4) For a similar specimen in an undrained test, initial application of a high enough back pressure can prevent the pore pressure falling below atmospheric as it tries to dilate, and therefore the measured pressure remains positive. In practice the pore pressure should not be allowed to fall below about 150 kPa.

(5) Any air bubbles remaining in the pore pressure and back pressure systems are eliminated, improving the response time of the former and avoiding the risk of air bubbles impeding drainage to the latter.

(6) Reliable measurements of permeability can be made on soils that are initially partially saturated if saturation is first achieved by applying a back pressure.

15.6.3 Solubility of Air in Water in the Soil Voids

ENTRY OF WATER FROM BACK PRESSURE SYSTEM

The introduction of additional air-free water under pressure into the voids of a partially saturated soil specimen increases the degree of saturation when equilibrium is established, because the pressure causes some of the air to be absorbed into solution. The theoretical additional pressure (back pressure) Δu_b required to increase the degree of saturation from an initial value S_0 to a final value S is given by the equation

$$\Delta u_b = [p_0] \frac{(S - S_0)(1 - H)}{1 - S(1 - H)} \tag{15.45}$$

(Lowe and Johnson, 1960) in which $[p_0]$ is initial absolute pressure, and H is Henry's coefficient of solubility (approximately $0.02 \, \text{cm}^3$ of air per cm^3 of water at $20°$ C) (Henry, 1803).

To obtain full saturation (i.e. $S = 1$), Equation (15.45) becomes

$$\Delta u_b = [p_0] \frac{1 - H}{H} (1 - S_0) \tag{15.46}$$

Putting $H = 0.02$,

$$\Delta u_b = 49[p_0](1 - S_0) \tag{15.47}$$

If the initial pressure $[p_0]$ is atmospheric, substituting one 'standard atmosphere' of $101.325 \, \text{kPa}$ in Equation (15.47) gives

$$u_b = 4965(1 - S_0) \, \text{kPa} \tag{15.48}$$

This relationship is shown graphically in Fig. 15.25, curve (a), and is summarised in Table 15.5. The two curves below curve (a) are for final saturation values of 99.5% and 99.0% (i.e. $S = 0.995$ and 0.990 in Equation (15.45)).

PRESSURISING AT CONSTANT WATER CONTENT

Theoretically the degree of saturation of the voids in a partially saturated specimen can be increased by raising the confining pressure so as to increase the air pressure without introducing additional water. The theoretical increase in pore air pressure, Δu_a, needed to achieve 100% saturation ($S = 1$) is given by the equation:

$$\Delta u_a = [p_a] \frac{1 - S_0}{S_0 H} \tag{15.49}$$

(Bishop and Eldin, 1950). Substituting for $[p_0]$ and H as above,

$$u_a = 5066 \left(\frac{1 - S_0}{S_0} \right) \text{kPa} \tag{15.50}$$

This relationship is included in Fig. 15.25, curve (b), and in Table 15.5. It is evident that an increase of air pressure alone, without the introduction of additional water, requires

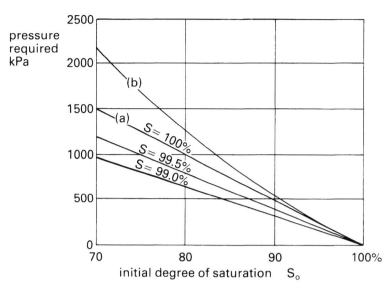

Fig. 15.25 *Pressures required for saturation of soil that is initially partially saturated: curve (a) with applied back pressure, curve (b) with confining pressure only*

Table 15.5. THEORETICAL PORE PRESSURES FOR SATURATION

Initial degree of saturation $S_0\%$	Theoretical pore pressure required	
	with back pressure kPa	confining pressure only kPa
100	0	
95	250	267
90	500	563
85	750	894
80	990	1266
75	1240	1690
70	1490	2170

appreciably higher pressures to give full saturation if S_0 is less than about 95%. When a partially saturated specimen is subjected to an increase in confining pressure only, the increase in the pore air pressure is likely to be less than the confining pressure increment, and very much less in soils that are stiff or have a 'cemented' structure. Saturation by increasing the confining pressure alone is therefore practicable only for relatively soft soils with a fairly high initial saturation.

15.6.4   Pressure and Time Required for Saturation

TRADITIONAL APPROACH

When attempting to saturate a specimen by the application of back pressure, the two factors of interest are:

(a) the pressure needed,

(b) the length of time required.

When using this procedure (described in Section 18.6.1, method (1)), the degree of saturation is monitored by observing pore pressure response to each confining pressure increment and calculating the value of B from Equation (15.4). The back pressure is increased in stages until a satisfactory B value is achieved. Traditionally a value of $B = 0.95$ is accepted as representing virtually full saturation.

Monitoring of pore pressure enables each stage to be held for as long as necessary to reach equilibrium. The two factors referred to above are thus established by trial, but sometimes a stage might be cut short prematurely for practical reasons if it would otherwise run to an excessively long time period. Some advance indication of the pressure and time likely to be needed for back pressure saturation of certain types of soil can be obtained from the data outlined below.

PRESSURE REQUIRED

The theoretical pressure required for saturation can be obtained from Equation (15.48), or from Fig. 15.25, which also indicates pressures required for degrees of final saturation of just below 100% derived from Equation (15.45). In practice the back pressures needed for saturation of undistributed specimens may not be quite as high as those indicated.

Provided that the pressures applied are enough to achieve the saturation objective, their values are immaterial. The effective stress required for the compression stage is obtained by consolidation and is equal to the difference between the confining pressure and cell pressure at the end of consolidation, irrespective of their actual values.

TIME REQUIRED

When pressurised air-free water is introduced into the void spaces in a specimen there is an immediate increase in the degree of saturation due to the compression of air in accordance with Boyle's law. If the pressure is maintained the degree of saturation increases further as air dissolves in the water but this process takes time owing to the slow rate of diffusion of small bubbles of air in confined spaces. The time element here is governed by diffusion, not by the effect of low soil permeability. The time factor was investigated by Black and Lee (1973), and some of their findings are outlined below. These data were derived by Black and Lee from tests on 71 mm diameter specimens of clean sand, although similar time effects have been observed in clay soils.

The time required for saturation under the appropriate back pressure (referred to above) depends on the initial degree of saturation of the specimen, and whether a degree of saturation of 100% is to be obtained or whether a slightly lower value is acceptable. Theoretical times for final saturation values of 99.0, 99.5 and 100% are plotted graphically in Fig. 15.26. The time required appears to be greatest when the initial saturation lies in the range 75%–85%. It decreases dramatically when the initial saturation exceeds 95%, and

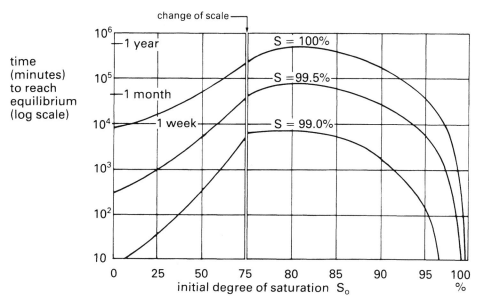

Fig. 15.26   *Time required for saturation under appropriate back pressure, related to initial degree of saturation (after Black and Lee, 1983)*

also decreases towards the dry end of the scale where the air voids are larger and interconnected and allow easier access for the penetration of inflowing water. There is a substantial saving of time if 99.5% or 99.0% saturation can be accepted.

In practice, the theoretical times to achieve 100% saturation often exceed one day, and can extend to several weeks. A compromise then has to be made on the grounds of practical expediency by accepting less than 100% saturation. Recommendations are suggested in Section 15.6.6.

15.6.5   Saturation Criteria

SOIL CATEGORIES

Instead of using the traditional arbitrary B value of about 0.95 as a criterion for saturation in all instances, it would be more realistic to relate the required B value to the properties of the soil. The relevant factor is whether a degree of saturation of less than 100% will have a significant effect on the pore pressure response, or whether at say 99.0% saturation the soil will behave as if fully saturated.

For soft soils the value of B at 100% saturation is close to 1.0, and a B value of 0.95 is obtained at about 97% saturation. Hence the commonly accepted 'saturation' requirement $(B = 0.95)$ is easily obtainable in soft soils, but this may not be high enough to justify the assumption of full saturation. In stiff soils the B value at saturation can be significantly less than 1.0, and could be little over 0.9 for very stiff materials. In these instances it is theoretically impossible to achieve $B = 1$ even at 100% saturation, and considerable well-intentioned time and effort could be wasted in the attempt.

Soils were divided into four categories by Black and Lee (1973) for the study of saturation effects, as follows:

Table 15.6. VALUES OF B FOR TYPICAL SOILS AT AND NEAR FULL SATURATION

Soil category	Degree of saturation		
	100%	99.5%	99.0%
soft	0.9998	0.992	0.986
medium	0.9988	0.963	0.930
stiff	0.9877	0.69	0.51
very stiff	0.913	0.20	0.10

(After Black and Lee, 1973)

Soft soils:	soft normally consolidated clays
Medium soils:	lightly overconsolidated clays compacted clays and silts
Stiff soils:	overconsolidated stiff clays average sands
Very stiff soils:	very stiff clays very dense sands soils consolidated to a high effective stress compacted clays with a stiff structure soils with a cementing agent, even if only very weak.

Most clays when subjected to small pressure increments show high stiffness; therefore the applied pressure increments should be of reasonable magnitude, e.g. 50 or 100 kPa.

Typical values of B at 100% saturation and just under, for each of the above categories of soil, are summarised in Table 15.6. The data are also presented graphically for initial degrees of saturation ranging from 85% to 100% in Fig. 15.27.

SUGGESTED CRITERIA

The data given above provide a basis for estimating a criterion for saturation of a particular soil that is realistic and relevant to pore pressure measurement in a triaxial test. For practical purposes a degree of saturation of less than 100% is acceptable if several successive equal increments of confining pressure give identical values of B within the range referred to above. The effective pressure should remain constant for each increment. If the pore pressure response increases with additional cell pressure increments the specimen is not saturated (Wissa, 1969).

Another check, if sensitive volume measurements can be made, is to observe carefully the tendency for water to flow into the specimen when the back pressure is increased. When the specimen is fully saturated the volume of inflow should be equal to the increase in volume of the specimen as measured on the cell pressure line (Chapter 18, Section 18.6.6).

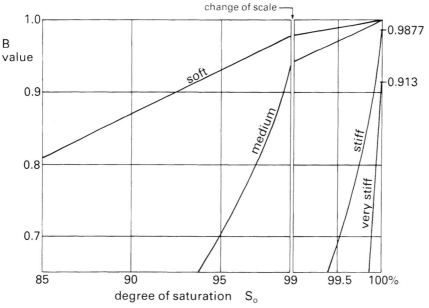

Fig. 15.27 *Typical values of pore pressure coefficient* B *related to degree of saturation and soil stiffness (after Black and Lee, 1973)*

15.6.6 Recommended Saturation Procedures

GENERAL COMMENT

Saturation by the application of back pressure is often accepted as normal practice for effective stress triaxial tests in which measurement of shear strength at failure is the main objective. However, this practice may not be suitable for all soils or in-situ conditions. Ideally the procedure used should attempt to follow the conditions imposed on the soil in the ground, but this is not always practicable.

SUGGESTED PROCEDURES

Some general guidelines on when and when not to use normal saturation procedures are outlined below.

(1) A degree of saturation of as close as possible to 100% is desirable for effective stress tests on the following soil types, and the procedure given in Section 18.6.1 is appropriate.

  (a) natural soils that will become saturated in the ground (e.g. under a dam);

  (b) compacted soils, compacted at about optimum moisture content;

  (c) any partially saturated soil in which changes in pore pressure before failure (i.e. at small strains) are significant.

(2) The saturation procedure referred to above is not necessary for effective stress tests

under the following conditions, but use of a back pressure on the drainage line is advantageous:

(a) soil that is virtually saturated initially (but a check on the B value should be made);

(b) drained tests from which only the shear strength parameters at failure (peak) are significant;

(c) soils compacted wet of optimum moisture content.

(3) For the following soil types, saturation should not be exactly as described in Section 18.6.1, but the procedure should be modified as suggested below when it is necessary to achieve or verify full saturation.

(a) Overconsolidated clay. The differential between confining pressure and applied back pressure might need to be greater than the suggested normal differential, in order to maintain an effective stress high enough to prevent swelling which could result in disturbance of the soil structure. Anisotropic stress application with K_0 greater than 1 may be more appropriate to represent the in-situ conditions.

(b) Soft normally-consolidated soil. The suggested pressure differential might cause premature consolidation, and a lower differential would be needed. This will require care in the measurement of the confining pressure and back pressure, so as to determine their difference accurately. Measurement of small differential pressure is discussed in Section 16.5.2.

(c) Residual soils. Some soils might need to be treated in the same way as overconsolidated clays to prevent them from excessive swelling.

(4) For soils of high permeability (e.g. sands), saturation can be effected relatively quickly by passing de-aired water through the specimen. This can be aided by the application of a vacuum, but care is needed to avoid disturbance to the soil structure.

(5) For stiff and very stiff soils, it is virtually impossible to achieve $B = 1$. Saturation should be checked by determining the value of B at two successive back pressures, or by observing the flow of water into the specimen as suggested in Section 15.6.5.

(6) Soils that have a relatively low initial degree of saturation should not be subjected to the cycles of effective stress imposed by a saturation process using incremental pressures, as this might progressively alter the soil structure. Cell pressure and back pressure should both be increased continuously, applying a differential pressure and a rate of pressure increase appropriate to the type of soil.

15.7 APPLICATIONS OF LABORATORY TESTS

Recommendations for the application of laboratory tests to the many different kinds of practical problems which can arise are beyond the scope of this book. However, some examples of the applications of many of the strength and compressibility tests described in this volume, and of some tests from Volume 2, are summarised in Table 15.7. This is intended only for general information and not as a comprehensive check-list of test requirements for each type of problem mentioned. Only the engineer who is fully aware of the in-situ conditions and restraints is competent to select the appropriate tests, and this requires judgement based on experience.

Table 15.7. APPLICATIONS OF LABORATORY TESTS TO FIELD PROBLEMS

Field problem	Critical period	Type of analysis	Parameters	Type of test	Comments
Foundations of structures (stability)					
on soft intact clay	end of construction	total stress	c_u, $\phi_u = 0$	QU or CU triaxial	numerous uncertainties:
on fissured clay			c_u, $\phi_u = 0$	QU or C-QU triaxial	errors may be self-compensating in some cases
(settlement)	immediate	elastic	E_u	elastic modulus	
	long term	consolidation —amount of settlement	m_v unloading curve) C_c, C_s, p'_c	oedometer consolidation	rate of settlement not reliable from lab. tests
Earth retaining structures	end of construction	total stress	c_u	QU triaxial	
	long term	effective stress	c', ϕ'	CU or CD triaxial	
	short or long term		K_0	zero lateral deformation	
Embankments: fill	during construction	effective stress	c_d, ϕ_d	CD (partly saturated)	partial consolidation represents stage construction;
	short or long term	total stress	c_u, $\phi_u = 0$	UU or CU triaxial	effect of water content on c_u;
		settlement	m_v, c_v	triaxial consolidation	isotropic or anisotropic
	long term	effective stress	c_d, ϕ_d	CD triaxial	
gravelly soils	long term	effective stress	c_d, ϕ_d	large shear-box (drained)	
foundation	during construction	effective stress	ϕ', $c' = 0$	CU with A values	
Natural slopes:					
first-time slides	long term	effective stress	c_d, ϕ_d	CD or CU triaxial	
previously failed slopes (including soils that have experienced solifluction)		residual strength	ϕ'_r	multi-reversal shear box or ring shear	
Cut slopes:	during construction	total stress	c_u	QU triaxial (extension and compression)	
first-time slides previously failed slopes (including solifluction)	long term	effective stress residual strength	c', ϕ' ϕ'_r	CU triaxial multi-reversal shear box or ring shear	
Earth dams	during construction	total stress	c_u	QU or CU triaxial	
	during construction	pore pressure effects	$\bar{B}$	special stress path test	

Table 15.7 — *continued*

Field problem	Critical period	Type of analysis	Parameters	Type of test	Comments
Earth dams drawdown, pervious soils	short term	effective stress	c', ϕ'	CU or CD triaxial	
		permeability	k	triaxial permeability	
impervious soils	short term	effective stress	c', ϕ'	CU triaxial (saturated)	
		pore pressure effects	$\bar{B}$	special test	
Tunnel linings	long term	total and effective stress	K_0	zero lateral deformation	
Temporary excavations: intact clay (base heave)	during construction	total stress		QU or CU triaxial extension (σ_v decreasing)	
fissured clay	during construction	effective stress	c', ϕ'	CU or CD triaxial	
Sheet pile retaining walls	immediate	total stress	c_u	QU or CU triaxial	
	long term	effective stress with pore pressure	c', ϕ'	CU or CD triaxial	
Settlement analysis in soft clays	short or long term	amount of settlement	C_c, p'_c	oedometer consolidation test	
Settlement analysis in laminated soils	medium to long term	rate of settlement	c_v, c_h	large hydraulic consolidation cell (Rowe cell)	
Settlement analysis in fissured soils	medium to long term	amount of settlement	E_u, E'	triaxial compression with local strain measurements: oedometer consolidation test	

REFERENCES

Annual Book of ASTM Standards, Volume 04.08: *Soil and Rock; Building Stones*. American Society for Testing and Materials, 1916 Race Street, Philadelphia, PA 19103, USA.

Atkinson, J. H. and Bransby, P. L. (1978). *The Mechanics of Soils*. McGraw-Hill, London.

Bishop, A. W. and Eldin, G. (1950). 'Undrained triaxial tests on saturated sands and their significance in the general theory of shear strength'. *Géotechnique*, 2:1:13.

Bishop, A. W. and Henkel, D. J. (1962). *The Measurement of Soil Properties in the Triaxial Test* (second edition). Edward Arnold, London.

Black, D. K. and Lee, K. L. (1973). 'Saturating laboratory samples by back pressure'. *J. Soil Mechanics & Foundation Division ASCE*, Vol. 99, No. SM1, Paper 9484, pp 75–93.

Blight, G. E. (1964). 'The effect of non-uniform pore pressures on laboratory measurements of the shear strength of soils'. *Symposium on Laboratory Shear Testing of Soils*, pp 173–184. ASTM Special Technical Publication No. 361. American Society for Testing and Materials, Philadelphia, USA.

BS 1377: Parts 1 to 8: 1990. *Methods of Test for Soils for Civil Engineering Purposes*. British Standards Institution, 389 Chiswick High Road, London W4 4AL.

Burland, J. B. (1961). Discussion. *Proc. 5th Int. Conf. Soil Mechanics and Foundation Engineering*, Paris, Vol. 3, pp 219–220.

Henkel, D. J. and Gilbert, G. D. (1954). 'The effect of the rubber membrane on the measured triaxial compression strength of clay samples'. *Géotechnique*, 3:1:20.

Henry, W. (1802). 'Experiments on the quantity of gases absorbed by water at different temperatures and under different pressures'. *Phil. Trans. Royal Society of London*, 1803, Paper III.

Lambe, T. W. and Whitman, R. V. (1979). *Soil Mechanics* (SI version). Wiley, New York.

Law, K. T. and Holtz, R. D. (1978). 'A note on Skempton's *A* parameter with rotation of principal stresses'. *Géotechnique*, 28:1:57.

Lee, K. L. and Black, D. K. (1972). 'Time to dissolve an air bubble in a drain line'. *J. Soil Mechanics & Foundation Engineering Div. ASCE*, Vol. 98, No. SM2, Paper 8728, pp 181–194.

Lowe, J. and Johnson, T. C. (1960). 'Use of back pressure to increase degree of saturation of triaxial test specimens'. *ASCE Research Conf. on Shear Strength of Cohesive Soils*. Boulder, Colorado, USA, pp 819–836.

Lyell, Sir Charles (1871). *Students' Elements of Geology*, pp 41–42. London.

Penman, A. D. M. (1953). 'Shear characteristics of a satuated silt, measured in triaxial compression'. *Géotechnique*, 3:8:312.

Pore Pressure and Suction in Soils (1960). Conference organised by the British National Society of ISSFME. Butterworths, London.

Rendulic, L. (1937). 'Ein Grundgesetz der Tonmechanik und sein experimenteller Beweis'. *Bauingenieur*, Vol. 18, pp 459–467.

Reynolds, O. (1886). 'Experiments showing dilatancy, a property of granular material'. *Proc. Royal Inst.*, Vol. 11. pp 354–363.

Roscoe, K. H., Schofield, A. N. and Wroth, C. P. (1958). 'On the yielding of soils'. *Géotechnique*, 8:1:22.

Schofield, A. N. and Wroth, C. P. (1968). *Critical State Soil Mechanics*. McGraw-Hill, London.

Scott, C. R. (1980). *An Introduction to Soil Mechanics and Foundations* (third edition). Applied Science Publishers.

Skempton, A. W. (1954). 'The pore pressure coefficients *A* and *B*', *Géotechnique*, 4:4:143.

Skempton, A. W. (1957). 'Discussion on planning and design of the new Hong Kong Airport'. *Proc. Inst. Civ. Eng.*, London, Vol. 7, pp 305–307.

Skempton, A. W. (1960). 'Significance of Terzaghi's concept of effective stress'. Contribution to *From Theory to Practice in Soil Mechanics*. John Wiley & Sons.

Skempton, A. W. (1964). 'Long-term stability of clay slopes'. Fourth Rankine Lecture. *Géotechnique*, 14:2:77.

Skempton, A. W. (1970). 'First-time slides in over-consolidated clays'. *Géotechnique*, 20:3:320.

Skempton, A. W. and Bishop, A. W. (1955). 'The gain in stability due to pore pressure dissipation in a soft clay foundation'. Trans. 5th Cong. Large Dams.

Taylor, D. W. (1944). Tenth progress report on shear strength to US engineers. Massachusetts Institute of Technology, USA.

Terzaghi, K. (1924). 'Die Theorie der hydrodynamischen Spannungserscheinungen und ihr erdbautechnisches Anwendungsgebiet'. *Proc. Int. Cong. App. Mech.*, Delft, pp 288–294.

Terzaghi, K. (1925a). *Erdbaumechanik auf bodenphysikalischer Grundlage*. Deuticke, Vienna.

Terzaghi, K. (1925b). 'Principles of soil mechanics'. *Engineering News-Record*, Vol. 95.

Terzaghi, K. (1936). 'The shearing resistance of saturated soils and the angle between the planes of shear'. *Proc. 1st Int. Conf. Soil Mechanics & Foundation Engineering*, Cambridge, Mass., USA, Vol. 1, pp 54–56.

Terzaghi, K. (1939). 'Soil mechanics — a new chapter in engineering science'. *J. Inst. Civ. Eng.*, London, Vol. 12, No. 7, pp 106–142.

Terzaghi, K. (1957). Presidential Address, Fourth Int. Conf. Soil Mechanics & Foundation Engineering London, Vol. 3, pp 55–58.

Wissa, A. E. (1969). 'Pore pressure measurement in saturated stiff soils'. *J. Soil Mechanics & Foundation Division ASCE*, Vol. 95, No. SM4, Paper 6670, pp 1063–1073.

Chapter 16

Test equipment

16.1 INTRODUCTION

16.1.1 Scope

For the purpose of this chapter, equipment required for effective stress tests is divided into the following categories. Descriptions of those items under each category that are more commonly used are outlined in the sections indicated.

(1) Cells or confining chambers in which specimens are mounted and tested (generally referred to as the 'cell'); Section 16.2.

(2) Means of applying regulated pressures to the specimens internally or externally ('pressure systems'); Section 16.3.

(3) Apparatus for applying axial force to the specimen (such as a 'loading frame'); Section 16.4.

(4) 'Conventional' measuring devices for determining the applied forces and pressures, and for measuring the resulting deformation and other changes imposed on the specimen; Section 16.5.

(5) Electronic instruments and their applications to soil testing; outlined in Section 16.6.

(6) Miscellaneous small accessories, tubing and fittings, materials, and tools; Section 16.7.

Reference is made in this chapter to a number of items which have been covered in Volume 2, especially those used for triaxial tests, and further details are given where necessary. The emphasis is on equipment for effective stress triaxial compression tests. Consolidation test equipment is listed for completeness but is described in more detail in Chapter 22. Historical notes outlining the development of some of the major items of equipment are included to illustrate progress made in the development of the test procedures.

Guidance on preparing and checking test apparatus is given in Section 16.8. Outline notes on general laboratory practice, and the calibration of measuring devices, are covered in Chapter 17, with references to Volume 2 where appropriate.

16.1.2 Conventional and Electronic Instruments

In this book procedures are generally presented in terms of conventional measuring instruments, e.g. micrometer dial gauges and pressure gauges. However, in many laboratories these are replaced by electronic devices which transmit data electrically via an electronic processing unit for display in digital form. The electric signals can also be recorded automatically for subsequent processing, either manually, or automatically by a programmed micro-computer. Electronic instrumentation is described separately in

Section 16.6, together with notes on the associated power supply and readout units. Automatic data-logging is also referred to, and the application of the micro-computer to automatic processing of data and control of many kinds of test is indicated.

Even where electronic instruments are used almost exclusively, the author believes that operators should first become familiar with test procedures and measurements using conventional instruments. The exception is the use of a pressure transducer for the measurement of pore water pressure, which is now universally recognised as standard practice.

16.1.3 Symbols

The test equipment and ancillary systems described in this book are generally in accordance with the requirements of BS 1377 : 1990 : Parts 6 and 8. They were derived from the apparatus originally developed by Professor Bishop at Imperial College, London (Bishop, 1960; Bishop and Henkel, 1962). Wherever possible the letter symbols assigned to valves are consistent with those that were used by Bishop and Henkel.

In the diagrams, certain items are represented by the symbols shown in Fig. 16.1

16.2 TEST CELLS

16.2.1 Types of Cell

The pressure vessel in which a test specimen is mounted for carrying out various kinds of

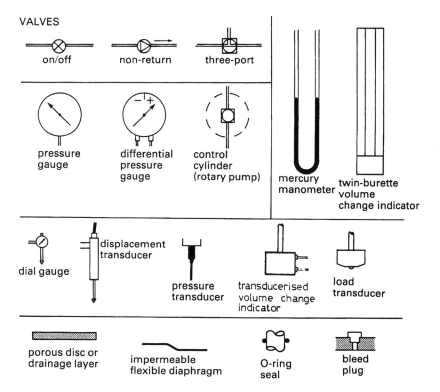

Fig. 16.1   *Symbolic representation of common items of equipment*

strength or compressibility tests is referred to as the 'cell'. It may also be called the 'pressure chamber' or 'chamber', especially in the USA.

Several types of cells are used for different types of effective stress tests on soils, as follows.

(1) Plain bearing piston-loading triaxial cells for carrying out strength tests on specimens subjected to a confining pressure, referred to as 'standard' triaxial cells. Several different sizes were referred to in Volume 2, Section 13.7.2.

(2) Triaxial cells fitted with linear ball bushings in the cell top to minimise piston friction and maintain alignment, as used in the USA.

(3) Triaxial cells fitted with a rotating bush to reduce piston friction ('rotating bush' cells), as developed by Imperial College, London.

(4) Special cells to facilitate tests in which no lateral yield is permitted ('K_0 cells').

(5) A special type of cell developed at Imperial College for a wide variety of purposes, especially stress-path tests (the 'Bishop–Wesley cell').

(6) Cells without a loading piston, for triaxial consolidation tests ('triaxial consolidation cells').

(7) Consolidation cells using a hydraulic loading system ('hydraulic consolidation cells' or 'Rowe cells'), as developed at Manchester University.

The main features of standard cells, type (1) and (2), are described in Section 16.2.2, which also includes reference to a cell of type (3). Cells of types (4) and (5) are more specialised and details are not given here. Consolidation cells, types (6) and (7), are covered in Section 16.2.4, but hydraulic consolidation cells of different sizes are described in greater detail in Chapter 22.

16.2.2 Standard Triaxial Cells

GENERAL FEATURES

Triaxial cells and their necessary fittings for various types of effective stress tests are described below. Special features relating to larger cells (i.e. for specimens of 100 mm diameter and upwards) are referred to separately.

Details of a typical cell are shown in cross-section Fig. 16.2. The main components are:

(1) Cell base

(2) Cell body and top

(3) Loading piston

(4) Loading caps (top and base)

COMPONENTS

(1) *Cell base*

The base is machined from corrosion-resistant metal or a sufficiently hard plastic material. Four outlet ports are normally required. The purpose of each outlet port, and the designation of the valve connected to it, are as follows (see Figs 16.2 and 16.3).

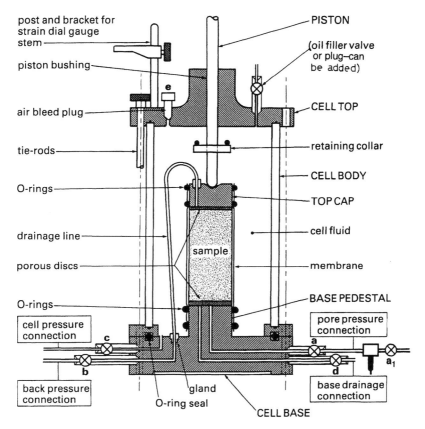

post and bracket for strain dial gauge stem
piston bushing
air bleed plug
tie-rods
O-rings
drainage line
porous discs
O-rings
cell pressure connection
back pressure connection
O-ring seal
gland
PISTON
(oil filler valve or plug—can be added)
CELL TOP
retaining collar
CELL BODY
TOP CAP
cell fluid
sample
membrane
BASE PEDESTAL
pore pressure connection
base drainage connection
CELL BASE

Fig. 16.2   *Details of a typical triaxial cell (the oil filler plug is not normally fitted to commercially available cells)*

(i)   Connection to base pedestal for the pore pressure measuring device (the transducer mounting block) (valve a).

(ii) Connection to sample top cap for the back pressure system for drainage and application of back pressure to the specimen (valve b).

(iii) Cell chamber connection for filling, pressurising and emptying the cell (valve c).

(iv) Second connection to base pedestal for drainage, or inflow of water in permeability tests (valve d). This valve is not mandatory in BS 1377 but here it is assumed to be included. Its absence would require amendments to certain procedures.

Some cells are fitted with an additional port for providing a second connection to the top cap, denoted by x in Fig. 16.3.

Each outlet terminates at a screwed socket on the edge of the cell base, into which can be fitted a valve, or a blanking plug if the line is not used. Each valve is connected to the appropriate pressure line. In addition, a valve is required on the side of the pore pressure transducer mounting block remote from valve a, between the block and the flushing system to which it is connected (valve a_1). Valve designations a, a_1, b, c, d given above are used consistently throughout this book. Details of suitable valves and other fittings are described in Section 16.7.2.

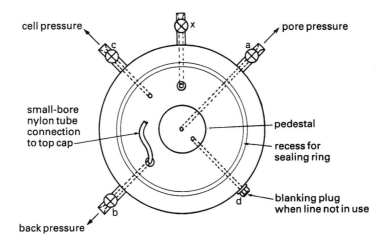

Fig. 16.3   *Triaxial cell base outlet ports*

The cell chamber port (iii) may be of a larger bore than the others to permit reasonably rapid filling and emptying of the cell. Ports (i), (ii) and (iv) leading to the base pedestal are of small bore in order to enclose the smallest practicable volume of water in the pore pressure measuring system (see Section 16.5.6).

The top surface of the pedestal may be either smooth, or machined with shallow radial and circular grooves. A grooved pedestal improves the drainage communication from the porous disc which is placed between the specimen and the small bore outlet, but a smooth surface makes for less likelihood of trapping air under the disc. If neither pore pressure measurement nor drainage is needed at the base, a solid perspex or aluminium alloy disc is placed on the pedestal instead of a porous disc.

When a porous disc of high air entry value (see Section 16.7.3) is needed, a more positive seal can be formed if the disc is of smaller diameter than the pedestal and bonded into a recess with epoxy resin. The arrangement for 38 mm diameter specimens is shown in Fig. 16.4 (a), and for large (e.g. 100 mm) diameter samples in Fig. 16.4 (b).

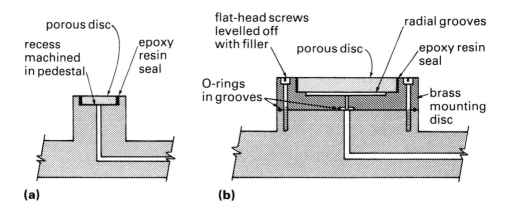

Fig. 16.4   *Triaxial cell base pedestals: (a) 38 mm diameter, (b) 100 mm diameter*

The base pedestal of some cells can be fitted with adaptors which enable one cell to be used for specimens of several diameters. Care must be taken not to trap air, or excess water, between the adaptor components when setting up. For tests requiring accurate measurements of pore pressure it is better to use a cell with an integral base pedestal that is compatible with the specimen diameter without having to use adaptors.

The cylindrical side of the pedestal should be polished smooth and free from scratches and deformities to ensure a watertight seal with the rubber membrane.

The back pressure port from valve b leads to a gland in the cell base (Fig. 16.2) which is connected by a length of small-bore nylon tubing to the specimen top loading cap (see (4) below). A similar arrangement is provided for a second top cap connection, if fitted.

(2) Cell body and top

Several types of cell body were described in Volume 2, Section 13.7.2, together with comments on their care and use.

A bracket fitted to the cell top forms a seating for the stem of the strain dial indicator or transducer (Fig. 16.2). The cell top is fitted with an air bleed plug. A useful modification is to drill and tap an additional hole for fitting a valve which can be connected to a pressurised oil supply. This enables oil to be inserted into the cell (see below), and to be replenished during a test if necessary.

The cell body expands when pressurised, but this volume change can be allowed for by calibration against pressure (see Section 17.4.4).

(3) Loading piston

The loading piston (also referred to as the 'ram' or 'plunger') is machined to a close tolerance in the bush in the cell top. The piston should slide down freely under its own weight when the cell is empty, and should allow negligible leakage of fluid from the cell when in use. For further details see Volume 2, Section 13.7.2. For effective stress tests, which are usually of long duration, a layer of castor oil 10–15 mm thick should be inserted to float on top of the water in the cell. This reduces both leakage past the piston and piston friction by providing lubrication. Measurement of leakage is described in Section 17.3.2, and the effect of piston friction is discussed in Section 17.4.5.

The piston is fitted with a fixed or adjustable collar to prevent it being forced out if the cell is pressurised when not in position in a load frame. A rubber O-ring on the collar forms a seal against leakage when in this position. Alternatively an external bracket can be used as a piston restraint, but the strain dial bracket should not be used for this purpose unless designed for the job.

The lower end of the piston may have a coned recess, or may be fitted with a hemispherical end, depending on the design of the top loading cap (see below). The piston diameter should not be less than about one-sixth of the test specimen diameter.

Some cells used in the USA are fitted with linear ball bushings to reduce piston friction.

(4) Loading cap

The force from the piston is transmitted to the specimen through the top loading cap, which is of lightweight corrosion-resistant rigid plastic or metal alloy. Several types of top cap are shown in Fig. 16.5. A solid cap (Fig. 16.5 (a)) is used for simple undrained tests.

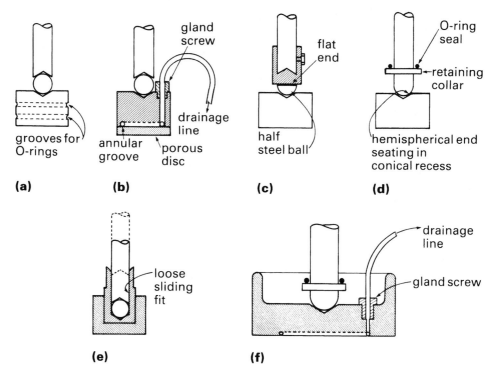

Fig. 16.5   *Triaxial sample top loading caps: (a) solid, (b) perforated with drainage line, (c) halved ball seating, (d) hemispherical ended piston seating, (e) provision for maintaining alignment during consolidation, (f) for large diameter sample*

Grooves are sometimes provided for O-rings. For effective stress tests a perforated cap, fitted with a gland for connecting a length of nylon tubing leading to the gland in the cell base, is used (Fig. 16.5 (b)). The underside of the cap may have a shallow circular groove connecting with the drainage hole, or it may be smooth. A porous disc is placed between the specimen and top cap. A second top drainage connection to the loading cap facilitates removal of air prior to permeability tests.

The tubing connecting the gland in the top cap to the back pressure inlet in the cell base (the drainage line) is usually of nylon, and should not be more than 2.5 mm diameter. It must be impermeable, and its expansion under pressure should not exceed 0.001 ml per metre length for every 1 kPa increase in internal pressure (i.e. 1 ml/m for 1000 kPa increase). Various types of tubing are listed in Section 16.7.1.

The illustrations in Fig. 16.5 (a) and (b) show a ball-bearing fitted between the coned end of the piston and a similar recess in the cap. This arrangement automatically corrects a small mis-alignment while loading is gradually applied, but may give an uncertain start to the stress/strain curve. This may be improved at small loads by using a halved steel ball and flat-ended piston (Fig. 16.5 (c)), as suggested by Bishop and Henkel, but it needs more care in aligning the specimen. The usual commercial compromise is the hemispherical-ended piston shown in Fig. 16.5 (d), which eliminates the need for a separate ball. The top cap should be capable of tilting by at least 6°, to prevent bending moments in the piston.

An arrangement which maintains alignment as the specimen is consolidated before the application of an axial force is shown in Fig. 16.5 (e). The tubular guide is a sliding fit on the piston. When first set up the piston is held in place so that it projects into the guide but is clear of the ball. After consolidation the piston is brought down into contact with the ball in the usual way.

A typical top cap for large diameter specimens (100 mm diameter and upwards) is shown in Fig. 16.5 (f), together with the usual type of hemispherical-ended piston. The cap must be deep enough to spread the centrally-applied load uniformly across the specimen area.

16.2.3 Special Triaxial Cells

ROTATING BUSH CELL

Piston friction can be virtually eliminated by using a cell top fitted with a rotating bush. The principle was developed at Imperial College (Bishop, Webb and Skinner, 1965) for use at cell pressures up to about 7000 kPa. The piston is of honed stainless steel and passes through a honed bronze bush which is continuously rotated at 2 rpm by a worm drive. Castor oil is used on top of the water in the cell to provide lubrication. The worm drive is connected to an electric motor with speed reduction gear by an extendable drive shaft fitted with a universal joint at each end. The cell and part of the drive shaft are shown in Fig. 16.6.

K_0 CELL

This type of triaxial cell is fitted with a piston of the same diameter as the specimen to ensure that volume changes measured on the cell pressure line are equal to the changes in

Fig. 16.6   Triaxial cell with rotating bush (courtesy Imperial College, London)

specimen volume. Connections to the cell, and other details, are similar to those on ordinary cells. Its use is not described here.

BISHOP–WESLEY STRESS PATH CELL

A hydraulically loaded triaxial cell was developed at Imperial College for readily reproducing a wide range of stress path conditions in the laboratory, and is described by Bishop and Wesley (1975). This cell and its use are beyond the scope of this volume.

16.2.4 Permeameter and Consolidation Cells

PERMEAMETER CELL

This type of cell, which can also be used for triaxial consolidation tests, is of similar construction to the triaxial cells described above but has no piston for axial loading. The only external total stress applied to the specimen is the isotropic confining pressure from the cell fluid. The cell accommodates specimens of 70 or 100 mm diameter, with heights of up to twice the diameter. It is sometimes referred to as a flexible-wall permeameter.

Base connection ports are similar to those provided in a normal cell, with two base pedestal drainage ports. The top cap is also fitted with two drainage ports, to facilitate de-airing and flushing.

A modular system designed for performing multiple permeability tests simultaneously, using cells of this type, is shown in Fig. 16.7.

HYDRAULIC CONSOLIDATION CELL

Hydraulically loaded consolidation cells (Rowe cells) are described in Chapter 22. The three usual sizes of cell are shown in Fig. 16.8. They are used for the determination of the consolidation properties of soils from tests in which pore pressures are measured, and back pressures applied if required, in ways similar to those used in effective stress triaxial tests.

Fig. 16.7   *Modular system for multiple permeability tests in triaxial permeability cells*

Fig. 16.8 *Hydraulic consolidation cells (Rowe cells) for nominal specimen diameters of 150 mm (dismantled), 250 mm and 75 mm*

16.3 PRESSURE SYSTEMS

16.3.1 Choice of Pressure Systems

Five types of system for applying and maintaining pressures for laboratory tests were described in Volume 2, Section 8.2.4. The three that are the most suitable for use in effective stress triaxial tests are:

Motorised compressed air system

Mercury pot system

Motorised oil-water system

The first is the most generally convenient and economical system for routine testing, and is usually referred to in this volume. For most applications either of the other two systems could be substituted. However, the mercury pot system is now little used, partly due to its complexity but mainly because of the potential health hazard inherent with the use of mercury, so it is not described here. The main features of the three types are compared in Table 16.1. Some further comments are given in Sections 16.3.2 to 16.3.4 below.

Pressures should be controllable and should remain stable to within the following limits:

Pressures not exceeding 200 kPa: within 1 kPa

Pressures exceeding 200 kPa: within 0.5%.

16.3.2 Motorised Compressed Air System

PRINCIPLE

One of the earliest applications of compressed air for maintaining a pressure supply for triaxial tests was referred to by Spence (1954), in which pressure regulation was effected in

Table 16.1. COMPARISON OF PRESSURE SYSTEMS

System	Pressure range normally available	Advantages	Disadvantages
Motorised compressed air (air bladder/water)	0–1000	versatile; economical; capable of extension to numerous regulated systems, subject to capacity of compressor; easy to regulate; automatic control possible	two compressors and stand-by generator desirable for long-term tests
Mercury pots (mercury/water)	0–1000, can be extended to 1700	highest accuracy and stability; visible indication that pressure is maintained; simple principle	limited volume capacity; high cost of mercury for each additional line; potential hazard from leaking mercury (for precautions see Volume 2, Section 8.4.4, and Volume 1, Section 1.6.7)
Motorised oil/water	0–1700, can be extended to 3500	self-contained	one unit needed for each pressure line

two stages. The single-stage pressure regulator valves commonly used today give a very stable source of constant pressure even if there are fluctuations in the supply line pressure.

The principle of the motorised compressed air system was described and illustrated in Volume 2, Section 8.2.4(2). The maximum working pressure is usually limited to 1000 kPa, which is high enough for most routine tests. When a higher pressure is necessary on a particular pressure line an oil-water pressure system can be substituted. Consumption of air is low and one compressor can serve numerous pressure regulators.

OPERATION

Bladder cells should be drained periodically and refilled with freshly de-aired water, because bladders can never be perfectly impermeable to air. At the same time the bladders should be inspected for wear, especially at the connection to the air inlet, and renewed if necessary. The connecting clip should be checked and replaced if showing signs of corrosion.

Condensation in compressed air distribution lines can cause malfunctioning of pressure regulator valves. This problem can be overcome by fitting an air refrigeration unit to cool the air as it leaves the compressor. Moisture held by the air condenses out as the air cools, and is removed. The dried cooled air does not release any remaining moisture as it subsequently warms up to the environmental temperature.

For long-term tests a standby air compressor is necessary to allow for the possibility of mechanical breakdown. The ideal arrangement is to have two identical compressors in the system, each with automatic starter operated by a pressure-control switch. During any one month one compressor is in active use and the other is on standby. Their roles are alternated monthly, enabling the inactive one to be serviced.

An automatically self-starting petrol or diesel generator is needed to supply electric power to the compressor in order to maintain pressures in the event of mains supply failure. It is also desirable to maintain the power supply to any compression machines which may be running. In practice this means maintaining the supply to all the laboratory power lines, and this factor will determine the necessary generator capacity. The batteries supplying the starter motor should be checked regularly for state of charge and acid level, and the standby system should be given a trial run at regular (e.g. monthly) intervals.

16.3.3 Motorised Oil–Water System

This type of system, for providing pressures up to 1700 kPa, was described in Volume 2, Section 8.2.4(4). A high-pressure unit, for pressures up to 3500 kPa, is also available.

Although one unit can provide a given pressure for up to three lines, two of these units are needed for effective stress tests because the pressures supplied to two lines must be independently controllable. However, Dr Penman at the Building Research Establishment developed a special modification of this apparatus to provide two independent pressures. A second dead-weight cylinder was added, supplied by excess oil from the first, and requiring only the one pump. A further modification was to substitute spring-loading for dead weights for use on board ship for off-shore testing. The spring stiffness was found to be very critical (Penman, 1984).

16.3.4 Digital Hydraulic Pressure Controller

This is a self-contained microprocessor-controlled hydraulic servo-system in which pressure is applied directly to water by means of the displacement of a piston on a cylinder similar in principle to the pore pressure control cylinder (Section 16.5.6). The original form of the system, known as the Automatic Programmable Triaxial Test (APTT), was described by Menzies and Sutton (1980). The screwed rod, fitting in a ballscrew to minimise friction, is rotated by a stepper motor and gearbox running on a linear bearing. One step of the motor, which is controlled by a microprocessor, causes a piston volume displacement of 1 mm³; thus the controller incorporates measurement of volume change. Pressure and volume change are displayed digitally on the control panel. The principle of operation is shown in Fig. 16.9.

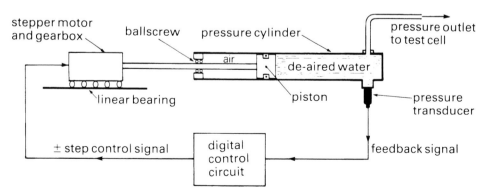

Fig. 16.9 *Principle of operation of digital hydraulic pressure controller (after Menzies and Sutton, 1980)*

For triaxial compression tests or Rowe cell consolidation tests a separate digital controller is needed for each pressure line. Units are available to give maximum pressures up to 2000 kPA.

The digital controller is a high-precision instrument, particularly suitable for research work. Its main applications are in conjunction with the Bishop–Wesley triaxial cell for special tests (e.g. stress-path tests) requiring feedback control linked to a computer.

16.3.5   Control and Distribution of Pressures

PRESSURE DISTRIBUTION PANELS

Water pressures generated by the constant pressure systems described in Sections 16.3.2 and 16.3.3 need to be controlled and distributed to the appropriate connections on the cell in which the specimen is to be tested. The connecting lines and cell ports must be flushed, de-aired and subjected to pressure checks before starting a test. Pressure distribution panels fitted with valves, gauges and other components provide for these functions. Essentially the panel provides a link between one constant pressure system and one test cell, and includes a pressure gauge and additional connections for other purposes, such as filling the cell from a water supply and draining it to waste. A larger panel with multiple inlets and outlets provides connections from several pressure systems to several cells, and uses wall space economically. Isolating valves enable the pressure gauge to be connected to each line in turn for setting and monitoring the pressure. Alternatively, a distribution panel can be fitted with multiple gauges, one for each pressure system to which it is permanently connected.

In addition to the pressure gauge or gauges fitted to the control panel, a pressure transducer of the appropriate range and readability may also be incorporated in each pressure line, normally at the cell port. Pressure transducers are essential when an automatic data-logging system is used.

CONTROL PANEL WITH PUMP

For flushing, de-airing and checking the cell connections, a hand-operated pump is required. A typical control panel fitted with a rotary hand pump is shown in Fig. 16.10, together with a low-pressure mercury manometer (see below). The pipework is concealed, as in most modern systems, but the connections are indicated schematically on the front panel.

The main components of the panel are as follows:

(1) Three outlets each with a valve, for connecting to a pressure source and to the test cell.

(2) Pressure gauge ('test' grade), usually 0–1200 or 0–1700 kPa.

(3) Rotary hand pump (also referred to as the control cylinder).

(4) Reservoir containing freshly de-aired water.

The arrangement is shown diagrammatically in Fig. 16.11.

A control panel of this type is suitable for the triaxial tests described in Chapter 18, and for the Rowe cell tests described in Chapter 22. One outlet (labelled (p)) on the panel is connected to the constant pressure system which provides back pressure. A second outlet (o) is connected to a volume change indicator and thence to the back pressure valve on the

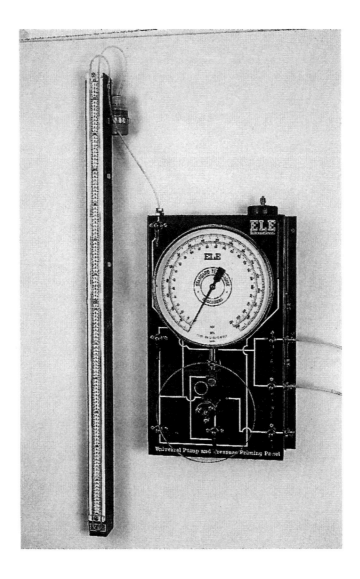

Fig. 16.10   *Pressure control panel, fitted with pressure gauge and rotary hand*
pump, and connected to mercury manometer

cell. The third outlet (k) is connected to the valve on the pore pressure transducer
mounting block. The control cylinder is then conveniently positioned for flushing, de-
airing and checking the back pressure and pore pressure lines.

The constant pressure system for providing cell pressure or diaphragm pressure is
connected through a separate gauge panel, which can be cross-connected to the control
panel. A typical installation of control panels and other items for a triaxial test is shown in
Fig. 16.12.

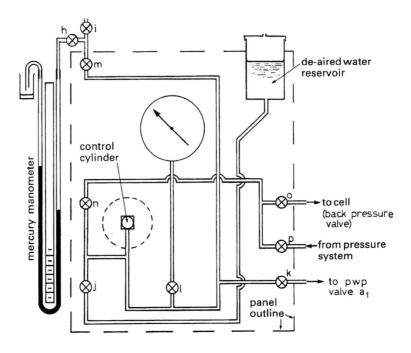

Fig. 16.11   Schematic diagram of pressure control panel shown in Fig. 16.10, with low-pressure manometer connected

Fig. 16.12   Layout of control panels, mercury manometer and volume-change indicator suitable for a triaxial test

Fig. 16.13   *Six-way pneumatic pressure control panel, connected to a pressure gauge and rotary hand pump panel*

A six-way pneumatic pressure distribution panel, suitable for three simultaneous triaxial tests, is shown in Fig. 16.13. This panel is used in conjunction with a control panel (Fig. 16.10).

CONTROL CYLINDER

The control cylinder is based on the original design by Bishop, which formed part of the early pore pressure measurement apparatus, and is shown diagrammatically in Fig. 16.14. The normal type is for pressures up to about 1700 kPa, while a cylinder of slightly smaller bore can provide higher pressures. It is a precision-made instrument for providing fine control of pressure, and must be watertight at low and high pressures.

MERCURY MANOMETER

For accurate measurement of pressures below 100 kPa a mercury manometer is an alternative to a low-range pressure gauge. It can be connected to the control panel at outlet (m) as shown in Fig. 16.11. The manometer was described in Volume 2, Section 8.3.5. A scale graduated in kPa is normally fitted, but older types may have a millimetres scale (1 mm difference equals 0.128 kPa). Provision of a mercury catch-pot is a necessary safety precaution because every operator at some time will forget to close the isolating valve when applying a pressure.

16.4   AXIAL LOADING SYSTEMS

16.4.1   Load Frame

The usual triaxial compression test is a 'controlled strain' test carried out in a motorised load frame. The motor drives the base platen at a predetermined rate of displacement (the platen speed), so that the specimen is deformed at a constant rate of strain. The resulting

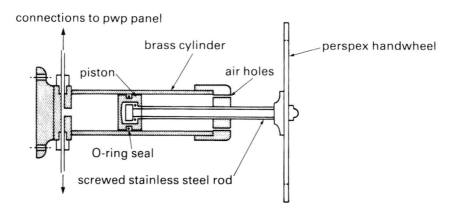

Fig. 16.14   Control cylinder (rotary hand pump) (after Bishop and Henkel)

axial force restraining the top of the specimen is observed at selected intervals of strain, or of time. Typical load frames for compression tests on soils were referred to in Volume 2, Section 8.2.3, Table 8.5. A general arrangement is shown in Fig. 16.15.

For effective stress triaxial tests, the duration of which may run into several days or even weeks, a drive unit giving displacement speeds down to about 0.0001 mm/minute is necessary, although a speed range of 0.5–0.001 mm/minute is sufficient for many tests. Modern machines are operated through an integral fully variable microprocessor-controlled drive unit, with key pad and LCD display. Steplessly variable displacement speed control down to 0.00001 mm/minute is possible. Rapid movement is also provided, for convenience when setting up and unloading.

The actual rate of displacement provided by any machine may vary with the applied force, and can be verified during a test by recording the time at which every set of readings is taken. The actual rate should not deviate by more than about $\pm 10\%$ of the required value. The rate of displacement of the lower end of the specimen relative to the upper end, from which the rate of strain is calculated, will be somewhat less than the platen speed because of the deformation of the force measuring device (which is very small for electronic force transducers).

16.4.2   Dead Weight Loading

In a 'stress-controlled' test the stress applied to the specimen is increased in increments at appropriate intervals by applying small increments of force, and the deformation resulting from each increment is observed. A convenient method of applying small stresses under controlled loading to a specimen in a triaxial cell is by means of a yoke hanger supporting slotted weights as indicated in Fig. 16.16. For higher stresses a beam supporting a balanced pair of load hangers may be more suitable.

The upward force on the cell piston from the pressure in the cell can be counteracted by fitting a separate pan to the hanger on which weights can be placed to exactly counterbalance the cell pressure; the main hanger weights then represent the net force applied to the specimen. The alternative is to take into account the mass of the hanger and weights, the cell pressure and the piston area when calculating the net applied force.

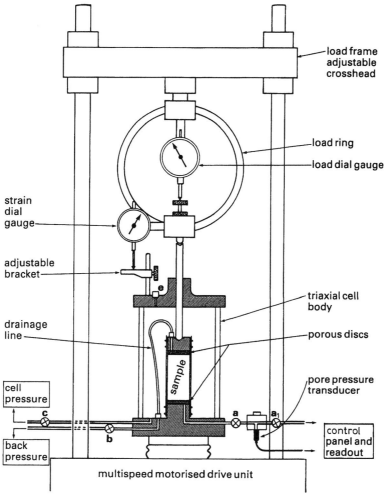

Fig. 16.15 *General arrangement of triaxial cell in load frame*

16.4.3 Lever Arm Loading

A lever-arm device can be used to apply larger compressive loads than can be conveniently applied by dead weights alone. The knife edges or bearings must be accurately positioned so that the precise level ratio is known. Weights should be applied particularly carefully because the effect of impact loading is magnified by the same factor as the lever ratio. The force applied to the specimen by the hangers and lever-arm alone need to be ascertained with care, either by measurement or by calculation. The principle is similar to that described for lever-arm loading of the shearbox, Volume 2, Section 12.5.5, stage 7(b).

16.4.4 Hydraulic Loading

For one-dimensional consolidation tests in which pore water pressures are measured, including tests on large diameter specimens, water pressure acting on a flexible diaphragm

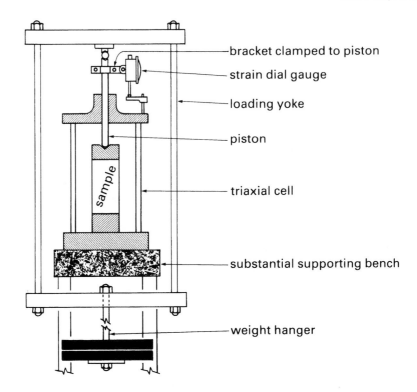

Fig. 16.16 Stress control for a triaxial test using dead weight loading

provides a loading system which offers many advantages over lever-arm loading. This principle is used in the hydraulic consolidation cell (the Rowe cell) and is described in detail in Chapter 22.

16.4.5 Pneumatic Loading

Regulated air pressure acting on a diaphragm, or in a triaxial cell chamber as frequently used in the USA, is an alternative method of using this principle. Unlike water pressure, an air pressure system cannot be used for measuring volume changes. Compressed air is hazardous, and should be used only in cells or vessels designed and tested for that purpose.

16.5 CONVENTIONAL MEASURING INSTRUMENTS

16.5.1 Categories of Instruments

Effective stress tests for both shear strength and compressibility require the use of many of the measuring instruments itemised in Volume 2, Section 8.2 together with additional devices and assemblies described in this section. Instruments are grouped according to the type of measurement for which they are used, as follows.

(1) Applied pressure

(2) Axial force

(3) Axial deformation

(4) Lateral deformation

(5) Pore water pressure

(6) Volume displacement

Only 'conventional' instruments are covered in this section, i.e. mechanical instruments such as pressure gauges and micrometer dial gauges which require manual observation and recording. This is to differentiate them from electronic instruments, whether used for automatic recording and processing or not, which are discussed in Section 16.6.

Calibration of all types of instrument is covered in Chapter 17.

16.5.2 Measurement of Pressure

PRESSURE GAUGES

Pressure gauges used in a soil laboratory were described in Volume 2, Section 8.2.1, and some were shown in Fig. 8.3. For most effective stress triaxial tests, gauges of 'test' grade, 250 mm diameter, reading 0–1000 or 1200 kPa, are desirable. Where higher pressures are used, gauges of the same standard but of the appropriate range are needed. Each pressure system should preferably have its own gauge permanently connected on line. The use, care and calibration of pressure gauges was covered in Section 8.3.4. Accurate calibration is particularly important, and calibration data should be displayed alongside every gauge. The accuracy and repeatability of gauges should be to within 0·5% of the full-scale reading within the range 10–90% of full-scale reading.

Application of vacuum, or partial vacuum, is often needed. A vacuum gauge can be of a smaller size because in most cases an indication of the degree of vacuum, not an accurate measurement, is sufficient.

PRESSURE DIFFERENCES

In many effective stress tests it is necessary to measure accurately a small difference between two relatively large pressures. Six ways of doing this are as follows.

(1) Reading the difference between two separate pressure gauges or transducers.

(2) Using a differential pressure gauge.

(3) Using two mercury manometers (up to 100 kPa only).

(4) Using a pressurised mercury indicator.

(5) Enclosing a pressure gauge in a pressure container.

(6) Using a differential transducer.

These methods are discussed below.

(1) *Separate gauge readings*

The gauges recording the two pressures should be very carefully calibrated against each other, at the same time as they are calibrated against a standard instrument. The same applies if two pressure transducers are used. Calibration data must be carefully applied to the readings.

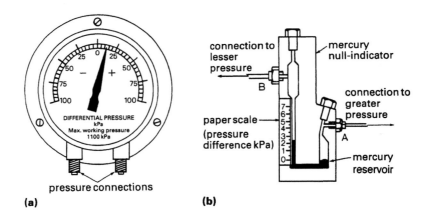

Fig. 16.17   Measurement of small pressure differences: (a) differential pressure
gauge, (b) pressurised mercury indicator

(2) Differential pressure gauge

This type of gauge, shown in Fig. 16.17 (a), has two connections, one for each pressure
line. It gives a direct reading of the difference in pressure, positive or negative, between the
two lines irrespective of the actual pressures. A pressure difference range of ± 25 or $50\,kPa$
is generally suitable. The overall safe working pressure should be not less than the
maximum available line pressure.

(3) Mercury manometer

Pressures up to $100\,kPa$ can be measured very accurately by using a mercury manometer
(Volume 2, Section 8.3.5). Pressure differences of equivalent accuracy can be determined if
each pressure line is connected to its own manometer. An ordinary mercury manometer
should not be connected between two pressure systems because the glass limbs are not
designed for pressures exceeding $100\,kPa$. However, if the mercury is contained in suitable
nylon tubing it can be pressurised from both limbs and used under pressures up to its safe
working pressure.

(4) Pressurised mercury indicator

The author had adapted a perspex mercury pore-pressure null indicator to measure
pressure differentials up to about $15\,kPa$ to an accuracy of $0.2\,kPa$. An early type of
indicator, about 200 mm long, was used, and could withstand pressures up to $1000\,kPa$.
The principle is shown in Fig. 16.17(b).

   The line at the higher pressure is connected to the mercury reservoir side at A, and the
other line to the top of the capillary tube section at B. Because of the very small bore of the
capillary the movement of the reservoir surface can be neglected over the range of use. A
paper scale with a zero mark at reservoir level is fixed to the side of the capillary. For a
'closed' manometer of this type every 8.1 mm represents a pressure difference of 1 kPa.
Markings at every $0.5\,kPa$ are enough for most purposes.

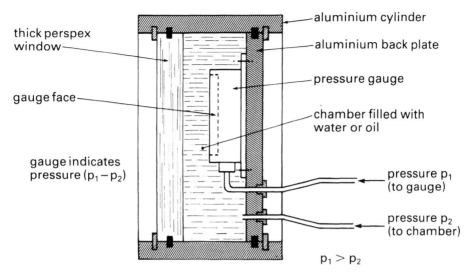

Fig. 16.18 *Measurement of pressure differences with an enclosed pressure gauge*
(after Lowe and Johnson, 1960)

Before opening the isolating valves the two pressures should be equal. Adjustment to the required differential is made by carefully increasing the pressure applied to A, or decreasing that at B. The valves should be closed when not in use. This device is very sensitive to small pressure fluctuations.

(5) Enclosed pressure gauge

An ordinary pressure gauge can be used for measuring differential pressure by enclosing it in a watertight container fitted with a perspex face for reading the dial (Lowe and Johnson, 1960). The container and gauge casing are filled with water or a suitable oil and connected to the lower of the two sources of pressure. A connection from the higher pressure is taken into the container and connected to the gauge in the usual way. The arrangement is shown in Fig. 16.18. The principle depends on the fact that the movement of the Bourdon tube operating the pointer is related to the difference between the internal and external pressures.

If oil is used as the surrounding fluid the effects of corrosion and deposits of water-dissolved salts on the gauge mechanism are avoided. An oil/water interchange vessel is needed.

The container should be properly designed and pressure-tested before fitting the pressure gauge. Care must be taken to prevent the pressure applied to the container exceeding that applied to the gauge.

(6) Differential pressure transducer (see Section 16.6.3)

16.5.3 Measurement of Axial Force

LOAD RINGS

Conventional load rings for triaxial compression tests were referred to in Volume 2, Section 8.2.1, and typical rings were shown in Fig. 8.2. The general arrangement of a triaxial cell and load ring mounted in a load frame is shown in Fig. 16.15.

The load ring should be selected with regard to the strength of the specimen to be tested, bearing in mind that compressive strength increases much more rapidly with increasing effective stress (especially in drained tests) than it does with increasing total stress in quick-undrained tests. The ring must be of more than enough capacity to cover the maximum force which the specimen can sustain, but with enough sensitivity to provide reasonable discrimination in the early stages of loading. Readings of force that are used for deriving soil parameters must lie within the calibration range of the ring. Properties of typical load rings were summarised in Volume 2, Table 8.3, and their care and use, and the importance of calibration, were discussed in Section 8.3.3.

In effective stress tests the use of a mean calibration factor might not be accurate enough. A calibration chart of the type shown in Fig. 8.32 should then be used for calculating the applied stress from each force reading. This is mandatory in BS 1377 : 1990 when the calibration factor lies outside the range 2% higher to 2% lower than the mid-scale calibration factor (see Section 17.2.4).

Good practice requires that the force measuring device should have an accuracy equivalent to Grade 2.0 of BS EN 10002–3, or better, i.e. the maximum deviation from linearity should not exceed 2% of the verification force.

16.5.4 Measurement of Axial Deformation

DIAL GAUGES

Micrometer dial gauges of the types described in Volume 2, Sections 8.2.1 and 8.3.2 are used for the measurement of axial deformation in triaxial tests and vertical settlement in consolidation tests. Gauges reading to 0.01 mm are generally required, with a travel of 12–50 mm depending on the size of specimen and magnitude of deformation expected.

The dial gauge is clamped to a bracket on the loading piston, or on the lower end of the load ring close to the point of contact with the piston. The stem of the gauge rests on an adjustable bracket supported by the triaxial cell, as indicated in Fig. 16.15. Movement of the cell relative to the piston is equal to the movement of the cell pedestal relative to the top cap, i.e. the change in length of the specimen.

VERNIER TELESCOPE

Axial deformation during isotropic consolidation, when there is no load applied to the piston to maintain contact with the specimen, is difficult to measure in the manner described above. Bishop and Henkel (1962) refer to the use of a vernier telescope or cathetometer mounted at a fixed point for observing the vertical movement of a scribe mark on the specimen top cap. Its application was mainly for precision measurements in research.

DIRECT MEASUREMENT ON SPECIMEN

Measurements made as described above (sometimes referred as 'end-cap' measurements) include any deformation due to bedding of the end caps and piston. These errors are small for specimens of compressible soil but can be appreciable for stiff soils. They can be eliminated if changes in distance between two fixed points on the specimen itself, such as the third-points, are measured. Four examples of ways of achieving this are with the use of:

(1) Demec strain gauge

(2) Submersible transducer(s)

(3) Electrolytic level

(4) 'Hall effect' transducers.

Method (1) is used mostly for uniaxial tests on concrete or rock cores, but can be used in a triaxial cell filled with oil as the pressure fluid. Methods (2), (3) and (4) were initially developed for research purposes, but procedures such as these are now considered to be the only reliable way of measuring small strains. Small strain measurement is not covered in this volume.

The Demec strain gauge consists of an invar bar with a fixed conical locating point at one end and pivoting point at the other (Fig. 16.19). The pivoting movement actuates a sensitive dial gauge, effectively doubling its sensitivity. To take a reading, the assembly is held so that the gauge points locate into holes on discs set into or bonded on to the surface of the specimen at precisely located positions along the axis. The relative movement of the axial gauge length defined by the target points is indicated by the dial gauge. This device is suitable for very stiff soils, rocks and concrete.

The use of submersible transducers is described in Section 16.6.5.

16.5.5   Measurement of Pore Water Pressure

CURRENT PRACTICE

The pressure of the pore water within the test specimen is normally measured by means of a pressure transducer, mounted as close as possible to the non-draining face of the specimen (see Fig. 16.15). Pressure transducers are described in Section 16.6.3.

MERCURY NULL INDICATOR — HISTÓRICAL

Before the advent of pressure transducers it was necessary to develop a special device for measuring pore pressure. The pressure of the relatively small volume of water in the pore spaces of a soil specimen cannot be measured directly by a pressure gauge or mercury

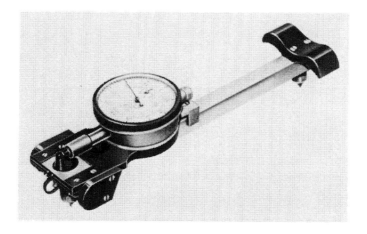

Fig. 16.19   Demec strain gauge

manometer because these devices require an inflow of water to actuate them. Flow of water from the specimen would appreciably change the magnitude of the pressure being measured, and would also introduce a time lag in the attainment of a steady reading. These difficulties were overcome by using the 'null method' which was first used for this purpose by Rendulic (1937). A true no-flow condition was ensured by maintaining a water–mercury interface at a constant level in a capillary tube connected to the base of the specimen, by adjustment of pressure with a manually operated screw plunger. This principle was developed at Imperial College, London, from 1956 and is described by Bishop and Henkel (1962) and by Bishop (1960).

The original stainless steel and glass 'null indicator' as designed by Bishop was difficult to set up and adjust. A simplified form of the device made of perspex was designed by T. G. Clark, and could be tilted to allow water to by-pass the mercury when de-airing. It was connected to the port in the triaxial cell base by a short length of small-bore rigid nylon tubing. Subsequent modifications at Imperial College and ELE led to the types of design shown in Fig. 16.20 which were widely used until the introduction of suitable pore pressure transducers.

The body of the type of null indicator shown in Fig. 16.20 (a) is of perspex blocks which are bonded together after drilling the bores. The capillary section is 1.5 mm diameter, and is connected to a mercury reservoir. For filling, de-airing and flushing the indicator can be rotated through 90° about the pivot bolt to the horizontal position, where it is clamped by a wing-nut. Mercury then drops clear of the connecting passages, as shown in Fig. 16.20 (b).

An improved type of null indicator, shown in Fig. 16.20(c), is made of a single acrylic block and can be mounted directly on to the pore pressure outlet from the triaxial cell. A double O-ring seal allows for rotation, eliminating the need for tubing between the mercury thread and the cell.

CONTROL PANEL

The layout of a control panel as originally designed by Bishop for use with the null indicator for the measurement of pore pressure in triaxial tests is shown diagrammatically in Fig. 16.21. The rotary hand pump (control cylinder) (described in Section 16.3.5) provided the back-up pressure for maintaining the null position of the mercury indicator.

The null indicator and its control panel were rendered obsolete for measurement of pore pressure by the introduction of pore pressure transducers, which require an insignificant movement of water for their operation. The manual method of measuring pore pressure is not described in this book. However, the rotary hand pump and connecting line to the cell were retained for flushing and de-airing the transducer mounting and base pedestal connection, as seen in Fig. 16.12 and shown diagrammatically in Fig. 18.4.

OTHER USES OF THE NULL INDICATOR

There are some situations in which the null indicator might be useful, three of which are referred to elsewhere. These are:

(a) checking for leaks in a pressure system (Section 18.3.2);

(b) measurement of very small volume changes (Section 20.3.4);

(c) measurement of small differences between two high pressures (Section 16.5.2 (4)).

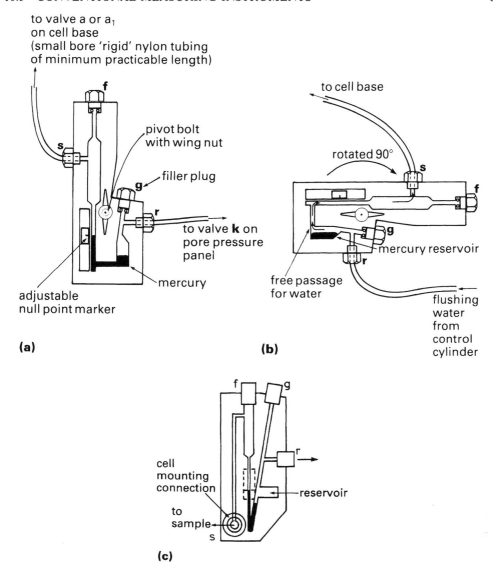

Fig. 16.20   *Commercial types of mercury null indicator: panel mounting type, (a) in working position; (b) tilted to allow for flushing; (c) modified design for mounting on cell outlet*

If a mercury null indicator is to be used at all, it must first be properly installed and prepared. A procedure for de-airing and checking is given in Section 16.8.6.

16.5.6   Measurement of Volume Change

HISTORICAL NOTE

The original methods devised by Bishop at Imperial College in 1956 for measuring a small flow of water under pressure out of a specimen or into the triaxial cell depended upon the use of mercury to provide an interface between two liquids. A similar apparatus, using

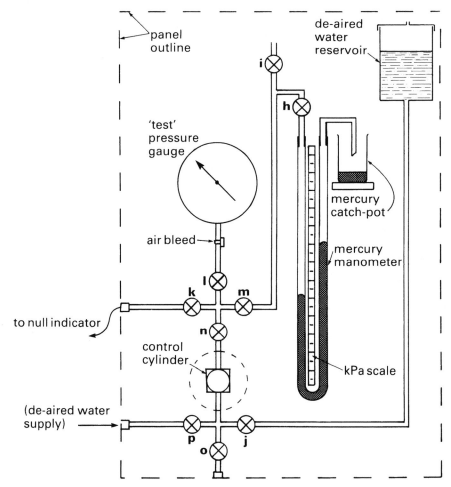

Fig. 16.21   *Control panel for mercury null indicator (as originally designed by Bishop)*

paraffin as the second liquid, was developed by Bishop and Donald (1961). The twin-burette unit used extensively today is based on the same principle.

TWIN-BURETTE VOLUME CHANGE INDICATOR

A typical twin-burette unit is shown in Fig. 16.22, and a diagrammatic section is given in Fig. 16.23. The apparatus consists of:

   2 glass burettes, each mounted within a clear acrylic tube.

   Valve system for flow reversal and by-passing the burettes.

   Paraffin containing red dye.

Two tubes are connected in series in order to facilitate reversing the direction of flow through the burettes when the cumulative flow exceeds the burette capacity. Flow is reversed by means of the lever-operated rotating plunger valve shown on the left in Fig. 16.23. This type of valve enables the direction of flow to be reversed almost

Fig. 16.22   *Twin-burette volume-change indicator*

instantaneously. A by-pass arrangement enables large volumes of water to be passed continuously into the cell system for flushing and de-airing. An alternative method of flow reversal is the use of two three-way valves connected as shown in Fig. 16.24, both of which must be operated at the same time.

Mounting the burettes within outer tubes avoids them being subjected to a net internal pressure, therefore their calibration is independent of pressure. It is also an important safety factor. Connections to the source of pressure and triaxial cell are made to the inner burettes.

Readings should be taken of the interface meniscus inside a burette, not in the outer tube. Recorded observations should indicate whether the left-hand or right-hand burette was read.

The capacity of the burettes should be related to the size and nature of the test specimen. For small specimens, or specimens of stiff soil, small diameter burettes should be used to obtain a suitable resolution of readings. For large or compressible specimens, and for permeability tests, sufficient capacity is desirable to avoid the need for multiple reversals of flow in the burettes. Maximum sizes and graduation intervals of burettes, as stated in BS 1377, are as follows.

Specimen diameter (mm)	Burette capacity (ml)	Graduation intervals (ml)
38	50	0.1
100	100	0.2

Larger burettes might be necessary for very compressible soils such as peats. Smaller burettes, e.g. reading to 0.05 ml, should be used where a higher degree of resolution is required.

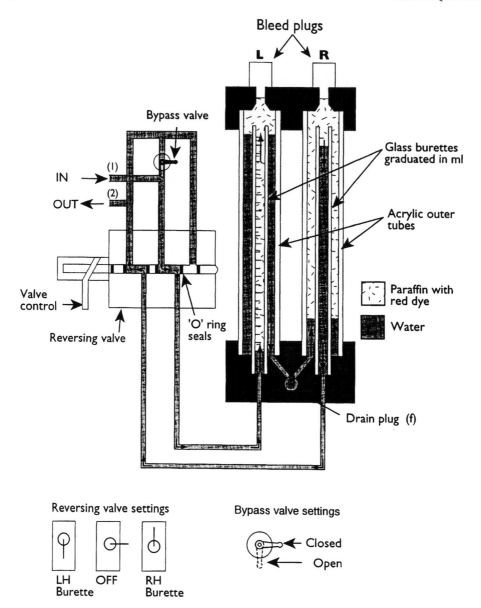

Fig. 16.23 *Principle of operation of twin-burette volume-change indicator*
(*courtesy ELE International Ltd*)

Where greater than normal precision of measurement of pressures is essential (e.g. in determining small differential pressures), the change in pressure caused by the movement of the interface between the paraffin and water should be taken into account. A change in relative level of 100 mm between the interfaces in a single tube would give a pressure change of about 0.2 kPa (assuming that the density of paraffin is 0.8 g/ml). This value is doubled for a twin-tube apparatus.

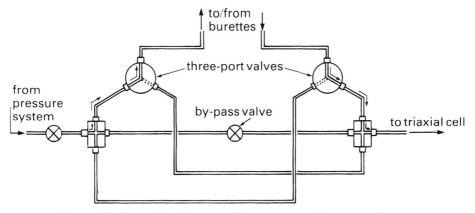

Fig. 16.24   *Reversal of direction of flow using two three-port valves*

Installation and filling of this type of volume change indicator are described in Section 16.8.3.

16.6   ELECTRONIC INSTRUMENTATION

16.6.1   General

SCOPE

In this section electronic instruments are described and some of their basic principles are outlined. More detailed information will usually be available from the manufacturer's operating instructions or handbooks.

The collective term 'electronic instruments' is used here to cover sensitive electronic devices (known as 'transducers') used for measuring certain physical quantities, and the signal-conditioning and read-out equipment used with them. Changes in applied physical conditions induce small changes in an electrical output signal, which are amplified electronically to provide a measure of those changes. Measurements can be shown in analogue or digital form, or as part of a screen display, and can be stored for subsequent computer processing.

HISTORICAL NOTE

The earliest electronic instruments gave a measurement of the modified electrical output on a voltmeter which then had to be converted to the appropriate units from calibration data. Modern systems automatically convert the electrical signal to give displays directly in engineering units (e.g. kPa; kN; mm).

During the last 20 years many of the conventional measuring instruments which have to be individually observed for taking and recording readings manually have been giving way to electronic instrumentation. This development not only allows data from several sources to be displayed at a central location, but also enables the data to be automatically recorded. A 'scanner' incorporating a timing device automatically selects each output channel in rapid sequence without the need for manual switching. One of the earliest developments which enabled the data to be processed by a micro-computer and the final test results to be printed out in both tabular and graphical form was described by Davison and Hills (1983).

IMPORTANCE OF CALIBRATION

There is a tendency to accept digitally displayed readings without question as being authentic and accurate, but their accuracy is no greater than the accuracy with which the measuring devices were calibrated. This applies whether or not conversion factors and calibration corrections are built into the electronic circuitry. Calibration of electronic equipment is just as important as the calibration of traditional instruments, but electronic instruments need to be checked and recalibrated more frequently than their conventional counterparts because they are subject to electrical noise, drift and other errors.

BASIC TYPES OF TRANSDUCERS

Four basic types of transducer are used in effective stress tests on soils, or incorporated in special measuring devices, namely

Pressure transducers

Load transducers

Displacement transducers

Volume change transducers

The principles of these devices, together with brief notes on their use and application, are outlined in Sections 16.6.3 to 16.6.6.

Power supply and ancillary equipment for transducers are described in Section 16.6.7. Automatic data-logging systems and computer-controlled testing are referred to in Sections 16.6.8 and 16.6.9. Some of the essential requirements and conditions for successful use of electronic instrumentation are outlined in Section 16.6.10.

ADVANTAGES OF ELECTRONIC INSTRUMENTATION

(1) Readings are displayed directly in engineering units, needing no conversion factors.

(2) Transducer outputs can be recorded on paper, disk or tape for subsequent analysis (particularly important when a test is left unattended).

(3) Digital display reduces the possibility of errors that can be made when reading gauges.

(4) One display unit can monitor all the data from a test, or data from several tests.

(5) Computer processing of test data can be carried out automatically.

(6) Control of tests by feedback from the computer is also feasible.

16.6.2 Definitions

Some of the terms used in electronic instruments are defined below.

A/D CONVERTER (or ADC) Analogue to digital converter. Electronic device converting a continuous electrical input to an equivalent digital output.

ADU Autonomous data-acquisition unit. A programmable interface unit between transducers and micro-computer which can capture transducer signals at predetermined intervals and store the data for subsequent processing.

AMPLIFIER Electronic device for magnifying a current or voltage.

CHANNEL An electrical pathway, usually connected to a transducer.

DATA ACQUISITION SYSTEM A system for gathering and treatment of information, usually measurements from transducers.

DATA LOGGER A device for gathering and storing data, usually from transducers.

DATA PROCESSOR Means of manipulating information.

DATA STORAGE Means of storing information.

DIGITAL DISPLAY Visual means of displaying actual numerical values.

ELECTRONIC INSTRUMENTATION In this context, a measurement system based on electronic components.

HARDWARE All the electronic and mechanical components that make up a computer system.

INTERFACE Means of allowing a computer to communicate with external devices.

LVDT Linear variable differential transformer. A device on which displacement transducers of high-linearity are based.

MICRO-COMPUTER A computer based on a 'microprocessor'.

OUTPUT SIGNAL Electrical value obtained from a measuring device.

PC Personal computer.

PERIPHERAL A device such as a printer or disk drive which is external to the computer.

PROGRAM A set of instructions to tell the computer to execute a sequence of tasks.

READ-OUT Means of displaying measurements.

SCANNER A component, usually of a data logger, which sequentially takes readings from a number of channels.

SIGNAL CONDITIONING UNIT Device for accepting a range of electrical inputs and producing electrical outputs within a common range.

SOFTWARE The programs required to operate the computer.

STABILISED POWER SUPPLY Electrical supply which remains constant despite changing imposed load or electrical input.

STRAIN GAUGE A system of resistances, bonded to or integral with a surface, whose resistance changes with deformation of that surface.

TRANSDUCER A device which converts a change in physical quantity to a corresponding change in electrical output.

VDU Visual display unit. A screen similar to a television screen on which instructions and data can be displayed in words, numbers or in graphical form.

16.6.3 Pressure Transducers

PRINCIPLE

A pressure transducer consists of a thin diaphragm on which electric strain gauge circuits are bonded or etched, mounted in a rigid cylindrical housing. A porous filter protects the diaphragm, and allows it to be influenced by the pressure of water. The resulting diaphragm deflection, though extremely small, gives rise to an out-of-balance voltage which is amplified and converted to a digital display in pressure units. The response time of the transducer depends on the volume of water required to make up for the small diaphragm movement, and the volume displacement per unit change of pressure is known as the 'compliance'. A typical 'stiff' pressure transducer of 1000 kPa capacity has a compliance of less than $2 \times 10^{-7} \, \text{cm}^3$ per kPa (i.e. $0.02 \, \text{mm}^3$ for a pressure change of 100 kPa), which is small enough to allow direct connection to a soil specimen for continuous measurement of pore water pressure. The magnitude of the displacement has negligible effect on pore pressure for specimens down to 38 mm diameter.

However, the presence of air in the system, or the use of a flexible connection between the specimen and the transducer, increases the compliance of the system considerably. A pore pressure transducer should therefore be mounted as close to the specimen as possible, either within the base pedestal or on the cell at the outlet from the cell base. Thorough de-airing of the system, and the absence of even the slightest leak, are essential if the very rapid response of a transducer to changes of pressure is to be relied upon.

The rigidity of the pore pressure system as a whole, in relation to the volume of the test specimen, is usually the important factor. In ASTM D4767 the rigidity requirement for the volume change of the enclosed system is given by the relationship

$$\frac{\Delta v}{V} < 3.2 \times 10^{-6} \times \Delta u$$

where Δv is the change in volume (mm^3) of the enclosed pore pressure measuring system (transducer and connecting parts) caused by a change in pore pressure of Δu kPa, and V is the total volume of the test specimen (mm^3).

The resolution of a pressure transducer is generally rather better than that of a good pressure gauge of equivalent range. If properly calibrated, its accuracy should be within 1% of indicated reading over the greater part of its range.

CALIBRATION

Every transducer must be calibrated against known pressures in order to determine the relationship between applied pressure and electrical output. Digital readout is normally displayed in kPa.

Calibration procedures are described in Section 17.2.3.

GENERAL APPLICATION

Pressure transducers have replaced gauges for applications such as measuring pore water pressure, and in automatic logging systems. However, pressure gauges are still desirable for continuous visual indications of pressures. A typical electric pressure transducer, fitted to a triaxial cell, was shown in Volume 2, Fig. 8.17.

Pressure transducers need to be readable to 1 kPa or better. They should always be mounted with the diaphragm facing upwards to facilitate de-airing. A transducer for measuring pore water pressure should be fitted on to a mounting and de-airing block which has four connecting ports, as shown in Fig. 16.25. The transducer is screwed into the lower connection (A), with the diaphragm upwards to reduce the possibility of trapping air. The two side ports (B and C) are connected to the source of pressure and the

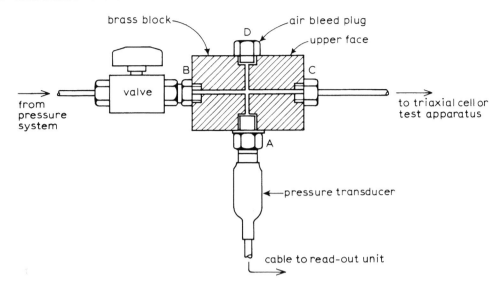

Fig. 16.25   *Mounting block for pressure transducer*

valve on the test apparatus respectively. The top port (D) is fitted with a plug and is used as an air bleed.

Three pressure transducers fitted to a triaxial cell, for measurement of pore pressure, back pressure and cell pressure, can be seen in Fig. 16.26.

MEASUREMENT OF PORE PRESSURE

Use of a pressure transducer is now standard practice for the measurement of pore water pressure. Essential requirements were discussed by Bishop and Green (1969). The transducer mounting block (Fig 16.25) is fitted directly to the triaxial or consolidation cell at the outlet port (valve a in Fig. 16.15) from the base pedestal, ensuring that the transducer is as close to the specimen as possible.

The other side of the block is connected to a second valve a_1, and thence to a control panel such as that shown in Fig. 16.10. The rotary hand pump (control cylinder) on the panel (Fig. 16.11) is used for de-airing the system as described in Section 16.8.3, and a mercury null indicator, if included on the panel, can be used for checking purposes. One panel can serve several transducers. The panel and control cylinder can also be used for checking the transducer calibration, with valve a closed and a_1 open. During a test, valve a_1 is kept closed and valve a is open.

DIFFERENTIAL PRESSURE TRANSDUCER

The most convenient and accurate method of measuring pressure differences is with a differential pressure transducer. It is based on the same principle as an ordinary pressure transducer but has two diaphragms, the circuits from which are electrically connected so that the difference between their outputs is measured. A differential transducer costs about the same as two transducers of similar capacity.

Fig. 16.26   *Triaxial cell for 100 mm diameter specimen, in load frame, fitted with pressure transducers for measuring pore pressure, back pressure and cell pressure* (*courtesy Sage Engineering Ltd*)

16.6.4   Force Transducers

PRINCIPLES

Three types of electrical force measuring instrument for use in place of conventional load rings with dial gauges were referred to in Volume 2, Section 8.2.6, namely:

(1) Ordinary load ring fitted with a displacement transducer which replaces or supplements the usual dial gauge (Fig. 8.14). Load rings of this type are available up to 500 kN capacity.

(2) Strain gauge force transducer (Fig. 8.15) in which deformation due to load is measured by electrical resistance strain gauges.

(3) Submersible force transducer fitted inside a triaxial cell chamber (Fig. 8.16). Compressive or tensile loads can be measured, and typical load capacities are 3 kN, 5 kN, 10 kN and 25 kN. Instruments of types (2) and (3) are collectively referred to as force transducers.

In a force transducer the applied force is transmitted to a metal web or diaphragm to which electrical resistance strain gauges are bonded. The resulting strains in the web cause changes in the electrical resistance of the strain gauges, and hence small voltage changes, which are amplified and converted to a digital display in force units (newtons or kilonewtons).

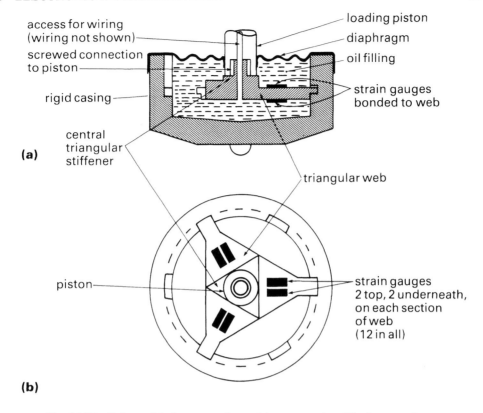

access for wiring
(wiring not shown)
screwed connection
to piston
rigid casing
central
triangular
stiffener

(a)

loading piston
diaphragm
oil filling
strain gauges
bonded to web

triangular web

piston

strain gauges
2 top, 2 underneath,
on each section
of web
(12 in all)

(b)

Fig. 16.27   *Submersible force transducer:* (*a*) *cross-section,* (*b*) *plan on web*

SUBMERSIBLE FORCE TRANSDUCER

The principle of the submersible force transducer shown in Volume 2, Fig. 8.16 is illustrated in Fig. 16.27. The chamber in which the triangular web is fitted is filled with oil and is sealed against the ingress of water by a diaphragm, which must be treated with care to avoid damage.

The force reading is not affected by changes in cell pressure. Cell pressures up to 1700 kPa can be used. The arrangement of the twelve strain gauges gives an output reading related only to the axial component of the applied force. Any transverse component is rejected, and the effect of eccentricity of loading is insignificant. The transducer can be used for measuring either compression or tension within its working range.

The outstanding advantage of the submersible transducer is the elimination of the effects of piston friction from the force measurements. An early device of this kind, using a displacement transducer to measure deflection, was described by Bishop and Green (1965). The type shown in Fig. 16.27 was designed at Imperial College specially for triaxial tests, and can be supplied with a triaxial cell as a complete assembly. The power supply cable is taken out through the hollow loading piston, and should have enough slack to allow for movement of the piston into the cell during a test.

CALIBRATION

A transducerised load ring or force transducer needs to be calibrated against an instrument complying with a recognised standard (see Chapter 17), in a manner similar to that used for an ordinary load ring of similar characteristics.

The accuracy and repeatability of a properly calibrated submersible load transducer should be better than that of a conventional load ring. A high quality commercial transducer can be expected to be accurate to within about 1%, and repeatable to within 2%, of indicated force, within the calibration range.

Calibration data from load transducers can be incorporated in the program of an automatic data-logging system.

16.6.5   Displacement Transducers

PRINCIPLE

The usual type of transducer for measuring displacement is known as a 'linear variable differential transformer' (LVDT). This device consists of electrical coils in a cylindrical casing, through the axis of which a metal rod (the 'armature') can slide. Movement of the armature changes the inductance of the windings, which is measured electrically and converted to a digital display in units of displacement (mm or μm). Transducers are obtainable for measuring ranges of displacement from a few millimetres to 600 mm.

CALIBRATION

Displacement transducers should be calibrated systematically, as indicated in Section 17.2.5. Their response is practically instantaneous. The accuracy and resolution of the LVDT type is similar to that of a comparable dial gauge, resolution generally being better than 1 part in 10,000. The last digit of a read-out might suggest a resolution 10 times greater than this, but that digit should be treated with caution.

GENERAL APPLICATIONS

Displacement transducers (LVDTs) are used for the measurement of axial deformation of specimens in triaxial and consolidation tests. Transducers are available that can be submerged in oil or water, and therefore used inside a triaxial cell. Miniature transducers can be fitted to other measuring devices such as lateral strain gauges and volume change transducers (Section 16.6.6).

AXIAL DISPLACEMENT

For the measurement of axial deformation in triaxial and consolidation tests, transducers with ranges of 15, 25, 50 and 100 mm are normally used in the same way as dial gauges.

A displacement transducer mounted on a bracket for attaching to a triaxial cell was shown in Volume 2, Fig. 8.13. The mounting bracket must be specially designed for the purpose so as to clamp the body firmly, but not so tightly as to cause distortion. Otherwise setting up is the same as for a dial gauge, and correct alignment is essential.

Use of submersible displacement transducers inside a triaxial cell enables the local axial strain of a gauged length of specimen to be measured. A pair of submersible displacement transducers are mounted diametrically opposite each other on a spring-loaded split collar located on a 'target' bonded on to the specimen. The transducer spindles rest on a second collar located at a precisely known distance from the first. This procedure was described by Brown and Snaith (1974). Small submersible transducers can alternatively be supported by stainless steel dowels inserted and sealed into the specimen.

The weight supported by the specimen can be minimised by using displacement transducers mounted in pairs on a post fixed to the base of the triaxial cell, as indicated in Fig. 16.28. Both collars then support only the transducer stems (Chamberlain, Cole and Johnson, 1979).

16.6.6   Volume Change and Flow Transducers

DEVELOPMENT

Automatic measurement and recording of volume change was the most difficult to accomplish of the measurements required for compression tests on soils, and was the last to be developed. Numerous methods have been tried, the earliest of which was probably that by Lewin (1971) at the Building Research Station, using a mercury null displacement unit. This device actuated an electric relay connected to a servo-mechanism which drove a piston to maintain equilibrium, and also operate a displacement transducer. Mercury-pot devices suspended from springs and actuating displacement transducers were used by Rowlands (1972), and Darley (1973). A spring lever fitted with strain gauges, supporting two mercury pots, was described by Klementev (1974). A device limited to low pressures

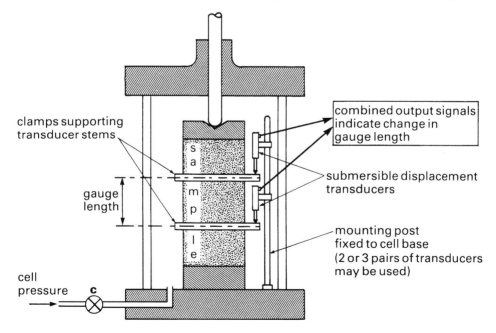

Fig. 16.28   *Measurement of axial deformation using pairs of submersible displacement transducers (after Chamberlain, Cole and Johnson, 1979)*

(up to 16 kPa) using small mercury pots and the output from an electric balance was described by Marchant and Schofield (1978). Another approach, based on variable capacitance due to the movement of a column of electrolyte, was reported by Sharpe (1978). The use of a rolling diaphragm was described by Menzies (1975).

ROLLING DIAPHRAGM TRANSDUCER

A commercial volume-change transducer based on the use of a rolling diaphragm is shown in Fig. 16.29. The principle of operation is illustrated in Fig. 16.30. A brass piston incorporating a Bellofram rolling seal in a pressure chamber is connected to the armature of a long-travel submersible displacement transducer, readings from which are converted to volumetric units (cm³). The displacement capacity is 80 cm³, but this can be extended by the use of reversing valves similar to those fitted to the conventional twin-burette gauge. The unit can be used at a pressure up to 1700 kPa. Its resolution is about 0.1 cm³.

IMPERIAL COLLEGE TRANSDUCER

A volume-change transducer of the type shown diagrammatically in Fig. 16.31 (a) was developed at Imperial College, London. This device consists of a hollow thick-walled brass cylinder containing a 'floating' piston attached to a Bellofram rolling seal at either end. Movement is measured by an externally mounted displacement transducer. Units of 50 cm³ and 100 cm³ capacity are available, reading to 0.01 and 0.02 cm³ respectively. They can be used at pressures up to 1400 kPa, but a minimum pressure of about 30 kPa is needed to expand the Bellofram seals.

Fig. 16.29 *Volume-change transducer, rolling diaphragm type*

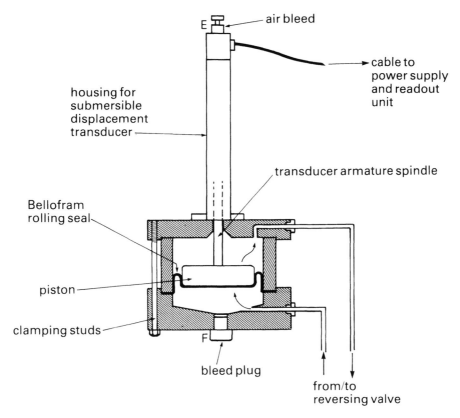

Fig. 16.30   *Principle of rolling diaphragm volume-change transducer*

A very small excess pressure is needed to support the mass of the piston. The error due to this effect can be eliminated by connecting the pressure gauge between the device and the triaxial cell, as shown in Fig. 16.31 (b). Changes in volume due to pressure changes are very small, and creep effects appear to be negligible. The gauge is not reversible, and at the end of travel of the piston the unit has to be isolated from the test cell so that the cylinder can be re-charged or drained, enabling the piston to be returned to the starting position.

One advantage of this type of transducer is that pressurised air can be applied to the lower end as an alternative to water. The unit then serves as a pressurised air/water interchange and a separate air/water bladder cell is not necessary. However, the continuous volumetric movement available is limited to the capacity of the transducer body.

FLOW PUMP APPARATUS

A new type of volume-change device, for use in permeability tests in which a constant rate of flow is maintained, was designed at the University of Newcastle upon Tyne by Araruna, Harwood and Clarke (1995). The device is shown diagrammatically in Fig. 16.32. It makes use of a flow pump, sensitive volume-change units and a differential pressure transducer.

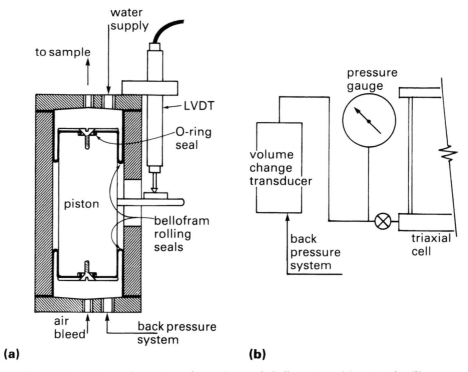

Fig. 16.31   *Volume-change transducer, Imperial College type:* (*a*) *principle,* (*b*)
layout of connection to triaxial cell (*courtesy Imperial College, London*)

The two flow tubes provide for a coarse setting for the consolidation stage, and a fine
setting with better resolution for permeability measurements. In operation a controlled
rate of flow, instead of a constant pressure difference, is applied to the specimen by means
of the flow pump. The system enables small hydraulic gradients and low rates of flow to be
measured with resolutions much better than can be achieved with conventional equipment.
Further details were given by Araruna, Clarke and Harwood in the Proceedings of the ICE
Site Investigation Conference (1995).

CALIBRATION

The calibration of volume change transducers is described in Section 17.2.6. Calibration
should be repeated under several pressures covering the normal working range.

16.6.7   Power Supply and Ancillary Units

In order to provide an output signal, electric transducers of all kinds need to be electrically
energised with a stabilised supply, typically 10 V d.c. The output voltage, which may be in
the range -5 V to $+5$ V, or 0 to 100 mV, depending on the type of transducer, must be
amplified and suitably conditioned if it is to actuate a digital display panel.

   The essential components of a transducer read-out system are as follows.

(a)

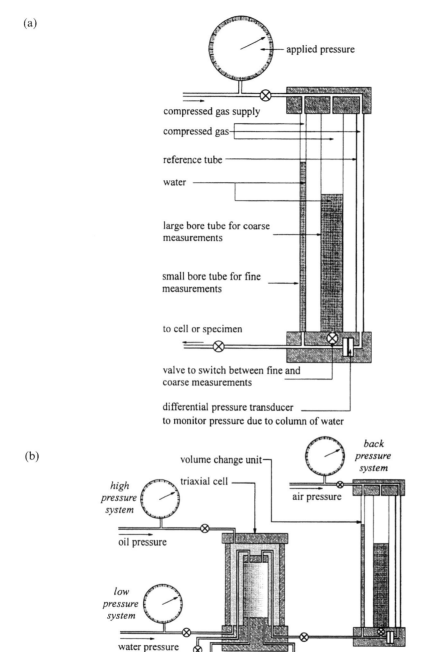

(b)

Fig. 16.32 *System developed at the University of Newcastle for maintaining constant rate of flow in permeability tests. (a) Details of volume-change unit (the height is about 800 mm). (b) Arrangement as used in triaxial permeability tests (courtesy Dept of Civil Engineering, University of Newcastle)*

Power supply unit:

  mains voltage input

  d.c. output to amplifier and other components

  stabilised low-voltage d.c. output to each transducer.

Low-level calibrated amplifier to step up and convert the output signal from each transducer to give readings in engineering units.

Switching system to connect each output channel to the display unit as required.

Analogue-to-digital converter and amplifier to provide digital signals.

Digital display panel.

All the above are incorporated in a typical 'signal conditioning unit' and the plug-in module connected to each transducer. One unit may serve several transducers, either of the same type (e.g. see Volume 2, Fig. 8.19), or of mixed types (e.g. Fig. 8.18). Both these examples are of 9-channel units. A 6-channel unit for monitoring reading from drained triaxial compression tests, displaying axial strain, axial load, cell pressure, back pressure, pore pressure and volume change, is shown in Fig. 16.33. Smaller units serve one, two or three channels.

A schematic representation of the system referred to above is shown in Fig. 16.34. The heavy dashed outline encloses the components of a typical commercial 'read-out' unit designed for soil testing.

A voltage stabiliser on the main input is essential for obtaining reliable data. This appliance smooths out mains voltage fluctuations which can give very erratic transducer outputs. The whole system should be one that is designed by specialists for the purpose of testing soils.

16.6.8   Automatic Logging and Data Processing

In addition to providing a display in digital form, test data from transducers can be recorded in several ways:

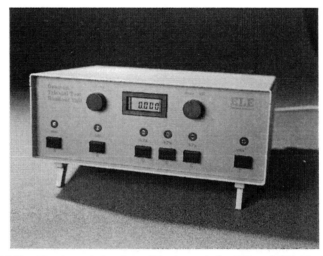

Fig. 16.33   *Six-channel signal conditioning unit for CU and CD triaxial tests*

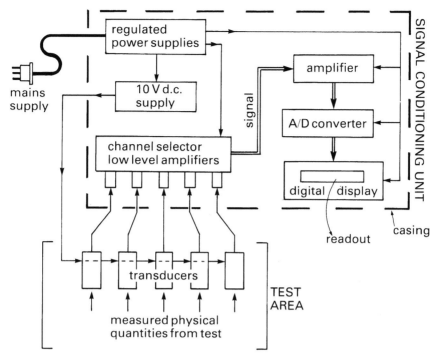

Fig. 16.34   *Schematic representation of signal conditioning unit*

Continuous print-out

Punched paper tape

Magnetic tape

Stored in computer memory or on disk for automatic processing

The first three methods are now virtually obsolete, and are not described here.

When several transducers are used at the same time, as in effective stress triaxial tests, a 'scanner' provides the switching facility which enables each transducer to be read in turn at predetermined time intervals. Electronic scanning is done so quickly that a set of readings can all be taken virtually at the same instant, and recorded. The system can be kept running to obtain readings when the laboratory is left unattended, such as overnight. Recorded data are fed into a computer for processing and thence to a printer or plotter for presentation of results. Data may also be abstracted manually when required, for producing graphs and performing calculations.

A modern automatic system which remains under the operator's control is the ELE DS6, shown in Fig. 16.35 linked to an effective stress triaxial test. The system comprises an interface unit, a desk-top computer with screen display, a printer/plotter and a disk recorder. The arrangement is shown schematically in Fig. 16.36.

The key to this system is the programmable interface unit for scanning the transducers, which enables the computer to be devoted to storage and analysis of test data. The computer initiates the programmed screen displays which guide the operator, and controls the printer/plotter which reproduces the results in both tabular and graphical form, ready for photocopying for insertion into a report. Once a test stage has been started, data are

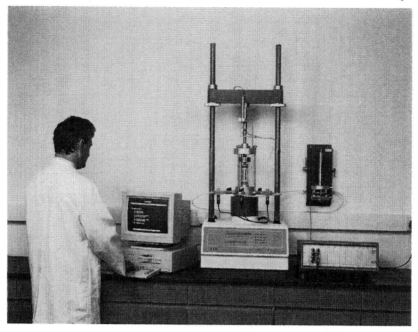

Fig. 16.35 *ELE DS6 and ADU automatic data-logging system linked to an effective stress triaxial test*

collected automatically without needing further attention until the stage has been completed.

A more flexible system is the ELE Autonomous Data-acquisition Unit (ADU). The interface unit which collects data has its own microprocessor and a large memory for storing test data. Collection of data can be programmed from most types of micro-computer by the user, in accordance with the test specification. The 'host 'computer can recall stored data for processing, display and printing out when required. A typical schematic arrangement for one ADU is shown in Fig. 16.37. Up to six units can be controlled by one computer, and each unit can collect data from a number of tests running simultaneously.

These systems are designed for use with standard test equipment (ordinary triaxial cells and Rowe consolidation cells), and can also be used with more advanced apparatus. An ADU linked to a Rowe consolidation cell is shown in Fig. 16.38.

16.6.9 Computer Controlled Testing

The application of micro-computers to effective stress testing allows many kinds of complex test procedures to be carried out automatically. This is made possible by using relays and stepper motors actuated by a process controller linked to the data-acquisition system via the computer. A stepper motor can be used for instance for operating a pressure regulator, and a self-contained motorised air pressure regulator has been developed for automatic control of triaxial cell and back pressures. This operates in response to signals fed back from the test observations of the computer in a closed-loop system. Programs fed into the computer can define the inter-relationship between the various factors measured during a test, including provision for following many types of stress path. Software

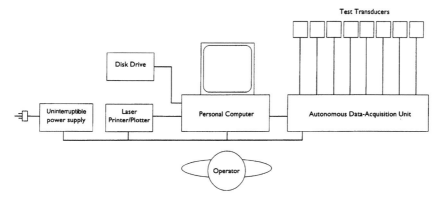

Fig. 16.36   Schematic arrangement of Datasystem components

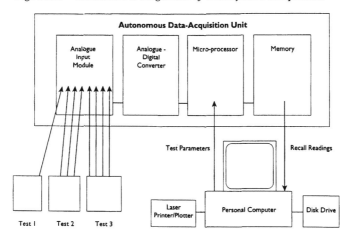

Fig. 16.37   Autonomous data-acquisition unit linked to a PC

Fig. 16.38   The ADU linked to a Rowe consolidation cell

packages are available for most of the common tests carried out on soils, including the triaxial and consolidation tests described in this book.

The ELE ADU system, as well as being an 'intelligent' data acquisition unit as described above, can be augmented by a process controller for automatic control of tests. A schematic layout is shown in Fig. 16.39.

Automatic systems should always allow for manual intervention to over-ride the automatic functions when necessary, so that the operator remains in full control of the whole procedure. Proper control can be maintained only if it is possible to obtain up-dated test data, in graphical and numerical form, at any stage while the test is still in progress.

16.6.10 General Requirements for Electronic Systems

Some of the essential factors on which the reliability of electronic equipment in general depends were outlined in Volume 2, Section 8.2.6. Further points are outlined below. Some may be obvious but none should be overlooked.

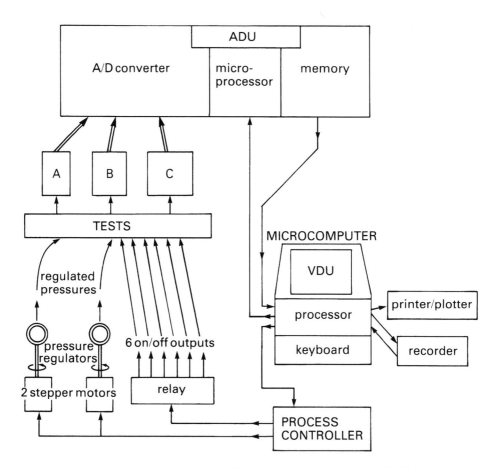

Fig. 16.39 *Schematic layout using the ADU for automatic control of laboratory tests*

POWER SUPPLY

The need for a stabilised voltage supply is referred to in Section 16.6.7. A further essential item for a computer-controlled system is an uninterruptible power supply unit which continues to supply power for a sufficient length of time in the event of mains supply failure, until a standby generator can be brought into operation. Interruption of power supply for only a few milliseconds can cause complete loss of the program and data stored in a computer. A single unit is available which provides both voltage regulation and standby power for up to one hour.

LAYOUT

Electric cables connecting transducers to the power supply/readout unit should be properly screened and laid carefully out of the way of normal laboratory activities, while allowing enough free movement for setting up and dismantling equipment between tests. Readout units should be conveniently positioned for easy observation and channel switching. Warm air vents should not be obstructed or placed where ventilation is impeded. The casing should not be used as a resting place for papers and books. Data-processing equipment should be kept in a clean area free from dust. When not in use the equipment should be covered.

CONNECTING UP

Transducers must be connected to the correct input sockets on the power supply unit, and every connection should be carefully double-checked. Plugs on connecting cables should be pushed firmly into their sockets, and secured in place if necessary.

EARTHING

For safety reasons the casing of a power supply/readout unit should be properly earthed. The manufacturer's instructions should be followed carefully in order to avoid an 'earth loop'. Earthing also cuts out electromagnetic interference from external sources, and prevents the escape of similar interference from the equipment.

ENVIRONMENTAL REQUIREMENTS

European and UK legislation includes the Electromagnetic Compatibility Directive (EMC). This requires that all electrical apparatus, including peripherals such as transducers, should be both immune to set levels of electromagnetic interference, and designed to prevent the emission of interference. Performance standards can be verified by an EMC test laboratory.

SWITCHING ON

The power supply/readout unit should be switched on at least 15 minutes before use, to allow both it and the transducers to warm up to their stable operating temperature. When in use for continuous testing the system should remain switched on overnight.

ADJUSTMENTS

The following adjustments to each channel to be used should be carried out in accordance with the manufacturer's instructions.

Setting of 'gain' control

Setting of zero reading

Calibration of transducer

Each transducer should then give a reliable read-out in engineering units.

VERIFICATION

Data-processing systems operate by reproducing all the manual processes of capturing readings, applying calibration constants, making various corrections, printing out results in the form of tabulated data and graphical plots, and in some cases calculating engineering parameters. Verification procedures should be carried out at prescribed intervals as part of the calibration programme to ensure that all these operations are performed correctly (see Section 17.2.7).

DESIGN AND MAINTENANCE

An electronic data-acquisition system for soil testing should be one that has been specially designed for the purpose, otherwise many difficulties will be encountered in its development and application.

For ordinary fault finding, self-checking diagnostic routines are usually available which enable the operator to trace a source of error and apply a correction. Any faults beyond the capability of these routines, and regular maintenance, should be attended to by suitably qualified and experienced personnel or entrusted to a specialist service organisation.

16.7   ACCESSORIES, MATERIALS AND TOOLS

16.7.1   Tubing

The types of tubing generally used for connecting the equipment described in this volume are summarised in Table 16.2.

Polyethylene tubing is impermeable to water but is not completely impermeable to air. This is of little consequence for pressure lines filled with de-aired water but migration of air from partially saturated soils could give errors in volume change measurements over a long period of time. Nylon tubing on the other hand is impervious to air but allows some migration of water when there is a difference in vapour pressure between inside and outside. A development at Imperial College was the use of nylon tubing closely fitted inside polyethlyene tubing. Difficulties in forming connections make this system too expensive for general commercial use, and it is not advised for routine testing.

Some polyethylene tubing is available in a variety of colours, and colour-coding of pressure lines makes for easy identification.

Table 16.2. TUBING FOR TRIAXIAL EQUIPMENT

Material	Outside diameter mm	Internal diameter mm	Maximum pressure kPa	Main use
Polyethylene	5	3	1700	general, for water
Nylon	3	1	8400	pore pressure null indicator
	4	2.5	8400	pressurised mercury manometer
	5	3.3	8400	oil/water pressure system
	6	4	3500	water distribution panel
	8	5.5	8400	general, alternative to polyethylene
	12.5	9.5	1700	compressed air
PVC	3	1	1700	drainage from small specimens
Rubber	16.5	6.5	0–100	vacuum
Copper	5	3.4	13,800	early type of null indicator

The flow capacity of a pipework system increases with increasing internal diameter of the tubing, but a coupling or fitting with a smaller bore will impose its own limit on the rate of flow.

16.7.2 Valves, Couplings and Fittings

VALVES

Valves used in pressure lines and pore pressure systems must be completely watertight under working pressures and must not displace water when turned on and off. The seals should be incompressible over the range of pressures to be used. Commercial sleeve-packed cocks (Klinger valves, type AB.10) are commonly used (Fig. 16.40 (a)). They should be lightly lubricated, and adjustment of the gland may be needed from time to time. A stock of spare packing sleeves should be kept for replacement when necessary. The position of the lever gives a clear indication whether the valve is on or off. After operating a valve the lever or knob should be set positively to either the open or the closed position, and not left somewhere in between.

Ball valves (Fig. 16.40 (c) and (d)) are rather more expensive but are reliable. The PTFE seals need no lubrication.

As well as on-off valves (with two ports), three-way valves (with three ports) are required in certain situations. For three-port valves, two flow paths are possible.

A special reversing valve used with a twin-burette volume change gauge is described in Section 16.5.7.

TYPES OF COUPLING

There is available a large variety of tube and pipe couplings for connecting tubing to valves, junctions and other fittings. For each type there may be several possible screw threads. Factors to bear in mind when making connections include:

Fig. 16.40   *Typical valves used with effective stress testing equipment: (left) ball valve RB166 (Automatic Valve Systems Ltd); (right) ball valve 64020713 (Legris Ltd) (bottom) Klinger cock AB10*

Suitability of coupling for the tubing

Compatibility of male and female threads on the coupling and other component

Correct fitting of coupling on tubing

Proper fitting of seals, which must be in good condition.

Four types of coupling used in soil laboratories in Britain, are described below; they are:

Coned metal couplings

Metal couplings for plastic tubing

Plastic couplings

Push-on connectors.

The first type was used with early apparatus but is now virtually obsolete.

METAL COUPLINGS FOR PLASTICS TUBING

This type of coupling consists of a connector with a male screw thread and a back-up ring (Fig. 16.41 (f)). The action of tightening the thread into a screwed socket clamps the back-up ring against a collar formed on the tubing, which makes the seal. The collar is formed by using a flaring tool (Fig. 16.41 (a)), as follows. A small source of heat (e.g. cigarette lighter or match) is needed.

(1) Thread the union nut and back-up ring on to the tube the right way round (Fig. 16.41(b)) before starting to form the flare.

(2) Soften the end of the tube by holding it vertically over the flame while rotating it (diagram (b)). The end will swell and bell out as it softens. Nylon needs a higher temperature than polyethylene; the amount of heating and softening is determined by trial.

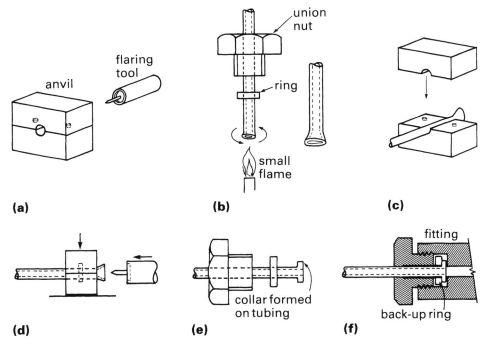

Fig. 16.41   *Metal coupling for plastic tubing: (a) anvil and flaring tool, (b) to (e) stages in forming a connection, (f) assembled coupling*

(3) Immediately lay the tube on the lower half of the anvil, with the softened end protruding about 5 mm (diagram (c)). Place the upper half of the anvil over the tube.

(4) Holding the anvil firmly on the bench, insert the spigot of the flaring tool into the bore of the tube, press and rotate the tool against the face of the anvil (diagram (d)). The softened plastic fills the recess in the end of the tool to form the end collar (diagram (e)).

(5) Assemble and tighten the coupling as shown in diagram (f).

PLASTICS COUPLING

A simple and effective method of fitting connections to either plastics or metal tubing is provided by Plasticon nylon fittings. These do not require flaring of the tube. A stainless steel grab washer grips the tube and retains it within the fitting, and is held in place by a thrust washer. A rubber O-ring seal compressed against the tube forms a seal. The fittings are designed for pressures up to about 1400 kPa. No tools are needed.

Details are shown in Fig. 16.42 (a). The connection is made as follows.

(1) Screw the cap assembly on to the body to half the thread length, without the rubber sealing ring.

(2) Insert the tube, the end of which should be cut off square, until it seats firmly at the end of the recess on the body (Fig. 16.42 (b)).

(3) Unscrew the cap and remove cap and tube from the body; the grab washer should hold the cap in place on the tube.

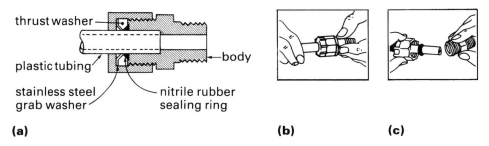

thrust washer

plastic tubing

body

stainless steel
grab washer

nitrile rubber
sealing ring

(a)   **(b)**   **(c)**

[courtesy of Pneumatic Engineering and Distribution Ltd, Sidcup, Kent]

Fig. 16.42   *'Plasticon' nylon coupling: (a) assembled coupling, (b) seating the tube,*
(c) fitting tube into coupling body

(4) Place the rubber sealing ring on the tube (Fig. 16.42 (c)).

(5) Replace the tube into the body and screw home the cap making it finger-tight. Do not use a spanner or other tool; excessive torque may split the nylon.

PUSH-ON CONNECTORS

This type of connector is now widely used. The fittings are of nickel-plated brass. The nylon or plastics tube is pushed on to a coned insert, to which it is firmly gripped when the retaining nut is screwed home. The nut need be only finger-tight. A typical connector is shown in Fig. 16.43.

16.7.3   Porous Discs

The porous discs which are fitted to the ends of triaxial test specimens must be of the same diameter as the specimen, with plane surfaces. They must be robust enough to withstand

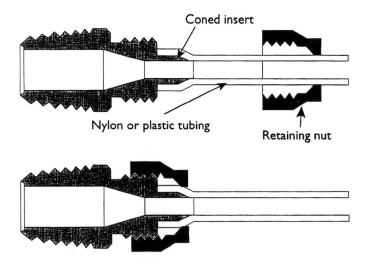

Coned insert

Nylon or plastic tubing

Retaining nut

Fig. 16.43   *Rapid push-on connector for flexible plastics tubing*

Table 16.3. TYPICAL PROPERTIES OF POROUS MEDIA

Material and designation	Porosity %	Water permeability m/s	Air entry value kPa
Ceramics:			
UNI A 80 KV		3×10^{-4}	('low')
UNI A 150 KV		5×10^{-7}	('medium')
Aerox 'Celloton' grade VI	46	3×10^{-8}	210 ('high')
Doulton grade P6A	23	2×10^{-9}	150
Kaolin dust, pressed and fired	39	4.5×10^{-10}	520
Filter paper:			
Whatman No. 54		1.7×10^{-5} at $\sigma' = 30$ kPa to	
Flow along length		3×10^{-7} at $\sigma' = 1000$ kPa	

compressive stresses up to about 5000 kPa without deformation, and allow free passage of water whether loaded or not. The water permeability characteristics of a disc depend upon the fineness of the material used, which may be sintered ceramic, bronze or stainless steel. The permeability should be substantially greater than that of the soil specimen.

More significant than permeability is the 'air entry value' of the porous medium. This is the maximum air pressure which the saturated, surface-dry disc can withstand before air breaks through the surface tension barrier into the void spaces. The smaller the voids (i.e. the finer the particles) the greater will be the air entry value. Two grades of porous disc are generally used, a normal disc of relatively low air entry value, and a 'fine' disc of high air entry value. The latter are necessary for measurements of pore water pressure in partially saturated soils. Typical properties of several grades are given in Table 16.3.

Porous discs are available for most standard specimen diameters. Discs of slightly smaller diameters are used when bonding into the cell base pedestal, as explained in Section 16.2.2, Fig. 16.4. Porous inserts of 5–10 mm diameter are used for bonding into the cell base when lubricated ends are used. Similar inserts are fitted into the base of the hydraulic loading consolidation cell (Chapter 22).

Before use, porous discs should be boiled for at least 10 minutes in distilled water to expel air from the voids, and then kept under de-aired water until required. Any remaining air is removed after fitting in place by the pressurising and flushing procedure.

After use, discs should be washed in distilled water, brushed with a natural or nylon bristle brush to remove adhering soil, and then boiled. A wire brush or steel wool should not be used. A clogged disc that does not allow free penetration of water should be discarded. A simple check for clogging is to try blowing through it. Properties of clogged discs were described by Baracos (1976).

16.7.4 Membranes, O-Rings and Stretchers

Examples of these and other accessories for triaxial tests are shown in Fig. 16.44. Details of rubber membranes used in triaxial tests, and a method for measuring the extension modulus, were given in Volume 2, Section 13.7.4.

The thickness of a few membranes from each batch received should be checked with a micrometer, so that the appropriate correction to be applied to the measured specimen strength can be determined (Section 17.4.2). Membranes should be stored in a cool dark place after lightly dusting with french chalk.

Before use a membrane should be carefully checked for flaws and leaks. A positive leak test can be made by sealing one end of the membrane to a solid end cap and the other end to a drainage top cap, using rubber O-rings. The membrane is inflated with air through the drainage lead while holding it under water. If air bubbles are visible before the membrane expands to about $1\frac{1}{2}$ times its initial diameter it should be rejected.

As a general rule a fresh membrane should be used for every effective stress triaxial test. Membranes should be soaked in water for at least 24 hours before use, in order to reduce absorption of water from the specimen and to lower the water permeability of the membrane. Two membranes, separated by a layer of silicone grease, should be used for tests of long duration.

Two rubber O-rings of the correct size should be used to seal the membrane to each of the sample end caps. Before use they should be carefully inspected for local weaknesses. After use they should be wiped clean, dry and free from grease, and stored as for membranes.

A procedure for making latex rubber membranes in the laboratory is described by Menzies and Phillips (1972). This is worthwhile only when membranes of non-standard shape or size are required.

A suction membrane stretcher (Volume 2, Fig. 13.59 and Fig. 16.44) is needed for each specimen diameter. In addition an O-ring placing tool is necessary for stretching and fitting O-rings to a top cap with a drainage connection. A split cylinder available for this purpose is included in Fig. 16.44 (see also Fig. 18.11).

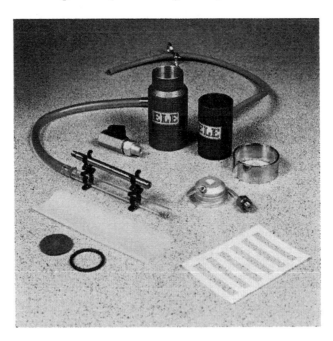

Fig. 16.44 Some miscellaneous accessories for triaxial tests: specimen split former; suction membrane holder; O-ring placing tool; valve; clamp-on burette; top cap with drainage lead; rubber membrane; porous disc; O-ring; filter paper drain

16.7.5 Drainage Materials

SIDE DRAINS FOR TRIAXIAL SPECIMENS

Side drains of filter paper are fitted around triaxial specimens of very low permeability in some tests in order to accelerate the rate of drainage. A continuous porous jacket would appreciably affect the measured compressive strength, and the accepted compromise is to fit drainage strips between the specimen and enclosing membrane.

The material used is a single layer of Whatman No. 54 filter paper, which does not soften in water. A sheet of filter paper is cut to the outline shown in Fig. 16.45 (a) for 38 mm and (b) for 100 mm diameter specimens. For specimens of other sizes, similar proportions are used. Cutting out is simplified if a template of thin metal is made to the same outline. Using this as a guide, several sheets can be cut at once with a sharp-pointed knife or scalpel. The filter strips should not cover more than 50% of the cylindrical surface

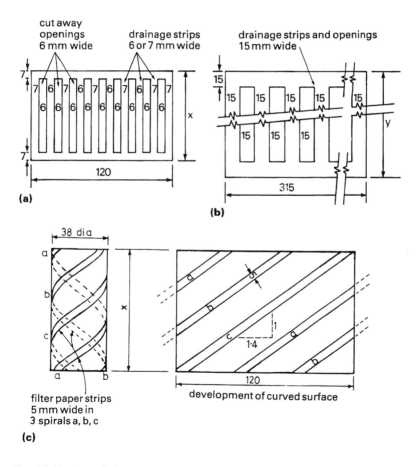

Fig. 16.45 *Details for making filter paper side drains for triaxial specimens:* (a) *38 mm diameter;* (b) *100 mm diameter:* ((c) *spiral drain strips for 38 mm diameter.* (*Dimensions* x *and* y *depend on whether or not the filters are to overlap the drainage discs*)

of the specimen. Drains of this pattern contribute to the measured strength of the specimen, and a correction must be applied (Section 17.4.3).

Side drains of spiral form were used by Gens (1982), and were found to need no correction to the measured strength, either in compression or in extension. Three spirals of filter paper, 5 mm wide, are wrapped around the specimen at an inclination of 1 (vertically) on 1·4 (horizontally), as shown in Fig. 16.45 (c). The drainage effect is equivalent to that of the conventional drains described above.

DRAINAGE LAYERS FOR CONSOLIDATION TESTS

For tests in hydraulic consolidation cells (Rowe cells), drainage to the periphery is provided for when required by fitting a layer of Vyon porous plastics material 1.5 mm thick (see Chapter 22). This is available in sheets, or in pre-cut lengths ready for fitting.

Similar material of 2 mm, 3 mm and other thicknesses is used as top and base drainage layers, porous inserts at pore pressure measuring points, and other purposes where a free-draining material is needed.

16.7.6 Liquid Materials

The fluids essential for several items of triaxial testing equipment are referred to below.

PARAFFIN

Paraffin (kerosene) with a specific gravity of about 0.8 is used in volume-change gauges of the glass burette (visual) type. Red dye is added to the paraffin to define clearly the meniscus between it and the water. The dye, Sudan powder, is dissolved in the paraffin at a concentration of about 0.02 grams per litre.

The insides of the glass burettes are rinsed with silicone water-repellant before filling to maintain a meniscus of uniform shape.

MERCURY

Small quantities of mercury are needed in a mercury manometer, and in a manual pore pressure null indicator. For all uses a purified (distilled or double-distilled) grade of mercury should be used.

It is particularly important that the mercury in the null indicator is clean and free from grease. The small amount (about 1 ml) required can be cleaned by squeezing it through a piece of chamois leather. Rubber gloves should be worn, and mercury should not be allowed to come into contact with the skin, or with gold rings and other valuables. Other precautions relating to the use of mercury are given in Volume 1, Section 1.6.7, and Volume 2 Sections 8.2.3(3) and 8.5.4. They should all be meticulously observed for safety and health reasons.

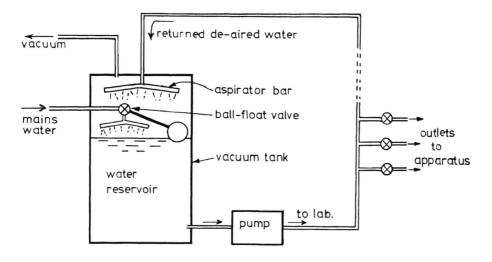

Fig. 16.46   *Schematic layout of a continuous circulation system for supplying de-aired water in a large laboratory* (*courtesy LTG Laboratories*)

DE-AIRED WATER

Water from which dissolved air has been removed must always be used in the pore pressure system, and for most tests in the back pressure line, cell chamber and cell pressure system. The traditional way of de-airing water for this purpose is by subjecting it to vacuum, as described in Volume 2, Section 10.6.2,(1). A typical vacuum de-airing arrangement was shown in Fig. 10.18.

Two types of self-contained de-airing apparatus were described in Volume 2, Section 10.6.2(2), and shown in Fig. 10.19. A system specially designed for a large laboratory is shown schematically in Fig. 16.46. De-aired water is continually circulated by pumping, and provides a supply which does not need to be interrupted.

The amount of air contained in water can be determined by measuring the dissolved oxygen (DO) content, and test kits are available for this purpose. A suggested upper limit for the DO content is 2 parts per million, but this is not mandatory in the amended BS 1377. The important requirement is to continue the de-airing process until no more bubbles are observed, while ensuring that the vacuum is maintained.

Distribution pipework from the de-airing tank can be led to a small reservoir serving each testing location. Water stored in the vacuum tank should be de-aired daily. Water in reservoirs to which vacuum cannot be applied should be run off and replaced with freshly de-aired water each day.

De-ionised water should not be used in triaxial cells. De-aired de-ionised water has been found to be very corrosive towards rubber seals and the like.

16.7.7   Miscellaneous Items

Other items required for the tests described in this volume, which were covered in Volumes 1 or 2, are listed in Table 16.4.

Table 16.4. MISCELLANEOUS EQUIPMENT FOR EFFECTIVE STRESS TESTS

Item	Section reference	Remarks
Sample preparation equipment	9.1.2	} in a separate room, preferably humidified
Trimming tools	9.1.2	
De-aired water apparatus	10.6.2	see also Section 16.7.6
Small tools	1.2.9	} pairs of spanners are especially important
	8.2.5	
General materials	8.2.5	
General laboratory apparatus	8.2.5	mainly for weighing and moisture content determination
Sundry pieces of:		
glassware	1.2.7	
hardware	1.2.8	selected items as appropriate
plastics ware	1.2.8	
Vacuum system	1.2.5(3)	electric pump and distribution line; or filter pump on water tap

16.8 PREPARATION AND CHECKING OF EQUIPMENT

16.8.1 Scope

General comments on the preparation of pressure systems and associated measuring devices are given in this section. They are intended to supplement the suppliers' detailed instructions, which should always be carefully followed. Initial checking procedures for these systems are also described. Attention is drawn to the importance of regular calibration.

Checking routines immediately prior to starting a test are given in Chapter 18, Section 18.3 (for triaxial tests), and Chapter 22 (for tests in the Rowe consolidation cell).

16.8.2 Control Panel

GENERAL

Procedures for installing and checking a control panel of the type shown in Fig. 16.10 (but without the mercury manometer) are described below. Similar procedures, as appropriate, apply to a gauge panel and to a multiple-outlet pressure distribution panel.

The panel must be mounted vertically and securely, and should be conveniently close to the test apparatus with which it is to be used. The pressure gauge should be about level with the middle of the test specimen.

Initially the control cylinder is fully wound in, and all valves are closed. If the mercury manometer is not connected, valve m opens to atmosphere. Valve designations are shown in Fig. 16.11. In the absence of a control cylinder, pressures can be applied from an appropriate constant pressure system.

FILLING AND DE-AIRING

The reservoir is filled with freshly de-aired water. A small quantity of liquid detergent added to the water will help to disperse and remove air bubbles adhering to the pipework and fittings during initial priming. This solution should be thoroughly flushed away and replaced by clean freshly de-aired water after completing the pressure checks and before using the system for a test.

The panel tubing is charged and primed as follows:

(1) Open valves j and m to allow water to feed under gravity from the reservoir through the pipework until the water level near valve m is level with that in the reservoir.

(2) Close valve m, and with valve j open wind out the control cylinder to its fullest extent (anti-clockwise) to fully charge it with water.

(3) Wind in the pump again to return the water to the reservoir.

(4) Repeat steps 2 and 3 until no air bubbles are observed in the water pumped back into the reservoir.

(5) Fill the control cylinder as in step 2, close valve j and open valve m.

(6) Operate the pump as in stages 3 and 4 until no air emerges from valve m, then close valve m.

(7) Recharge the control cylinder as in step 2.

(8) Open valve k, flush water through until no air emerges, then close valve k.

(9) Repeat steps 7 and 8 to flush the line to valve o, then the line to valve p.

(10) Open valve 1 and the bleed screw immediately below the pressure gauge.

(11) Wind in the pump steadily until water, without air bubbles, emerges from the bleed screw, then tighten it (hand-tight; do not use a spanner). Close valve 1. Recharge the cylinder if necessary.

(12) Check the pipework for bubbles of air. If necessary continue flushing until all air is removed from the system. Removal of air may be made easier by connecting vacuum tubing from a water tap filter-pump to valve p or o, with valve j closed, drawing de-aired water in through valve k. If a motorised vacuum pump is used a sufficiently large water trap should be included in the vacuum line.

(13) Recharge the control cylinder.

PRESSURE CHECK

(14) With all valves closed except 1, wind in the pump to pressurise the system to the maximum working pressure, as indicated by the pressure gauge. Leave under pressure for several hours, or overnight.

(15) If there is only a small drop in pressure it may be due to expansion of the tubing and solution of remaining bubbles of air. Release the pressure and flush the system with freshly de-aired water, by discharging from valve m. Leave under pressure again and repeat until the system is completely air-free, as confirmed by no fall in gauge pressure.

(16) A large pressure drop after step 14 indicates a leak in the system, which must be traced either by observation or by isolating and pressurising each section in turn. The whole system must be made free from leaks.

(17) On completion of the pressure check, release the pressure slowly by winding out the hand pump or by steadily reducing the pressure supply. Avoid a sudden drop in pressure, which might damage the pressure gauge or affect its calibration.

The panel should now be free of air and leaks, and ready for use. Flushing, de-airing and checking of the connecting lines to the test cell are described as part of the pre-test procedures (Section 18.3 for triaxial tests; Section 22.3 for Rowe cell tests).

If a mercury null indicator is to be used, it is prepared and checked as described in Section 16.8.7.

MERCURY MANOMETER

Mercury is a potentially hazardous substance, and the precaution given in Volume 1, Section 1.6.7, and referred to in Volume 2, Section 8.5.4 and Section 16.7.6, should be carefully observed. Additional items needed for charging the mercury manometer, if it is required, are:

About 200 mm length of flexible tubing to fit on manometer tube

Hypodermic syringe, or small plastics wash-bottle

Mercury — about 20 ml (nearly 300 g) for a manometer of 5 mm bore.

Two additional valves (h and i in Fig. 16.11), and a tee connection, if not already provided.

(1) Fill both limbs of the manometer with de-aired water from the control cylinder after opening valves m and h. Open valve i until the emerging water is free of air bubbles, then close it.

(2) Charge the syringe or wash-bottle with mercury.

(3) Connect the length of tubing to the open (left-hand in Fig. 16.11) arm of the manometer U-tube.

(4) Connect the other end of the tubing to the syringe or washbottle containing the mercury.

(5) Open valves i and h and gently pour in the mercury. Allow it to displace water until the mercury level in both limbs is at the 50 kPa mark. Close valves h and i.

(6) Remove the temporary tubing, and fit a length of tubing from the open limb to the mercury catchpot as shown in Fig. 16.11. The end of the tubing must remain open to atmosphere, and the tubing should be secured in place so that a sudden spurt of mercury under pressure will not cause it to jump out of the catchpot.

(7) If the mercury thread is broken by 'bubbles' of water, they can be displaced by rapidly jiggling the mercury up and down using the control cylinder, with valves i and l closed and valves m and h open.

READING THE MANOMETER

The manometer is used for making accurate measurements of pressures up to about 100 kPa, and for reading negative pressures. Valve h must be kept closed when the manometer is not in use, and opened only when the pressure in the system is known to be below 100 kPa.

A reading is taken by recording the levels of the mercury surface in both limbs of the manometer when they are steady. The difference between the two readings is the pressure, in kPa, if the scales are graduated in units of kPa. If the scales are in millimetres and centimetres, the difference in millimetres is multiplied by 0.128 to obtain the pressure in kPa. The difference is calculated as $(r_o - r_s)$, where r_o is the reading on the limb open to atmosphere, in order to get the correct sign (positive or negative pressure). Examples are shown in Fig. 16.47.

If the manometer is used in conjunction with a mercury null indicator, the initial zero setting of the thread of mercury in the indicator will give a small positive reading, such as that shown in Fig. 16.47 (a). This reading is then subtracted from subsequent readings and this correction (if applicable) is included in Fig. 16.47 (b), (c) and (d).

PORE PRESSURE SYSTEM

Checking procedures for the pore pressure system (initial checks, and routine checks between tests) are described in Section 18.3.2 for triaxial tests, and Section 22.3.4 for Rowe cell tests.

16.8.3 Volume Change Indicator

COMPONENTS

The arrangement of a twin-tube volume change indicator is shown in Fig. 16.23 (Section 16.5.7). The unit includes two graduated burettes inside outer tubes, reversing valve, by-pass valve, and three air bleed screw plugs.

Installation of the unit, or replacement of a burette, should be carried out strictly in accordance with the manufacturer's instructions. The glass burettes are fragile and should be handled with great care. Joints and mating surfaces may be lightly coated with silicone grease before starting assembly.

The fluids for charging the burettes are de-aired water and high-grade paraffin (kerosene), coloured by adding a little dye (see Section 16.7.6). A 100 ml capacity gauge requires about 250 ml of paraffin, which is initially placed in a beaker.

FILLING THE BURETTES

The inlet connection (1) in Fig. 16.23 is connected to the control panel (Fig. 16.11) at valve o. The control cylinder is fully charged with freshly de-aired water. All valves are initially closed. The filling procedure is as follows:

(1) Set the reversing valve to connect the left-hand burette to the inlet (1) (referred to as the left-hand position). Remove the bleed plug L at the top of the left-hand burette.

(2) Open valve o on the control panel and pump water into the left-hand burette. Recharge the pump if necessary, and continue pumping until the burette and outer acrylic tube are completely filled with water. Tap the burette tube to encourage air bubbles to rise to the top and escape through the opening at L.

(3) Replace and tighten bleed plug L.

(4) Change the reversing valve to the right-hand position, which connects the right-hand burette to the inlet (1), and remove bleed plug R.

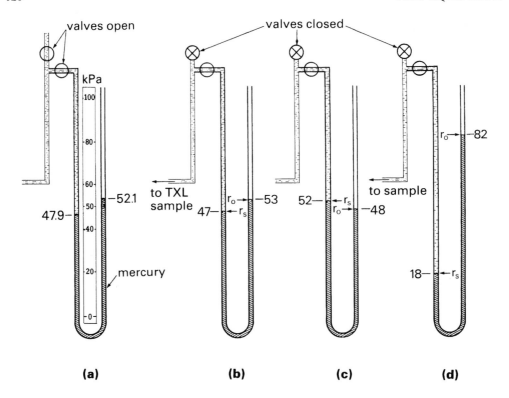

Fig. 16.47   *Examples of use of mercury manometer for manual measurement of small pressures:*

(*a*) *zero setting*

$$(r_0 - r_s) = 52.1 - 47.9 = 4.2 \, \text{kPa}$$

(b) *small positive reading*

$$(r_0 - r_s) = 53 - 47 = +6$$
$$u_0 = 6 - 4.2 = 1.8 \, \text{kPa}$$

(c) *small negative reading*

$$(r_0 - r_s) = 48 - 52 = -4$$
$$u_0 = -4 - 4.2 = -8.2 \, \text{kPa}$$

(d) *subsequent positive reading*

$$(r_0 - r_s) = 82 - 18 = 64$$
$$u = 64 - 4.2 = 59.8 \, \text{kPa}$$

(5) Fill this burette as in step 2, but using opening R, and then replace and tighten bleed plug R.

(6) Loosen the drain plug F and pump water into the burette until no air emerges from F. Retighten the plug.

PRESSURE CHECK

(7) Set the reversing valve to either the left- or right-hand position. Connect the outlet (2) to an appropriate valve on the test cell, ensuring that the connecting line is filled with water and free of air, pumping de-aired water through as the connection is tightened. Close the valve on the cell.

(8) Operate the pump on the control panel to apply a pressure of about 700 kPa to the burettes. Release any air bubbles and check for leaks, which should be rectified. Connections should be carefully tightened. Some adjustment of the pump might be necessary initially to maintain the pressure, due to expansion of the components.

(9) Maintain the pressure for 24 hours. Leakage will be indicated by a fall in pressure. Any leakages should be rectified and the pressure check should then be resumed.

(10) When the pressure check confirms the absence of leaks, reduce the pressure slowly by operating the pump. Close all valves and set the reversing valve to its central position.

CHARGING WITH PARAFFIN

(11) Disconnect the outlet line from the cell valve and place the end in a suitable container (a large beaker or bucket) below the bottom level of the burettes.

(12) Remove the bleed plug L and set the reversing valve to the right-hand position, enabling water to drain from the left-hand burette into the container.

(13) When the water level in the left-hand burette has fallen to the 100 ml mark, return the reversing valve to the central position.

(14) Using a small funnel, pour the coloured paraffin through the bleed aperture L until the burette and the space above it to the top of the aperture is filled. Replace and tighten the bleed plug L, ensuring that no air is trapped.

(15) Remove bleed plug R, set the reversing valve to the left-hand position and repeat steps 12 to 14 for the right-hand burette.

(16) Set the reversing valve to the left-hand position. Place a beaker under the drain plug F and loosen the plug to open it slightly.

(17) Recharge the control cylinder and pump water slowly into the left-hand burette, displacing paraffin into the outer tube. Excess water will bleed from F.

(18) When the paraffin/water interface in the left-hand burette reaches the zero mark, retighten the bleed plug F and set the reversing valve to the central position. The right-hand fluid interface should be at or near the 100 ml mark.

The unit is then ready for use. The inlet (1) is connected to the pressure system, without entrapping air, e.g. via valves o and p on the control panel (Fig. 16.11). The outlet (2) is connected to the appropriate valve on the test cell. Open the by-pass valve so that the line can be flushed and slightly pressurised when making the connections. Close the by-pass valve. If water is expected to drain out of the specimen due to consolidation, set the reversing valve to the left-hand position before starting a test. Set it to the right-hand position if the specimen is likely to take in water, such as in a saturation stage.

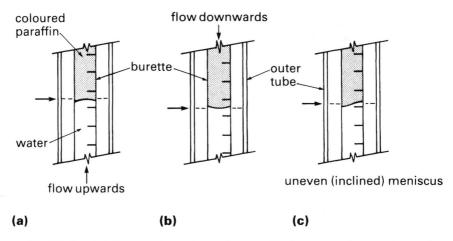

Fig. 16.48   *Reading a volume-change burette:* (*a*) *flow upwards — read top of meniscus,* (*b*) *flow downwards — read bottom of meniscus,* (*c*) *uneven meniscus — read the mean value*

USE OF VOLUME-CHANGE INDICATOR

The burettes are not subjected to a net internal pressure and it is therefore not necessary to calibrate them against pressure. Readings can usually be estimated to within half a division of the scale markings, e.g. a burette marked every 0.2 ml can be read to the nearest 0.1 ml.

The main source of inaccuracy is the possible variation in shape of the interface meniscus between the water and paraffin. For this reason it is important to clean the tubes occasionally to remove grease and other deposits. The configuration of the meniscus usually becomes inverted when the direction of flow in the tubes is reversed. This is allowed for by recording the reading of the inverted meniscus immediately after reversal to provide a new datum for subsequent readings. Meniscus reading procedures are indicated in Fig. 16.48.

The paraffin/water interface must not be allowed to pass the 100 ml mark in either burette. Volume changes in excess of 100 ml can be measured by reversing the direction of flow.

Burettes are emptied by opening bleed plugs L and R (Fig. 16.23, Section 16.5.7) in turn and draining the outer tubes through bleed valve F. The inner burettes are drained similarly, by changing over the reversing valve.

Clean the burette units by rinsing with warm water. Some detergents may adversely affect the acrylic plastic tubes.

Take care to reassemble correctly. Mark each component of every joint before disconnection. Handle with great care. Check the O-ring seals and replace if necessary.

VOLUME CHANGE TRANSDUCER

The automatic volume change transducer is filled, de-aired and initialised in accordance with the manufacturer's instructions. The procedure outlined below relates to the type of gauge shown in Figs 16.29 and 16.30 (Section 16.6.6) connected to the reversing valve assembly in Fig. 16.23.

(1) Set the three-port valve (designated below as A) to the straight-through horizontal position and reversing valve lever (designated B) to the middle (off) position.

(2) Allow de-aired water to flow in and displace air. A pressure of 20 kPa, or the head of water in an overhead de-aired water storage tank, should be enough.

(3) Turn valve A to the vertical position and lever B to the 'piston-down' position.

(4) Open the air bleed E (Fig. 16.30) and expel air from the upper chamber.

(5) Turn the whole unit upside-down.

(6) Move lever B to the 'piston-up' position (as viewed normally).

(7) Open bleed plug F and expel air from the lower chamber.

(8) Return lever B to the central 'off' position.

(9) Turn the unit right way up again.

(10) Check that the system is free from air bubbles.

(11) If bubbles are present, pressurise the unit for a few hours under a pressure of 700 kPa with the lever B in either the 'up' or the 'down' position, then flush through with freshly de-aired water.

(12) After connecting to the readout unit and allowing time for warm-up, pressurise the system to about 20 kPa.

(13) Connect the gauge into the system by turning valve A to the vertical position, and set lever B to the 'down' position. Observe the movement of the Bellofram piston.

(14) When the piston reaches its lowest position, reverse the direction of flow by moving B to the 'up' position, and allow the piston to move upwards by about 2 mm.

(15) Move lever B to the 'off' position, and set the readout to zero.

(16) The unit is then ready for calibration.

16.8.4 Cell Pressure and Back Pressure Systems

Requirements are less onerous than for the pore pressure system, and adequate checks can be made using the pressure gauge and volume-change gauge connected into each system. The systems should be pressurised and left under pressure for several hours. After the initial expansion there should be no visible movement in the volume gauge readings if the system is watertight. If a leak is indicated it should be located, if necessary by isolating sections of the line in turn, and eliminated.

Before starting a test, freshly de-aired water should be drawn into the constant pressure systems and flushed through the pipework connections to the test apparatus. Pressure vessels (air-water cells) should be charged with de-aired water as appropriate to meet the immediate requirements of each system. For instance if a triaxial cell is to be filled from the system, the maximum practicable volume of water should be available. To allow for drainage of water from a specimen there should be enough capacity available in the back pressure system to take in the water.

Checking procedures as required by BS 1377 for pressure systems are described in Sections 18.3.1–18.3.4 (systems used in triaxial compression tests) and Sections 22.3.4–22.3.5

(systems used in Rowe cell consolidation tests). 'Complete' checks are necessary for new or refurbished equipment, and at intervals not exceeding three months. Before starting each test, 'routine' checks are required.

16.8.5 Pressure Transducers

A pressure transducer is fitted into a mounting block as shown in Fig. 16.25 (Section 16.6.3) with the diaphragm upwards. A hypodermic syringe can be used to introduce de-aired water into the space immediately above the diaphragm without entrapping a bubble of air. Care is needed to avoid damaging the diaphragm. The block is connected into the system and filled with de-aired water. Visible air bubbles are removed through the bleed port D. With the plug D closed the system is pressurised for several hours to check for leaks, then flushed with freshly de-aired water. In the pore pressure system the transducer block is pressurised and flushed as described in Section 18.3.2. The transducer readings can then be calibrated against the pressure gauge.

   Pressure should not be applied to or released from a transducer suddenly, and rapid fluctuations of pressure should be avoided, otherwise damage may be caused to the diaphragm. For the same reason, suction should be applied carefully, and only if the transducer is designed to withstand negative pressures.

16.8.6 Mercury Null Indicator

PRINCIPLE

If a mercury null indicator is connected to the control panel for checking purposes, the following procedures should be followed to ensure that it is free from air and leaks. These procedures relate to Fig. 16.20 (a) and (b), Section 16.3.5, and to Fig. 16.49. The valve to which the null indicator is connected is referred to below as valve a, which could be valve a, or a_1, shown in Fig. 16.15.

DE-AIRING

(1) Connect the union joint at valve k on the control panel (Fig. 16.11) to the pressure inlet r on the null indicator (Fig. 16.20 (a)). Valve a is detached from the cell and closed, and immersed in a beaker of water.

(2) Open the bleed screws on the null indicator (f and g, Fig. 16.20 (a)), and slowly wind in the pump until water displaces all the air from the right-hand chamber. Close bleed screw g.

(3) Open valve a under water and continue pumping until air is completely expelled from the null indicator and tubing. Close bleed screw f and valve a, and reconnect valve a to the cell base.

CHARGING WITH MERCURY

See Section 16.8.2 regarding precautions to be taken when handling mercury.
   A syringe with hypodermic needle is needed, or a wash bottle with a capillary plastic tube, charged with mercury. About 1 ml (14 g) will be used.

(1) Remove bleed screw g and open f (Fig. 16.20 (a)) with the null indicator vertical.

(2) Insert the syringe needle or wash-bottle tube into the right-hand chamber at g and introduce the mercury, tapping the indicator to help the mercury through, until the reservoir is charged as in Fig. 16.20 (a).

(3) Replace and tighten bleed screw g without introducing any air. Use a wash-bottle of de-aired water to completely fill the bleed-screw recess first.

(4) Tighten bleed screw f as soon as excess water has been forced out, and ensure that no air is trapped inside.

(5) Turn the indicator to the horizontal position. Tap it to enable the mercury to fall into the reservoir chamber (Fig. 16.20 (b) and Fig. 16.49). A very small back-winding of the control cylinder may help.

(6) If air is visible, flush through with de-aired water from the control cylinder with bleed screw f open to release the air. A small bubble lodged in the capillary tube below f may be impossible to expel without turning the indicator upright, but do not allow the mercury level to climb closer than about 5 mm below the outlet in the left-hand chamber.

(7) Further traces of air, or bubbles which refuse to move, can be dissolved by pressurising for several hours as far as valve a, then flushing through with freshly de-aired water.

(8) Immerse the cell base, to which the null indicator is attached via valve a, in a tray containing de-aired water, deep enough to ensure that the aperture in the base pedestal is submerged. If complete immersion is not practicable, use a wash-bottle to maintain a layer of de-aired water in the base pedestal enclosed within a short length of rubber membrane, as shown at w in Fig. 16.49. Check that the connections at valve a are tight.

(9) Reduce the pressure in the system to zero and turn the null indicator to the upright position.

(10) Open valves k, m and h, and slightly open valve a. Carefully adjust the mercury in the capillary tube with the control cylinder to a convenient level a little above the mercury reservoir level.

(11) Mark this level with the adjustable datum marker, and read the levels of the limbs of the mercury manometer. Record this as the initial 'zero pressure' datum for checking the system.

(12) When the system is not in use the null indicator should be left in the horizontal position (Fig. 16.20 (b)). Immediately before use the device should be flushed and de-aired as described above.

CHECKING

The installation as far as valve a is checked as follows. All valves are initially closed. The null indicator is set as in step 11 above. The control cylinder is fully charged with de-aired water.

(1) Open valves k and l. Use the control pump to increase the pressure to the maximum working value, and observe the change in level of the mercury in the null indicator. If it moves no more than about 15 mm under a pressure of 1000 kPa and remains steady, and the pressure does not fall after initial small adjustments, the system as far as valve a should be free from leaks and air bubbles.

126

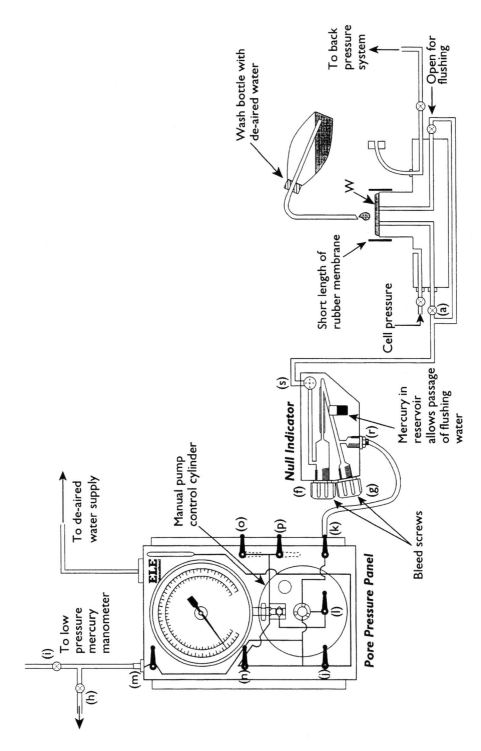

Fig. 16.49   Arrangement for flushing pore pressure line with mercury null indicator

(2) When the pressure is reduced to the initial manometer reading the mercury should return to its original level.

(3) If the mercury thread shows excessive or continuing movement when pressure is applied in step 1, it could be due to one or more reasons, which can be deduced as follows.

(a) Air in the system between the mercury thread and valve a: The rate of rise of the mercury thread decreases with time until the air is dissolved. A similar delay occurs on reducing the pressure, but the mercury does not return to its initial level.

(b) Expansion in the connecting tubing or valve sleeves: The rise of mercury continues for a short time, and the movement is recoverable on reducing the pressure.

(c) Leakage at valve a or in the connecting tube or joints: The mercury level rises at a constant rate and continual adjustment of the control cylinder is needed to maintain pressure. When pressure is reduced the mercury does not return to its original level. If a negative pressure is applied by winding back the control cylinder, air may be drawn in through the leak.

(d) Leakage between control cylinder and null indicator: Pressure cannot be maintained except by continued pumping, and the mercury level returns to its initial mark when the pressure falls off.

(4) The above faults can be dealt with as follows.

(a) If the presence of air is suspected, the system should be pressurised again for a few hours and then flushed through with freshly de-aired water, as described in step 7 above.

(b) A very small degree of flexibility in the system is unavoidable. If expansion is excessive (i.e. if there is more than about 15 mm movement of the mercury thread under a pressure of 1000 kPa) a shorter or more rigid length of tubing should be fitted between the null indicator and the cell.

(c) and (d) The source of leakage must be traced and rectified. Leakage may be due to the following:

    loose connection

    badly fitting connection

    loose gland on valve

    worn gland

    perforated or split tubing.

CLEANING

If there is evidence of dirt or grease in the capillary of the null indicator it may be possible to clean it without dismantling it, as follows. All valves should be closed initially, and the control pump wound in about half way, the pressure in the system being zero.

(a) Remove the bleed plug f from the null indicator and connect a short length of plastics tubing to the outlet. Slowly wind in the pump while the connection is being made to prevent air being drawn in.

(b) Place the other end of the tubing under the surface of a solution of mild detergent in warm water in a beaker.

(c) Turn the null indicator to the horizontal position so that all the mercury falls into the reservoir, allowing passage of water over it (Fig. 16.20 (b)).

(d) Open valve k (Fig. 16.49), and draw detergent solution into the indicator by slowly winding back the pump.

(e) After a few minutes force out the detergent and flush with cold de-aired water.

(f) Repeat steps (d) and (e) several times if necessary.

(g) Replace bleed plug f without entrapping air, and reset the zero datum as described above.

(h) If the mercury itself contains grease or dirt the thread in the capillary tube may be discontinuous and a sensitive balance is difficult to achieve. The mercury should be drained out and cleaned, as described in Section 16.7.6. The null indicator should be washed with warm detergent solution and rinsed with cold de-aired water, and then recharged with clean mercury as described above.

OPERATION

The small diameter of the mercury thread enables very small movements of water to be observed within a pressurised system. This is useful for detecting leakage, presence of air, or very small rates of flow (see Section 16.5.6).

When using the null indicator to determine pressure, the surface of the mercury thread must be at the level of the datum mark when observing the pressure gauge. This is achieved by careful adjustment of the control cylinder, the rule being as follows:

If mercury starts moving up: decrease pressure by winding out (anticlockwise)

If mercury starts moving down: increase pressure by winding in (clockwise).

The null indicator is not normally used for measuring the pore pressure in a soil specimen mounted on the cell base, but if the need should arise the procedure is outlined as follows. First pressurise the system to the anticipated level using the control cylinder. Open valves l, k, a_1, and finally a (Fig. 16.11). As valve a is opened, keep one hand on the control cylinder to make a rapid response if necessary. The mercury thread should not be allowed to deviate by more than a few millimetres from the datum level when connected to the specimen, otherwise water will move out of or into the soil. The condition of no drainage will apply as long as the mercury thread remains at the datum level.

16.8.7 Calibration

Regular calibration of all measuring devices is an essential factor in maintaining a high standard of reliability in the laboratory, and must not be neglected. Calibration data should always be readily available for reference, or displayed alongside fixed items such as pressure gauges. The date when the next calibration is due should be clearly shown.

Calibration of measuring instruments such as dial gauges, load rings, and pressure gauges is described in Volume 2, Section 8.4, and of rubber membranes in Section 13.7.4. Calibration of other instruments and apparatus referred to in this volume is covered in Chapter 17.

REFERENCES

Araruna, J. T., Harwood, A. H. and Clarke, B. G. (1995). 'A practical, economical and precise volume change measurement device'. Technical note, *Géotechnique* 45:3:541–544.

Araruna, J. T., Clarke, B. G. and Harwood, A. H. (1995). 'Quick, accurate, consistent measurements of permeability of clays'. *Proc. International Conference on Advances in Site Investigation Practice*, ICE, London, March 1995, pp 840–850.

Baracos, A. (1976). 'Clogged filter discs'. Technical note, *Géotechnique*, 26:4:634.

Bishop, A. W. (1960). 'The measurement of pore pressure in the triaxial test'. *Conference on Pore Pressure and Suction in Soils*, London, March 1960 (Butterworths, London).

Bishop, A. W. and Donald, I. B. (1961). 'The experimental study of partly saturated soil in the triaxial apparatus', *Proc. 5th Int. Conference on Soil Mechanics & Foundation Eng.*, Paris, Vol. 1, pp 13–21.

Bishop, A. W. and Green, G. E. (1965). 'The influence of end restraint on the compression strength of a cohesionless soil'. *Géotechnique*, 15:3:243.

Bishop, A. W. and Green, G. E. (1969). 'Pore pressure measurement in the laboratory'. Speciality session, Part II, *7th Int. Conf. on Soil Mechanics & Foundation Eng.*, Mexico.

Bishop, A. W., Green. G. E. and Skinner, A. E. (1973). 'Strength and deformation measurement of soils'. *Proc. 8th. Int. Conf. on Soil Mechanics & Foundation Engineering*, Moscow.

Bishop, A. W. and Henkel, D. J. (1953). 'A constant-pressure control for the triaxial compression test'. *Géotechnique*, 3:8:339–344.

Bishop, A. W. and Henkel, D. J. (1962). *The Measurement of Soil Properties in the Triaxial Test* (second edition). Edward Arnold, London (out of print).

Bishop, A. W., Webb, D. L. and Skinner, A. E. (1965). 'Triaxial tests on soil at elevated cell pressures'. *Proc. 6th Int. Conf. Soil Mechanics & Foundation Eng.*, Montreal, Vol. 1, pp 170–174.

BS 1377 : Part 6 and Part 8 : 1990, *British Standard Methods of Test for Soils for Civil Engineering Purposes*. British Standards Institution, London.

BS EN 10002, *Tensile Testing of Metallic Materials. Part 3 : Calibration of Force Proving Instruments Used for the Verification of Uniaxial Testing Machines*. BSI, London.

Darley, P. (1973). 'Apparatus for measuring volume change suitable for automatic logging'. *Géotechnique*, 23:1:140–141.

Davison, L. R. and Hills, J. M. (1983). 'A data acquisition system for soil mechanics'. *Ground Engineering*, Vol. 16, No. 1, January 1983, pp 15–17.

Gens, A. (1982). 'Stress–strain and strength characteristics of a low plasticity clay'. PhD Thesis, Imperial College of Science & Technology, University of London.

Klementev, I. (1974). 'Lever-type apparatus for electrically measuring volume change'. *Géotechnique*, 24:4:670–671.

Lewin, P. I. (1971). 'Use of servo mechanisms for volume change measurement and K_0 consolidation'. *Géotechnique*, 21:3:259–262.

Lowe, J. and Johnson, T. C. (1960). 'Use of back pressure to increase degree of saturation of triaxial test specimens'. *Proc. Research Conf. on Shear Strength of Cohesive Soils*, pp 819–836. Boulder, Colorado. ASCE.

Marchant, J. A. and Schofield, C. P. (1978). 'A combined constant pressure and volume change apparatus for triaxial test at low pressures'. *Géotechnique*, 28:3:351–353.

Menzies, B. K. (1975). 'A device for measuring volume change'. *Géotechnique*, 25:1:132–133.

Menzies, B. K. and Phillips, A. B. (1972). 'On the making of rubber membranes'. *Géotechnique*, 22:1:153–155.

Menzies, B. K. and Sutton, H. (1980). 'A control system for programming stress paths in the triaxial cell'. *Ground Engineering*, Vol. 13, No 1, January 1980, pp 22–23, 31.

Penman, A. D. M. (1984). Private communication to author.

Rendulic, L. (1937). 'Ein Grundgesetz der Tonmechanik und sein experimenteller Beweis'. *Bauingenieur*, Vol. 18, pp 459–467.

Rowlands, G. O. (1972). 'Apparatus for measuring volume change suitable for automatic logging'. *Géotechnique*, 22:3:526–535.

Sharp, P. (1978). 'A device for automatic measurement of volume change'. *Géotechnique*, 28:3:348–350.

Spencer, E. (1954). 'A constant pressure control for the triaxial compression test'. *Géotechnique*, 4:2:89.

Chapter 17

Calibrations, corrections and general practice

17.1 INTRODUCTION

17.1.1 Definitions and Scope

This chapter deals with the calibration and correction procedures necessary for the correct evaluation of data from the tests described in this volume. These procedures are aimed at reducing or eliminating many of the effects of errors which can occur in laboratory measurement and analysis.

Some degree of uncertainty is inherent in all measurements. The objective of calibration is to ensure the uncertainty in any measurement can be quantified.

In the context of this book, 'calibration' and 'corrections' are defined as follows.

CALIBRATION

(1) (for measuring instruments). Determination of the relationship between an observed reading (or transmitted electrical signal) and the physical quantity being measured.

(2) (for other apparatus). Determination of the effect of a physical change or process on other measured quantities (e.g. the change in volume of a cell due to an internal pressure change).

(3) (for certain accessories). Determination of the physical properties of a piece of apparatus which has an influence on the measurements made during a test, so that that influence may be allowed for.

(4) Determination or verification of fixed measurements which influence the test results (e.g. those which determine the dimensions of the test specimen).

CORRECTIONS

(1) Application of calibration data or calculations to observed test data to eliminate inaccuracies in the results.

(2) Adjustment of test data to make allowance for various physical factors inherent in the apparatus or in the test method.

The processes described fall into three main categories, the first two consisting of experimental procedures, the third being the application of these and other techniques to test data. They are presented as follows.

Calibration of measuring instruments, especially electronic devices, including effects dependent on the laboratory environment (Section 17.2). Calibration of some of these instruments was described in Volume 1, Section 1.7, and Volume 2, Section 8.4.

Methods of measuring and calibrating test equipment (triaxial cells and consolidation cells) and certain accessories (Section 17.3).

Application of corrections to readings taken during triaxial compression tests (Section 17.4) and hydraulic cell (Rowe cell) consolidation tests (Section 17.5), based on calibration data generally.

Some measurements, such as linear dimensions for calculating cross-sectional area, once determined seldom need to be checked, except to make allowance for wear. Other calibrations need to be rechecked from time to time to maintain accuracy.

The chapter concludes with comments on general laboratory practice (Section 17.6) covering safety, the laboratory environment, sources of error and the need for critical appraisal of test data and results.

17.1.2 Importance of Calibration

Calibration of instruments and equipment is an essential factor in maintaining a high standard of reliability in the laboratory. Recalibration at regular intervals is equally important. Calibration data should always be readily available for reference, or displayed alongside fixed instruments such as pressure gauges. They should show the date when the next calibration is due. Detailed records should be maintained and kept on file for future reference.

Suggested maximum intervals between recalibrations for the items used most often are indicated below. Additional regular checks will be necessary for items used intensively. Items that have been misused, or which are suspected of giving erroneous results, should be recalibrated immediately.

Suggested calibration frequencies

Dial indicators (Section 8.3.2 and 8.4.4)	One year
Load rings (Section 8.3.3 and 8.4.3)	One year
Pressure gauges (Section 8.3.4 and 8.4.4)	6 months
Pressure transducers (Section 17.2.3)	6 months*
Force transducers (Section 17.2.4)	One year*
Displacement transducers (Section 17.2.5)	One year*
Volume change transducers (Section 17.2.6)	One year*
Burette volume change indicators	2 years
Triaxial cells (Section 17.3.1 and 2)	One year
Rowe cells and diaphragms (Sections 17.3.3–6)	One year
Rubber membranes (Section 17.3.9)	Samples from each batch received

These intervals between recalibrations relate to working instruments and are based on the requirements of BS 1377 : Part 1 : 1990, Clause 4.4. More recently it has been recommended that measuring devices which make use of electronic instrumentation (marked * above) should be recalibrated at intervals shorter than those specified for equivalent mechanical instruments (see Section 17.2.1).

Recalibration intervals for laboratory reference standards are given in Volume 2, Section 8.4.7, based on Clause 4.3 of BS 1377: Part 1.

Calibration of measuring instruments should be traceable back to national standards of measurement. This means that the laboratory's working instruments should be calibrated against its reference instruments, which in turn through a traceable chain of calibration steps are referred to instruments held by the national standards organisation (in the UK, the National Physical Laboratory (NPL)). Every calibration step should be carried out precisely to the appropriate technical requirements, by suitably trained and experienced personnel.

Guidance on calibration systems and procedures is given in BS EN 30012:1992, which supersedes BS 5781. General criteria for the technical competence of testing laboratories, including calibration requirements, are set out in European Standard EN 45001.

Calibration procedures should be fully documented, and all calibration readings and observations should be recorded and retained. Properly documented calibration records are one of the essential requirements for recognised accreditation of a testing laboratory. The accreditation body in the UK is NAMAS, which is part of the United Kingdom Accreditation Service (UKAS).

If checks reveal that an instrument or apparatus has ceased to function correctly, or recalibration reveals that an instrument no longer performs within its calibration limits, all test data produced since the previous check or calibration should be treated as suspect. Recipients of such data should be informed.

17.1.3   Significance of Test Corrections

After allowing for the calibration of the measuring instruments there are numerous factors which have to be taken into account when processing the test data if a sufficiently accurate graphical plot or test result is to be produced. These are covered in Sections 17.4 and 17.5, and are related to the behaviour of the test specimens or the characteristics of various parts of the apparatus.

Some corrections are important enough to be applied regularly as a matter of routine. Others are of less significance and may be required only where the highest accuracy is essential, such as in research. The corrections necessary for routine application are listed in Section 17.4.7.

Where there are varying opinions on the application of a correction, several are outlined and the author's compromise is suggested.

17.2   CALIBRATION OF MEASURING INSTRUMENTS

17.2.1   Principles

SCOPE

Calibration of conventional measuring instruments such as dial gauges, load rings and pressure gauges was described in Volume 2, Section 8.3, and is not repeated here. Many of the principles given there apply also to the calibration of electronic devices and systems, discussed generally in Section 17.2.2. Calibration of electric transducers of various types is described in Sections 17.2.3 to 17.2.5.

Included in the category of measuring instruments are volume-change indicators fitted with transducers (Section 17.2.6).

ELECTRONIC INSTRUMENTS

Conventional measuring instruments, which are essentially mechanical in operation, are relatively robust, and it is usually fairly obvious if they cease to function properly. Electronic instruments can be less reliable, they are subject to electrical noise, drift and other errors, and it can be difficult to detect when they are no longer working properly as long as they give a reading. Intervals between checks and recalibrations for electronic instruments should therefore be shorter than those suggested in Section 17.1.2 for conventional instruments (J. H. Atkinson, pers. comm.). In many cases intervals not exceeding 3 months might be more appropriate.

Calibration of electronic instruments should include checks for the effects of electrical 'noise' and drift. Electrical noise can be significantly reduced by including a voltage stabiliser on the power supply line. Checks for drift should ideally determine the extent of drift of the output signal, while the measured quantity remains constant, over a period not less than the duration of a typical test. It is practicable to do this in a large laboratory, where there are sufficient circuits to allow one or more to be dedicated in turn to long-term checks. The evidence thus obtained enables the limits of errors due to electronic drift to be expressed with confidence.

17.2.2 Calibration of Transducers

PRINCIPLES

A transducer with its connecting leads and electrical readout unit should be regarded as a single system for calibration purposes. Any change in a part of the system will invalidate the calibration. However, it is possible to maintain the calibration of a transducer on several systems simultaneously, once each system has been calibrated.

The principles of good calibration practice for any type of electric transducer can be summarised as follows.

(1) The reference instrument used for calibration should itself be accurately calibrated by measurements that are traceable back to a national standard, as recommended in Section 17.1.2.

(2) After setting up the transducer and making the electrical connections, the electric circuits are energised and allowed to warm up for long enough to ensure that readings are stable.

(3) With the transducer unloaded or at the datum position, output signals are set to zero or to a suitable datum value.

(4) Three fairly rapid cycles of loading and unloading from zero to maximum are normally applied.

(5) Outputs are then carefully reset to zero.

(6) Three cycles covering the full range of the transducer are applied successively.

(7) During each half-cycle at least 10 comparative readings are observed and recorded.

(8) Average readings from the three cycles are tabulated, or plotted as a graph, against the known applied values.

(9) The temperature at the time of calibration is recorded.

(10) The observed data are processed as outlined below.

ZERO DATUM

Transducers for measuring force and pressure have a clearly defined 'fixed zero' point, corresponding to zero load or atmospheric pressure respectively, at which their reading would normally be zero. On the other hand measurement of displacement depends upon the position of an arbitrarily chosen datum point, a 'floating zero'. For a displacement transducer, whether used for measuring linear displacement such as axial deformation, or fitted to the diaphragm of a volume change indicator to measure volume changes, the zero datum is determined by the position of the transducer at the start of the test or calibration.

CALIBRATION EQUATION

Calibration data for a transducer can be used in the same way as for any other calibrated instrument, i.e. the correct value corresponding to an observed reading is read off from a table or calibration curve. However, when test data are fed directly into a computer for analysis and printing out of results, the computer must be given the calibration data beforehand, in the form of an equation, to enable it to make the necessary corrections automatically. Transducers are normally designed to give a practically linear relationship between the physical quantity being measured (q) and the reading (r), of the form

$$q = b(r - i) \tag{17.1}$$

where b and i are the calibration constants, b being the rate of change of q with respect to r, and i the initial value for a particular test.

The derivation of a calibration line and its equation, from a set of observe data, is illustrated in Fig. 17.1.

The initial zero position (fixed or arbitrary) establishes the value of i, and the slope of the calibration line is equal to b. Several factors need to be taken into account in the determination of b, as outlined below.

RANDOM ERRORS

Random errors due to electrical 'noise' can be reduced to acceptable limits by taking a large enough number of readings within a few milliseconds at each calibration point. If for instance 24 readings are taken, the 4 largest and 4 smallest are discarded and the average of the remaining 16 is taken as the accepted reading. This statistical procedure has been shown to reduce random errors to less than 0.02% of the full scale reading, and provides the basis for a reliable linear regression analysis.

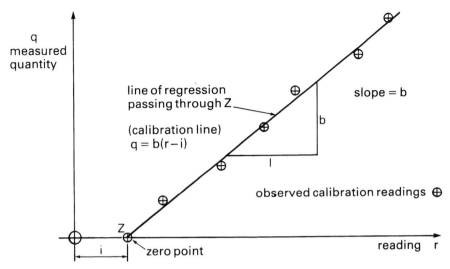

Fig. 17.1   *Calibration line derived from a set of observed data*

SYSTEMATIC ERRORS

No transducer gives a true linear response over its full range, and the working range is therefore restricted to reduce the deviation from linearity to within acceptable limits. Deviations from linearity are greatest near each end of the transducer's range, but use can be made of the near-linear middle section. A large number of calibration points are determined within this range, and a form of linear regression analysis is used to derive the slope, b, of the 'line of best fit' (Equation (17.1)) which may or may not pass through the zero point.

By carrying out three complete cycles of calibration measurements (as specified in BS EN 10002) the repeatability of the transducer can be determined as a percentage deviation from indicated value. The mean line from the three cycles is used for calculating the calibration constants.

A linear calibration is usually acceptable provided that the maximum deviation from linearity does not exceed a certain limit, depending on the type of transducer. The maximum deviation, expressed as a percentage of indicated value, is quoted as the maximum percentage error.

NON-LINEAR CALIBRATION

In cases where some calibration points are scattered from a linear relationship to an unacceptable extent, a more complex relationship may be necessary. This could take the form of two or more lines, or some kind of curve. A quadratic curve fitted through the origin needs two constants to define it. More complex curves require a greater number of constants, but for most purposes a linear or quadratic curve is adequate. Use of a quadratic expression can be avoided if b itself is treated as a variable. An example is the use of a variable 'ring factor' for a load ring, as illustrated in Volume 2, Fig. 8.32.

ENVIRONMENTAL EFFECTS

The effects of environmental changes on transducer calibrations should be investigated. The most likely variations from normal laboratory conditions are:

(1) Ambient temperature of the whole system significantly different from the calibration laboratory temperature (e.g. in tropical conditions).

(2) Temperature changes in the transducer only.

(3) Excessive increase in temperature of the electronic power supply and readout system (e.g. over-heating due to inadequate ventilation).

(4) Exposure to high humidity.

(5) Interference from stray electromagnetic fields.

(6) Unstabilised mains supply voltage.

Errors due to these effects can be reduced considerably if the problems are recognised so that appropriate remedial measures can be taken and any necessary additional calibrations made. In any case, all electronic instruments should be calibrated in the laboratory environment in which they are to be used, immediately after installation and periodically thereafter.

APPLICATION

A fully automatic data-acquisition system such as the ELE DS6 (Fig. 16.35, Section 16.6.8) allows for the entry of calibration data, from which the calibration constants are automatically calculated and stored for use in the test program. This enables signals from transducers during a test to be corrected before being displayed or printed in engineering units.

EXAMPLE CALIBRATION

An example of a set of data obtained automatically from a computer calibration program for a force transducer is given in Fig. 17.2(a). Other types of transducer can give similar data in terms of the appropriate units.

The output signal (bits) for each input value (true load, kN) is plotted in Fig. 17.2(b), and the line of regression is drawn through the set of points. Its slope is calculated by the computer program, and is stored in memory for subsequent use in the automatic display of load in kN. The extent to which the observed calibration points deviate from the calibration line is shown by the 'percentages of indicated load' in the right-hand column of the print-out, Fig. 17.2(a).

17.2.3 Pressure Transducers

A pressure transducer should be calibrated against a calibrated reference standard such as a dead-weight gauge tester, or in series with a 250 mm diameter 'test' pressure gauge that has already been accurately calibrated over the same range as the transducer. In the latter case, gauge and transducer are connected to a source of pressure such as a stable constant-pressure system or a screw control cylinder. The system is flushed with de-aired water, and

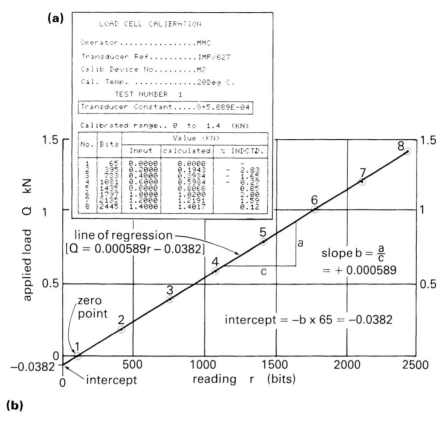

(a)

```
LOAD CELL CALIBRATION

Operator................MMC

Transducer Ref..........IMP/627

Calib Device No.........M2

Cal. Temp. ............20Deg C.

     TEST NUMBER  1

Transducer Constant.....6+5.889E-04

Calibrated range.. 0  to  1.4  (KN)
```

No.	Bits	Value (KN)		% INDCTD.
		Input	calculated	
1	65	0.0000	0.0000	–
2	395	0.2000	0.1943	– 2.83
3	733	0.4000	0.3934	– 1.65
4	1081	0.6000	0.5934	– 0.37
5	1435	0.8000	0.8068	0.85
6	1797	1.0000	1.0200	2.00
7	2135	1.2000	1.2191	1.59
8	2445	1.4000	1.4017	0.12

line of regression —
$[Q = 0.000589r - 0.0382]$

slope $b = \dfrac{a}{c}$
$= + 0.000589$

intercept $= -b \times 65 = -0.0382$

zero point

(b)

Fig. 17.2   Typical calibration data for a load transducer: (a) printed data, (b) graphical plot and line of regression

precautions are taken against leaks and entrapped air, as described in Section 16.8.5. The transducer should be at the same level as the gauge tester or reference gauge.

Calibration is carried out, at a known temperature, by applying three cycles of pressure increase and decrease in suitable stages, in a manner similar to that described in Volume 2, Section 8.4.4. The calibration data may be presented as a table or graphical chart for use when readings are being observed manually, as in Fig. 8.34. With a computer-controlled data acquisition system the data are processed automatically, as described in Section 17.2.2 above. A calibration can be assumed to be linear if the maximum deviation does not exceed 1% of indicated pressure.

17.2.4   Force Transducers

A transducerised load ring or force transducer should be calibrated at a known temperature against a load cell or proving ring complying with Grade 1.0 of BS EN 10002–3, as described in Volume 2, Section 8.4.3. Three load/unload cycles should be applied over the full range of the transducer.

Calibration curves similar to those shown in Volume 2, Figs. 8.31 and 8.32 can be obtained for manual observations. A mathematical relationship can be derived by, and used in, a computer-controlled system.

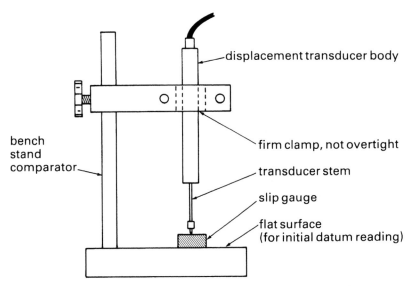

Fig. 17.3   Arrangement for calibrating a displacement transducer

Force transducers should be recalibrated at intervals not exceeding one year, or more often if in constant use, either by a calibration house certificated by a national calibration service (UKAS in the UK) or against the laboratory's own reference standard. The reference instrument should be recalibrated at intervals not exceeding two years, as described in Section 8.4.7 of Volume 2.

17.2.5   Displacement Transducers

A displacement transducer is calibrated by mounting it vertically in a rigid bench stand comparator, and observing readings when certificated slip gauges of accurately known thickness are inserted under the tip of the transducer stem (Fig. 17.3). The calibration temperature should be recorded.

The resolution of displacement transducers depends upon their range of movement, and is typically better than 1 part in 10,000. For example a transducer with a travel of 10 mm can discriminate down to about 0.001 mm, but a digital readout to the next decimal place should be treated with caution.

Linearity of calibration can be assumed if the calibration readings deviate from the mean calibration line by no more than 0.1% of the reading at the upper limit of calibration. This requirement is illustrated diagrammatically in Fig. 17.4, where the deviations have been intentionally exaggerated. In practice the assessment of compliance must be checked arithmetically because these deviations would be too small to be seen on a graphical plot of this kind.

17.2.6   Volume Change Transducers

Each volume-change transducer should be individually calibrated. The unit is connected between a paraffin burette indicator of appropriate accuracy and a control cylinder, with a

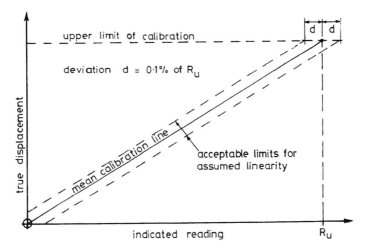

Fig. 17.4 *Limits within which linearity of calibration can be assumed to be acceptable (deviation from mean line greatly exaggerated)*

connection to a de-aired water supply, such as in the arrangement shown in Fig. 17.5. The whole system must be thoroughly de-aired and leak-free. The transducer should first be pressurised to the maximum working value, and operated over its full range a few times to ensure that the diaphragm or Bellofram seals are properly seated.

The control cylinder is operated to provide a movement of water into or out of the constant pressure system. Transducer readings are compared with readings of the burette over the range of piston travel, in both directions. For very precise calibration of the Bellofram unit a small single-tube paraffin burette should be used. Emptying or recharging of the burette, using the control cylinder, is necessary every time its capacity is reached.

The piston movement of the single diaphragm type must be kept within the stated range of the unit, otherwise the diaphragm might be damaged. The effect of reversal should be investigated so that any 'lost volume' due to backlash of the diaphragm can be determined. In a computer-based data acquisition system this error can be allowed for automatically if it has been measured and the data entered into the software.

17.2.7 Data-logging and Processing Systems

In addition to calibrating the separate measuring instruments of a data-logging system, overall verification checks should be carried out on the system as a whole at prescribed intervals, for instance of not less than six months, as part of the calibration programme. These checks should cover the whole process, from collection of readings to the printed output data and plots. The aim is to ensure that all calculations and corrections are properly performed and that the results are correctly presented and interpreted.

17.3 CALIBRATION OF CELLS AND ACCESSORIES

In this section the methods used for calibration of cells and other equipment are described. Sections 17.3.1 and 17.3.2 relate to triaxial cells; Sections 17.3.3 to 17.3.6 to Rowe

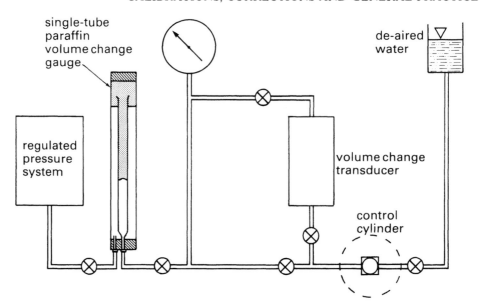

Fig. 17.5 *Arrangement for calibration of volume-change transducer*

consolidation cells; and Sections 17.3.7 to 17.3.10 deal with ancillary items relevant to both types of apparatus.

The applications of these calibrations are covered in Sections 17.4 and 17.5.

17.3.1 Triaxial Cell Deformations

VOLUME CHANGE RELATED TO PRESSURE

When the pressure in a triaxial cell is increased the cell walls expand. This introduces an error in the sample volume change measurements made on the cell pressure line which must be allowed for. The volume change/pressure relationship for a triaxial cell is determined as follows. Reference is made to Fig. 16.2, Section 16.2.2.

(1) Connect the cell to a constant pressure line, incorporating a volume-change gauge, at valve c (Fig. 16.2). The cell clamping studs or nuts should be tightened with a controlled torque; variations in bolt tension can affect the calibration.

(2) Fill the cell completely with de-aired water, making sure that no air is entrapped. Flush water through the connecting ports in the cell base, to displace air, and close the valves and blanking plugs.

(3) Applying enough pressure (say 20 kPa) to hold the piston up at its fullest extent, with the retaining collar against the cell top, or with the upper end of the piston restrained against the crosshead of a load frame (not against a load ring, which will deflect under increasing pressure).

(4) Read the volume-change indicator when steady.

(5) Close valve c and increase the line pressure to 100 kPa. Observe the volume change indicator.

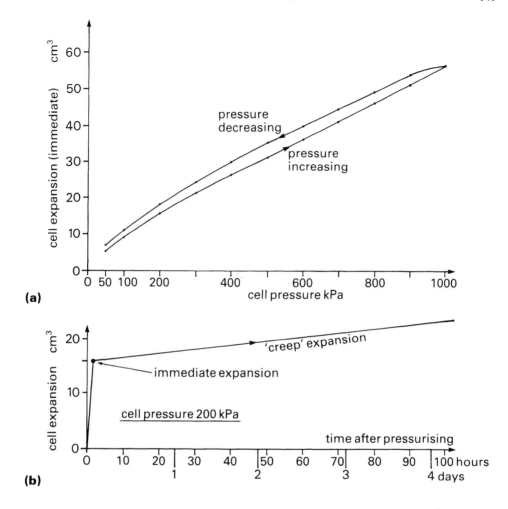

Fig. 17.6   *Example of triaxial cell volume changes with:* (a) *pressure changes,* (b) *time*

(6) When the indicator reading is steady, open valve c to pressurise the cell.

(7) Record the volume change indicator reading after a fixed interval of time, say 5 minutes.

(8) Repeat steps 5, 6 and 7 at pressure increments of 100 kPa up to the maximum working pressure of the cell.

(9) Reduce the cell pressure in decrements of 100 kPa and take readings as for pressure increments.

(10) Record the temperature of the cell during calibration.

(11) Plot a graph of volume change (related to the initial pressure as datum) for each reading against cell pressure.

An example of a pressure calibration curve for a cell taking specimens up to 50 mm diameter is shown in Fig. 17.6 (a). A bigger cell would give larger volume changes.

VOLUME CHANGE RELATED TO TIME

The volume change relationship with time, under several different pressures covering the working range, should also be determined. After applying each selected pressure and taking the immediate readings described above, further volume change indicator readings are taken during the ensuing 2–3 days. Curves of cell expansion against time are then plotted for each pressure. These give the expansion of the cell with time (the 'creep' calibration), an example of which is shown in Fig. 17.6 (b).

VERTICAL DEFORMATION RELATED TO PRESSURE

The dial gauge or transducer used for vertical strain measurements is usually supported by the cell. Deformation of the cell due to internal pressure change may cause a false strain reading, which can be determined by pressurising the cell containing a rigid metal cylinder fitted in place of a soil specimen. This correction is applicable only to tests in which the cell pressure is changed while axial strain is being measured, and does not apply when the cell pressure remains constant during a compression or extension test.

17.3.2 Leakage from Pressure Systems and Cells

A thorough check for leaks in the triaxial pressure systems can be made by setting up a metal 'specimen' in the cell, with the drainage valve closed. Suitable pressures are applied and maintained for a period not less than the test duration. No change should be observed in the readings of the volume change indicator or the pore pressure transducer. Details of pre-test checking procedures for pressure systems are given in Section 18.3.

Loss of water from a triaxial cell under pressure can occur in five ways.

(1) Through the piston bushing.
(2) Past the seals between the cell body and the base or top.
(3) Through the sample membrane and bindings.
(4) Absorption into and migration through the cell walls.
(5) From valves, blanking plugs and pressure line connections.

(1) *Piston leakage*

This is the most common source of leakage, and accounts for most of the losses from a triaxial cell. A newly machined piston should be virtually watertight, but leakage inevitably develops with wear. Leakage can be reduced, and piston lubrication increased, by floating a layer of castor oil on top of the water in the cell. If the external collar around the piston has a recess the rate of leakage can be determined by soaking up the oil which collects during a known time on pieces of weighed filter paper or blotting paper. If the density of the oil is known the volume collected can be calculated after weighing the oil-soaked papers. This procedure cannot be used with leaking water because of evaporation losses. The rate of leakage at the appropriate cell pressure can be used to correct measured volume changes on the cell pressure line.

(2) *Seals*

If the O-ring seals are not seated properly before tightening the clamping bolts, leakage will be obvious and the cell should be re-assembled. Long-term leakage is very unlikely when properly assembled with sealing rings of good condition.

(3) *Membrane and bindings*

This possibility is discussed under 'rubber membrane' in Section 17.4.2 (8).

(4) *Absorption into cell walls*

Transparent acrylic plastics from which cell bodies are usually made can absorb and transmit water to an extent which depends upon temperature, pressure and humidity gradient. Technical data on these properties can be obtained from ICI Limited, Plastics Division. However, these effects can be neglected for most practical purposes.

(5) *Valves and couplings*

Pressure checks before test will reveal any obvious leaks due to faulty seals or valve seatings. Before starting a long-term test in which volume change measurements are critical, a pressure test should be made to assess whether the magnitude of leakage is significant.

If all seals, couplings and valves are in good condition, for most purposes only the piston bushing leakage need be allowed for, but the possibility of membrane leakage might also have to be considered.

17.3.3 Rowe Cell Measurements

The following dimensions of Rowe consolidation cells should be carefully measured and recorded for future use. Reference letters refer to Fig. 17.7 (these cells are described in Section 22.1.4).

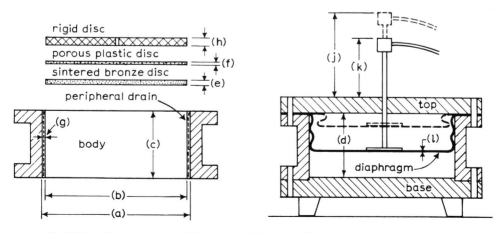

Fig 17.7   *Measurements of Rowe consolidation cell:* (*a*) *cell body and internal fittings,* (*b*) *assembled cell*

(1) Internal diameter of cell body (a), and of cell with the porous peripheral drain fitted (b).

(2) Overall height of cell body (c).

(3) Internal depth from top flange to base (d), when body is bolted to the base.

(4) Thickness of sintered bronze porous disc (e), porous plastic disc (f), and porous plastic peripheral drain material (g).

(5) Thickness of rigid loading disc (h).

(6) Mass of cell body with normal attachments, which should be listed.

(7) Mass of cell body assembled on base including bolts and attachments (listed).

(8) Mass of rigid disc, and of each porous disc (dry).

(9) Mass of a typical porous liner.

(10) Distance of upper face of drainage spindle above cell top when the diaphragm is at the upper limit (j) and at the lower limit (k) of its extension.

(11) Thickness of diaphragm (l).

Each cell should be individually measured, and clearly numbered so that the correct data (including calibrations from Sections 17.3.4 to 17.3.6) can be identified easily. The body, cover and base of a cell should all bear the same number.

17.3.4 Rowe Cell Diaphragm Load

The force exerted by the diaphragm may be less than that calculated from the hydraulic pressure and cross-sectional area of the cell (Shields, 1976). This is due to stiffness of the diaphragm and side friction, and is more significant in a small cell than in a large one. The difference between actual and calculated force can vary with both the applied pressure and the diaphragm extension, and is greatest at low pressures. In small cells at low diaphragm pressures the relative differences can be very large, and repeatable calibrations are difficult to obtain and therefore unreliable. The effect of side friction can be reduced to some extent (when a peripheral drain is not fitted) by applying a film of lubricant, such as silicone grease or petroleum jelly, to the inside wall of the cell behind the diaphragm.

The actual pressure transmitted to the specimen is equal to the actual diaphragm force divided by the specimen area, and can be determined as follows. (This procedure is as specified in Clause 3.2.5.2 of BS 1377 : Part 6 : 1990.)

(1) Remove the settlement dial gauge supporting brackets from the cell top (see Fig. 22.1, Section 22.1.4).

(2) Fit the cell top to the body and fill the space above the diaphragm with water without entrapping air. Close valve C, valve D and bleed plug E.

(3) Place the assembly (without the cell base) upside-down on the platen of a triaxial load frame, as indicated in Fig. 17.8. Place the rigid circular plate on the diaphragm.

(4) Insert suitable spacers between the plate and a load ring mounted in the load frame. For stability an extension stem passing through a stabiliser bearing, as fitted for CBR tests, is desirable. Without an arrangement of this kind there is no restraint on the plate to keep it flat.

(5) Adjust the water content of the diaphragm to make its extension about the same as when in contact with a sample of the recommended initial length. Connect valve C to a pressure system fitted with an accurately calibrated pressure gauge or transducer.

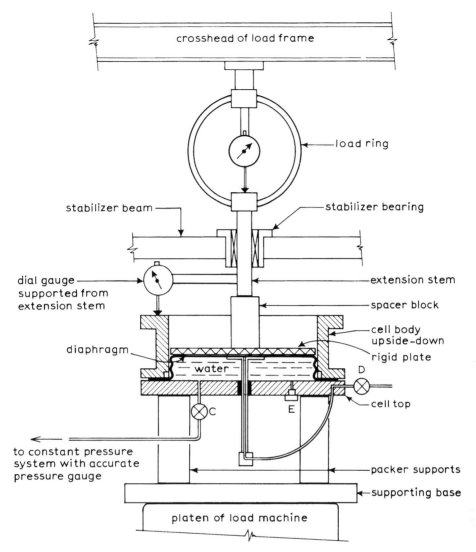

Fig. 17.8   Arrangement for calibration of load transmitted by diaphragm of Rowe cell

(6) Clamp a dial gauge in position to measure the movement of the plate relative to the cell body flange. A suitable arrangement is indicated in Fig. 17.8.

(7) Set the pressure system to an initial seating value (e.g. 10 kPa) and open valve C. Record the load ring reading and the vertical movement of the plate (i.e. the diaphragm), which should correspond to the deflection of the load ring.

(8) Repeat stage 7 with increased values of diaphragm pressure (e.g. 25, 50, 100 kPa, etc.) up to the working limit of the load ring or the maximum working pressure of the diaphragm. For low pressures a sensitive load ring should be used, and a ring of higher capacity substituted for a second stage of calibration covering high pressures.

(9) Repeat stages 7 and 8 with different initial diaphragm extensions, over the range of permissible extension.

(10) Calculate the following from the test data:

$$\text{Upward force on ring} = (\text{dial division}) \times (\text{ring factor}) \text{ newtons}$$
$$= P \text{ newtons}$$
$$\text{Downward force on diaphragm} = (P + mg) \text{ newtons}$$

where m is the mass of plate and spacers etc. supported by it (kg) and g is 9.81 m/s^2.

$$\text{True pressure on specimen area} = \sigma = \frac{P + 9.81m}{A} \times 1000 \text{ kN/m}^2 \qquad (17.2)$$

where A is the circular area of cell (mm^2).

$$\text{Pressure difference } \delta p = p_d - \sigma \qquad (17.3)$$

where p_d is the applied diaphragm pressure and δp is the diaphragm pressure correction.
(11) Plot values of δp (y-axis) against true pressure σ (x-axis) for each extension of the diaphragm.

The above calibration can be carried out for three conditions:

with peripheral drains fitted;

without peripheral drain, diaphragm not lubricated;

without drain, with diaphragm and cell wall greased.

Examples of calibration curves are of the form shown in Fig. 17.9. The significance of diaphragm lubrication (Fig. 17.9 (b)) should be noted. For any required test pressure on a specimen the graph gives the amount to be added to the indicated diaphragm pressure to give the required pressure on the sample.

The above calibration automatically allows for the small non-pressurised area under the drainage stem. This accounts for about 1.2, 0.3 and 0.1% of the specimen area for the 75, 150 and 250 mm diameter cells respectively, which in themselves are negligible compared with the overall diaphragm calibration corrections.

The calibration should be repeated at regular intervals to allow for any changes which take place in the characteristics of the diaphragm with use and with time.

17.3.5 Rowe Cell Deformation Under Pressure

Distortion of large Rowe cells when subjected to high internal pressures could influence the measured settlements. This effect can be assessed by supporting the diaphragm on a rigid disc inside the cell (as in Fig. 17.10 (a)), pressurising both sides of the diaphragm equally in increments and recording the resulting changes in the settlement dial gauge or transducer readings. A calibration curve of deflection against diaphragm pressure can then be drawn to enable the correction appropriate to any given total pressure to be applied, in a manner similar to that used for a conventional oedometer press (Volume 2, Section 14.5.6 (5)). The effect is small and this correction is not called for in BS 1377.

17.3.6 Rowe Cell Rim Drain

The volume of water trapped between the side of the diaphragm and the cell wall, and the time required for it to escape, can be estimated approximately as follows.

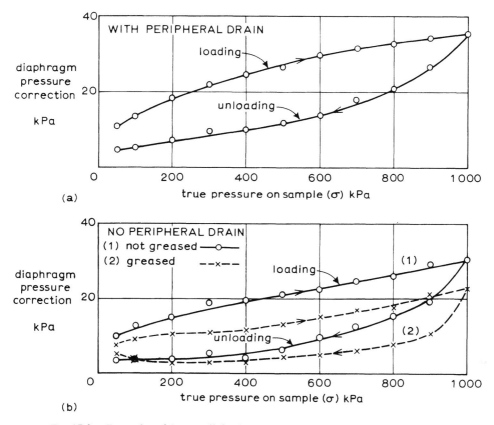

Fig. 17.9   Examples of Rowe cell diaphragm calibration curves: (a) with peripheral drain fitted, (b) without peripheral drain, showing effect of greasing the cell walls

(1) Bolt the cell body to its base, and place a rigid dummy specimen, such as the rigid loading plate supported on three spacer blocks, in the cell (see Fig. 17.10 (a)). The top of the plate should be at the same level as the top surface of a normal specimen at the start of a test.

(2) Fit a strip of porous plastic material about 20 mm wide over the rim drainage outlet, to form a drainage 'wick' reaching down to the top of the plate (Fig. 17.10 (b)). It can be held in place by a wire clip of similar metal to the cell body. Alternatively a liner of the same material may be fitted all around the cell wall behind the diaphragm.

(3) Fill the cell with water.

(4) Fit the cell top, without trapping any air under the diaphragm, and bolt it to the cell body.

(5) Fill the space above the diaphragm with water from a header tank at about 1 m above bench level, through valve C, with the air bleed E open. Close E when the air has been displaced.

(6) Push the drainage spindle firmly on to the rigid disc with valve D open, to remove any water from between diaphragm and disc, then close valve D. Keep valve C open to maintain a small seating pressure on the diaphragm.

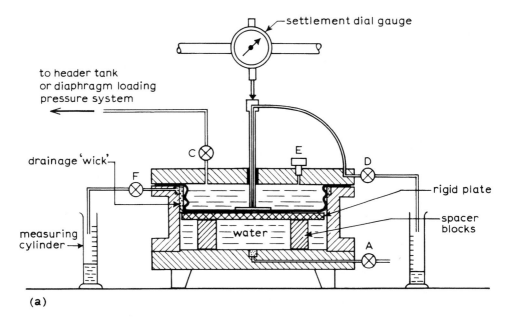

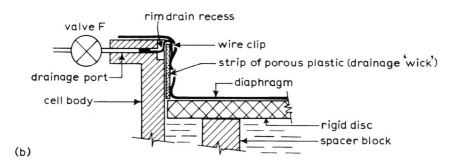

(b)

Fig. 17.10   Measurement of volume of water escaping from behind Rowe cell diaphragm: (a) general arrangement, (b) provision of drainage to release water trapped between convolutions

(7) Open valve F and collect the water that drains out in a measuring cylinder. Record the volume of water, and the time until flow virtually ceases.

(8) Repeat stage 7 under several higher diaphragm pressures (e.g. 25, 50, 100 kPa, etc) until the volume of water emerging from valve F is negligible. Ensure that water from behind the diaphragm convolutions does not lift the diaphragm from the rigid disc, if necessary by locking the spindle in place.

If the drainage strip referred to in step 2 is not fitted, drainage may be impeded by the folds of the convoluted diaphragm touching the cell wall. The above procedure may then take an hour or more, and should be timed for future guidance.

The recorded volume of expelled water provide an indication of the volume of surplus water that should be drained off from behind the diaphragm at any effective pressure before starting each drained stage of a consolidation test.

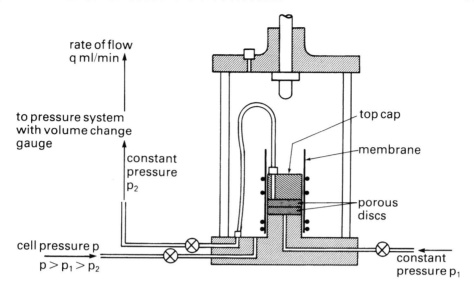

Fig. 17.11   *Arrangement for measuring head losses in triaxial cell connections and pipelines*

17.3.7   Pipeline Losses in Cell Connections

CAUSE OF LOSSES

When carrying out a permeability test in a triaxial cell or Rowe cell, there are inevitable losses of pressure head due to friction and turbulence in the connecting tubing, porous discs, and constrictions in valves, tube connectors and cell ports. These losses may be insignificant when testing clays of low permeability, where the rate of flow is very small. But for faster rates of flow, such as occur with silts, the head losses in the pipework and connections (referred to here as 'pipeline losses') may be as much as or greater than the pressure difference across the specimen itself. They must then be taken into account in the permeability calculations.

LOSSES IN THE TRIAXIAL CELL

The following procedure is used for determining the pipeline losses in a triaxial cell.

(1) Place the two saturated porous discs that are to be used in the test one on top of the other on the cell pedestal, covered by the loading cap, and surrounded by a rubber membrane (Fig. 17.11). Flow conditions then correspond to those during a test with a specimen present.

(2) Fill the cell with de-aired water. Throughout the test the cell pressure should always exceed the pressure applied to the porous discs.

(3) Connect a pressure system to the cell outlet from the base pedestal, and a second system incorporating a volume change indicator to the top cap outlet.

(4) Ensure that the whole system is de-aired and leak-free, and full of de-aired water. Both pressure systems are at the same pressure initially.

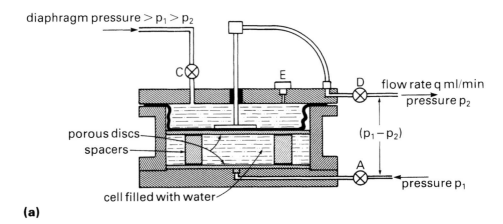

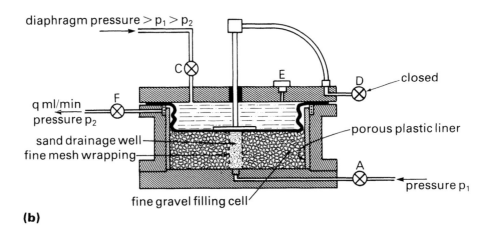

*Fig. 17.12   Arrangement for measuring head losses in Rowe cell: (a) with vertical
flow, (b) with radial flow*

(5) Increase the base pressure very slightly. The pressure difference $(p_1 - p_2)$ kPa should be small initially, and should be measured accurately with a manometer or differential gauge.

(6) Observe the rate of flow of water over a period of time, and when it becomes steady record it ($q \, cm^3$ per minute).

(7) Repeat steps 5 and 6 at several higher differential pressures up to about 100 kPa, or until the maximum practicable rate of flow is reached.

(8) Plot the recorded data as a graph of rate of flow (q) against pressure difference $(p_1 - p_2)$. The use of this graph is explained in Section 17.4.6.

LOSSES IN ROWE CELL

For vertical flow in the Rowe cell, one porous disc is placed on the cell base and the other is supported on spacers at a suitable height to support the diaphragm (Fig. 17.12 (a)). The cell is completely filled with de-aired water. The diaphragm pressure is maintained slightly in excess of the water pressure applied inside the cell to ensure that the diaphragm is seated flat on the top disc. A seal around the circumference of the lower disc is not necessary.

The calibration procedure and plotting is similar to that described above for the triaxial cell. Use of the calibration curve is explained in Section 17.4.6.

If it is considered that the presence of the spacers impedes the flow of water, the test can be repeated with twice the number of spacers, or spacers covering twice the area. If the two rates of flow differ at a given pressure, this difference is added to the original rate of flow to give the value corresponding to no spacers.

To obtain a calibration for radial flow, the cell is fitted with the peripheral drain of porous plastic, and the central drainage well of fine sand is wrapped with mesh fine enough to retain its finest particles. The cell is filled with fine uniform gravel to support the drainage well, and then with de-aired water (Fig. 17.12 (b)). The rate of flow calibration is similar to the above.

17.3.8 Porous Discs

PERMEABILITY

The water permeability of a porous disc may need to be known for comparison with other discs; as a means of assessing the extent to which its voids are clogged by soil particles; or to verify that it is appreciably greater than the permeability of the test specimen. The permeability of porous discs used in triaxial or Rowe cells can be determined by measuring rates of flow under known pressure differences using the procedure outlined in Section 17.3.7.

Measurements are first made with the disc mounted between two layers of material having a much higher permeability, such as glass fibre, as indicated in Fig. 17.13 (a). A second set of measurements is made with the test disc removed and the two interposed layers in contact, to determine the total head losses to be deducted in the disc permeability calculations. The method of calculation is as follows.

A graphical plot of rate of flow against pressure difference is made for each set of readings, represented by curves OD and OA respectively in Fig. 17.13 (b). For an appropriate rate of flow, q_m ml/minute, the difference between the two curves (δp) kPa represents the pressure drop across the disc itself. If its diameter and thickness are denoted by D mm and t mm respectively, the area of cross-section A mm^2 is equal to $\pi D^2/4$ and the permeability of the disk k_D is calculated from the equation

$$k_D = \frac{q_m t}{60A \times 102(\delta p)} \text{ m/s} \qquad (17.4)$$

AIR ENTRY VALUE

Air entry value is usually more significant than permeability for porous discs used at pore pressure measuring points. The value for a triaxial cell disc can be determined as follows.

(1) Connect the base pedestal of the triaxial cell to the control panel, as in Fig. 18.4, Section 18.2.3. The panel is connected to a mercury manometer. Ensure that the system is completely air-free and leak-free.

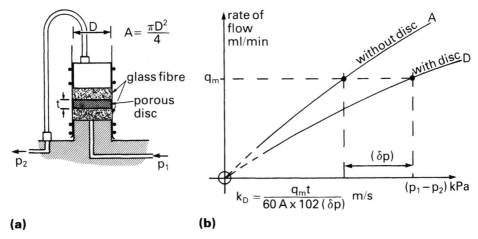

Fig. 17.13   *Measurement of water permeability of porous discs: (a) arrangement,*
(b) curves of flow rate related to pressure difference

(2) Moisten the pedestal with de-aired water and place the saturated porous disc on it without entrapping air.

(3) Seal the outside edge of the disc to the pedestal, either with a sealant or with a rubber membrane clamped to the disc with an O-ring, or as in Fig. 16.4 (Section 16.2.2).

(4) Use the control cylinder to push de-aired water through the disc to ensure that it is saturated.

(5) Soak up free water from the surface of the disc so that it becomes surface-dry but still saturated.

(6) Apply steadily increasing negative pressure to the disc with the control cylinder, measured by the mercury manometer, until air breaks through the surface tension barrier and no additional suction can be sustained.

(7) Record the maximum negative pressure sustained as the air-entry value of the disc.

MEASUREMENTS

Additional useful data to have recorded are the diameter and thickness of each disc, and the mass both dry and fully saturated. The latter is used if a 'moisture balance' check on the triaxial sample is needed at the end of a test.

17.3.9   Membranes and Side Drains

MEMBRANES

A procedure for the determination of the elastic modulus of rubber membranes, from which the barrelling correction to the measured compressive strength of a soil specimen can be estimated, was described in Volume 2, Section 13.7.4. Further published data on the barrelling correction, and on other aspects of the effect of rubber membranes, are reviewed in Section 17.4.2, together with suggested methods of application.

SIDE DRAINS

A correction for the presence of conventional filter-paper side drains, additional to the membrane correction, for a barrelling type of failure was determined by Bishop and Henkel (1962). Details are given in Section 17.4.3, together with a comment on spiral drains and recommendations for slip plane corrections.

17.4 CORRECTIONS TO TRIAXIAL TEST DATA

Details of corrections which are applied to observed data from triaxial tests are given in this section. Those that are invariably applied as part of the routine procedure are listed in Section 17.4.7. Others are additional refinements which are restricted to certain applications such as research work.

17.4.1 Area Corrections

BARRELLING

The correction to apply for the increasing area due to axial strain (the 'barrelling' correction) in an undrained test is the same as that described in Volume 2, Section 13.3.7. The corrected deviator stress $(\sigma_1 - \sigma_3)$ is given by the equation

$$(\sigma_1 - \sigma_3) = \frac{P}{A_0} \frac{(100 - \varepsilon\%)}{100} \tag{13.7}$$

Using the conventional units, applied axial force is P newtons, initial area of the consolidated specimen is A_c mm^2, and axial strain is $\varepsilon\%$, then

$$(\sigma_1 - \sigma_3) = \frac{P}{A_c} \frac{(100 - \varepsilon\%)}{100} \times 1000 \, \text{kPa} \tag{17.5}$$

In a drained test this expression has to be modified to take account of the change in volume ($\Delta V \, \text{cm}^3$) due to drainage (positive if water is draining out of the sample). The volumetric strain is equal to $\Delta V / V_c$ where $V_c \, \text{cm}^3$ is the volume at the start of compression, i.e. immediately after consolidation.

The corrected area A is given by the equation

$$A = \frac{1 - \dfrac{\Delta V}{V_c}}{1 - \dfrac{\varepsilon\%}{100}} A_c \tag{17.6}$$

Equation (17.5) then becomes

$$(\sigma_1 - \sigma_3) = \frac{P}{A_c} \frac{(100 - \varepsilon\%)}{100} \times \frac{1000}{1 - \dfrac{\Delta V}{V_c}} \, \text{kPa}$$

i.e.

$$(\sigma_1 - \sigma_3) = \frac{P(100 - \varepsilon\%)}{A_c\left(1 - \dfrac{\Delta V}{V_c}\right)} \times 10 \, \text{kPa} \tag{17.7}$$

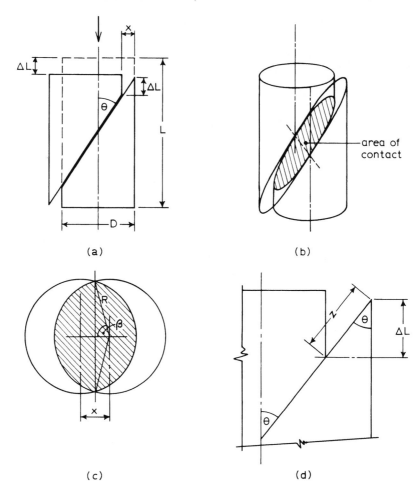

Fig. 17.14   *Area correction due to single-plane slip: (a) mechanism of slip, (b) area
of contact between the two portions of the specimen, (c) projected area of contact,
(d) displacement along slip surface related to vertical deformation*

SINGLE-PLANE SLIP

When failure occurs by slipping along a surface of shear the effective plan area, used
for calculating axial stress, decreases as movement takes place. The mechanism is
illustrated in Fig. 17.14 (a). The overlapping elliptical surfaces of sliding (Fig. 17.14 (b))
can be projected vertically to give a double segment area in plan, as shown in
Fig. 17.14 (c).

Each half of the shaded area of overlap in Fig. 17.14 (c) is a segment of a circle of radius
$R = D/2$ subtending an angle 2β (radians) at the centre. The area of each segment is equal
to

$$\tfrac{1}{2}R^2(2\beta - \sin 2\beta)$$

The plan area of overlap, A_s, is therefore given by the equation

$$A_s = 2 \times \frac{1}{2} \left(\frac{D}{2} \right)^2 (2\beta - \sin 2\beta)$$

$$= \tfrac{1}{2} D^2 (\beta - \sin \beta \cos \beta) \tag{17.8}$$

The angle β depends on the horizontal displacement x, i.e.

$$\cos \beta = \frac{x/2}{D/2} = \frac{x}{D}$$

From Fig. 17.14 (a), strain $\varepsilon_s = \Delta L/L$ and $x = \Delta L \tan \theta = \varepsilon_s L \tan \theta$.

Therefore
$$\cos \beta = \varepsilon_s \frac{L}{D} \tan \theta \tag{17.9}$$

where θ is the inclination of the slip surface relative to the specimen axis, and ε_s is the axial strain measured from the start of slip.

If the $L:D$ ratio of the specimen is $2:1$, Equation (17.9) can be written

$$\cos \beta = 2\varepsilon_s \tan \theta \tag{17.10}$$

The above expressions were given in a different form by Chandler (1966), who used the inclination of the slip plane relative to the horizontal, denoted by α, where $\alpha = 90° - \theta$. His graphs relate the plane area of contact to axial compression for values of α of 60°, 55°, and 45° for a 1.5 inch (38 mm) diameter specimen.

To enable the correction to be applicable to specimens of any diameter it is convenient to use the area ratio A_s/A, where A is the area at the start of slip ($= \pi D^2/4$), and A_s is the contact area at a certain axial displacement during slip. The area ratio is given by the equation

$$\frac{A_s}{A} = \frac{2}{\pi} (\beta - \sin \beta \cos \beta) \tag{17.11}$$

The slip area factor, f_s, by which values of P/A must be multiplied to obtain the corrected deviator stress, is the reciprocal of the above, i.e.

$$f_s = \frac{\pi}{2(\beta - \sin \beta \cos \beta)} \tag{17.12}$$

(β is expressed in radians).

Values of f_s for axial strains up to 40%, and for values of θ from 25° to 45° are shown graphically in Fig. 17.15. These curves are applicable to specimens of any diameter provided that the $L:D$ ratio is $2:1$. For specimens of a different ratio, the same graphs may be used if the strain is first multiplied by $L/2D$. The dashed curve in Fig. 17.15 is for a 'corner to corner' slip plane in a specimen of $2:1$ ratio (i.e. $\theta = \tan^{-1} 0.5 = 26.6°$).

If a slip along a surface of shear does not begin from zero strain, the area A at which it does begin should be calculated from the barrelling equation

$$A = \frac{100}{(100 - \varepsilon\%)} \times A_c \qquad \text{(from Equation (13.6))}$$

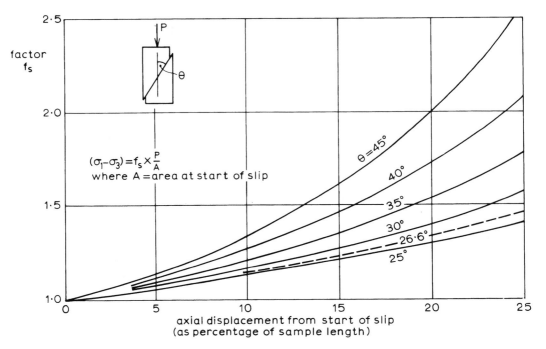

Fig. 17.15 *Slip area factor* f_s *related to axial displacement and inclination of shear surface* (*for* $L/D = 2$)

The strain ε_s is measured from the point at which slip begins. It is often difficult to observe just when this occurs, but it can usually be estimated more easily from measurements of slip in the specimen at the end of the test. If the displacement z along the slip surface is measured (Fig. 17.14 (d)), the longitudinal movement due to slip, ΔL, is equal to $z\cos\theta$. Subtracting this from the total recorded axial deformation gives the deformation at the start of slip.

An approximate correction factor derived by the author is

$$f_s = \left[1 + \left(0.06\theta \times \frac{\varepsilon_s\%}{100}\right)\right] \qquad (17.13)$$

where θ is the angle referred to above, measured in degrees, and $\varepsilon\%$ is the strain (%) measured from start to slip. This approximation is reasonable if ε_s does not exceed 15% and θ lies between 27° and 35°.

Although the term 'strain' has been used above for mathematical convenience, it does not have the same meaning after the start of slip. The 'magnitude of displacement', which is not dependent on specimen length, is more relevant.

17.4.2 Membrane Corrections

Membrane corrections are discussed under the following headings

(1) Barrelling

(2) Slip plane

(3) Membrane penetration effect on volume change

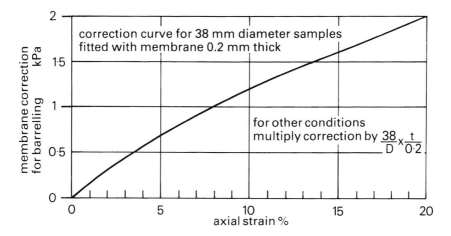

Fig. 17.16   *Membrane correction curve (barrelling failure) for 38 mm diameter specimens*

(4) Membrane penetration effect on pore pressure

(5) Membrane stretching in shear plane failure

(6) Trapped air

(7) Air permeability

(8) Water permeability

(9) Alternatives to the use of membranes

(1) *Barrelling*

The membrane correction to be applied to the measured deviator stress with a barrelling type of failure is as described in Volume 2, Section 13.3.8. The correction curve (Fig. 13.17) is shown here in Fig. 17.16. This curve is a compromise derived by the author from data published by Wesley (1975), Sandroni (1977) and Gens (1982), and has been incorporated into BS 1377:Part 8:1990 (Fig. 4).

The correction curve applies to a specimen of 38 mm diameter fitted with a membrane 0.2 mm thick. The correction is multiplied by $38/D \times t/0.2$ to allow for specimens of any other diameter (D mm) or membranes of any other thickness (t mm). For large diameter specimens, and for stiff soils, the correction is unlikely to be significant.

(2) *Slip plane*

In effective stress triaxial tests in which failure occurs on a surface of shear a more elaborate correction for the effect of the membrane than that given above is justified. Several workers who have investigated the membrane effect in slip plane deformation (e.g. Chandler, 1966; Blight, 1967; La Rochelle, 1967; Symons, 1967; Symons and Cross, 1968; Balkir and Marsh, 1974) have shown that the restraint provided by the membrane is dependent on cell pressure. The published tests were carried out on dummy specimens of

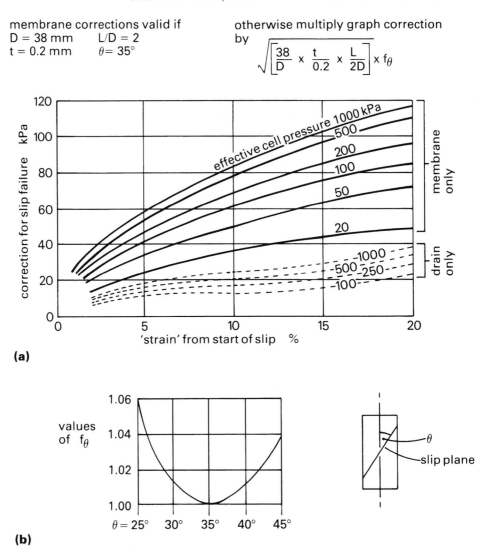

membrane corrections valid if   otherwise multiply graph correction
D = 38 mm   L/D = 2   by
t = 0.2 mm θ = 35° $\sqrt{\left[\dfrac{38}{D} \times \dfrac{t}{0.2} \times \dfrac{L}{2D}\right]} \times f_\theta$

(a)

(b)

Fig. 17.17 *Membrane and drain corrections for single-plane slip: (a) correction
curves and general equation, (b) values of factor f_θ for various inclinations of slip
plane (after La Rochelle, 1967)*

Plasticine or Perspex with pre-formed sliding surfaces which were well lubricated or fitted
with ball bearings, and results are not all compatible with one another.

The author suggests the use of the set of correction curves based on equations of La
Rochelle (1967), given in Fig. 17.17 (a). These relate to a slip plane inclined at 35° to the
axis of a 38 mm diameter specimen 76 mm long fitted with a rubber membrane 0.2 mm
thick. For any other specimen diameter (*D* mm), length (*L* mm), and rubber thickness
(*t* mm) the values derived from the curves should be multiplied by

$$\sqrt{\dfrac{38}{D} \times \dfrac{t}{0.2} \times \dfrac{L}{2D}}$$

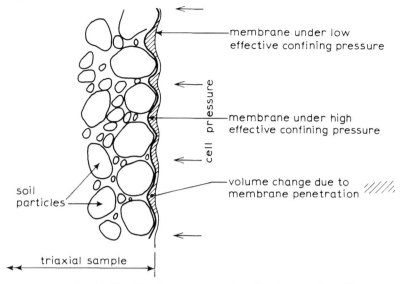

Fig. 17.18   *Membrane penetration effect in granular soils*

Cell pressures marked on the curves relate to the effective confining pressure, i.e. (cell pressure minus p.w.p.). 'Strain' refers to the displacement after the start of slip on the surface, which may be easiest to determine from measurements made on the specimen after test (see Section 17.4.1). The inclination of the slip plane is not a critical factor but it may be allowed for by multiplying by the correction factor f_θ derived from the graph in Fig. 17.17 (b).

Corrections for slip plane failure are not given in BS 1377.

(3)   *Membrane penetration effect on volume change*

In a triaxial test on a granular soil a volume change measurement made as a result of an increase in confining pressure will be influenced by the penetration of the membrane enclosing the specimen into the voids between particles at the interface. This is known as the membrane penetration effect, illustrated in Fig. 17.18. It can affect volume change measurements in both the back pressure line and the cell pressure line. The effect has been shown to depend mainly on particle size, and to a much lesser extent on state of packing (i.e. density) and particle shape, as well as on membrane thickness and stiffness. It applies to materials of medium sand size and upwards, having a 50% particle size, D_{50}, exceeding 0.1 mm, and is of greatest significance with large diameter specimens. The effect is negligible for fine-grained soils.

Newland and Allely (1959) studied the effect using lead shot, and Roscoe, Schofield and Thurairajah (1963) investigated it using Ottawa sand. A graphical method for estimating the effect in sandy silts, based on the 50% size D_{50}, was given by Frydman, Zeitlen and Alpan (1973). Corrections of the same order of magnitude were derived theoretically by Poulos (1964) using measured elastic properties of the membrane material. More recent investigations were made by Molenkamp and Luger (1981). This correction is unlikely to be significant in routine testing.

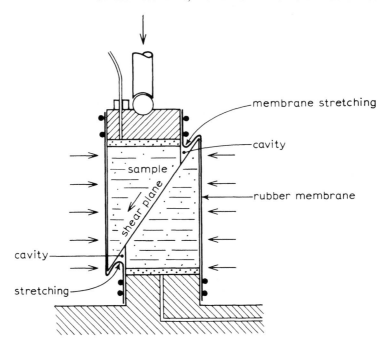

Fig. 17.19   *Stretching of rubber membrane due to single-plane slip*

(4)   *Membrane penetration effect on pore pressure*

The membrane penetration effect illustrated in Fig. 17.18 introduces a degree of flexibility into the pore pressure measuring system when testing granular soils. It allows movement of pore water which affects the magnitude of the pore pressure measured in an undrained test, and therefore influences the pore pressure parameters and effective stresses. However, the effective strength envelope and shear strength parameters are not likely to be affected.

The influence of membrane penetration on pore pressure was investigated by Lade and Hernandez (1977), using uniformly graded sands and glass beads. Details of their theoretical analysis and experimental work are of interest mainly for research. The effect of membrane flexibility is greatest for specimens made up of large size particles. For soft soils with high compressibilities membrane flexibility can be neglected in routine tests.

(5)   *Membrane stretching in shear plane failure*

When slip occurs along a surface of failure, the 'wedging' effect of the relative movement of the two portions of the specimen theoretically stretches the rubber membrane, tending to increase the volume between specimen and membrane (Fig. 17.19). In a drained test water could be drawn into the space thus formed. In a slow undrained test a negative pore pressure may develop, reducing the pore pressure in the specimen as a whole and increasing the effective stress. The shear strength may then continue to increase with further axial strain. This effect was investigated by Chandler (1968). However, the strength of overconsolidated clay which shows an appreciable drop in strength when a slip plane develops is not likely to be influenced by this effect. In any case local deformation of a real

Table 17.1 VOLUME CORRECTIONS FOR AIR TRAPPED UNDER
MEMBRANE ($L:D$ ratio $2:1$)

Specimen diameter mm	Volume correction cm³	
	trimmed specimen	compacted specimen
38	0.2	1
50	0.4	2
70	1.1	5
100	3.1	16
150	11.0	53

(After Bishop and Henkel, 1962)

soil ensures that the amount of membrane stretching is less than that indicated by the idealised condition in Fig. 17.19, and this effect is usually neglected.

(6) *Trapped air*

The possibility of trapping air between the membrane and the specimen cannot be avoided, no matter how carefully the membrane is fitted. Flushing the side of the specimen with water is not acceptable because it introduces an equally uncertain volume of water. In any case this procedure cannot be used with partially saturated or overconsolidated soils. Application of back pressure during the saturation stage eventually forces air into solution and eliminates the formation of bubbles in the back pressure line, provided that the back pressure is maintained.

The volume of trapped air is not likely to be less than 0.2% of the specimen volume (Bishop and Henkel, 1962). For a 100 mm diameter compacted specimen the volume may be as much as 1%. Corrections for the most usual diameters of specimen, of 2:1 ratio, based on these data, are summarised in Table 17.1

These corrections can be considered as bedding errors which disappear under high pressures. The volume (v) of trapped air can be calculated from the following equation, but for small specimens the correction is usually neglected.

$$v = \frac{\dfrac{\Delta V}{V_0}}{\dfrac{101}{101 + p} - 1} - n_0 \left(\frac{100 - 0.98 S_0}{100} \right) V_0 \text{ cm}^3 \qquad (17.14)$$

This equation relates to undrained conditions in which the pore pressure is increased due to the application of a confining pressure.

If the back pressure is raised enough to result in full saturation, the initial volume of entrapped air is given by the equation

$$v = \Delta V - n_0 \left(\frac{100 - S_0}{100} \right) V_0 \text{ cm}^3 \qquad (17.15)$$

In Equations (17.14) and (17.15), n_0 is the initial porosity of specimen ($= e_0/1 + e_0$), S_0 is the initial degree of saturation (%), V_0 is the initial specimen volume (cm^3), ΔV is the change in volume of air in the pore fluid (cm^3) (equal to specimen volume change as measured on cell pressure line), and p is the final pore pressure (kPa). The atmospheric pressure is taken as 101 kPa.

(7) Air permeability

The permeability of a rubber membrane to air is quite high. Migration of air through the membrane can cause errors in volume change and pore pressure measurements in the specimen if the water in the cell contains dissolved air, and is the main reason for using de-aired water as the cell fluid in long-term tests on saturated soils. Conversely, air in a partially saturated soil can diffuse through the membrane into the cell if it is filled with de-aired water. Use of two or more membranes makes no difference because the air permeability of the rubber is of the same order as that of water.

(8) Water permeability

A rubber membrane is not completely impervious to water, but if the water in the cell is free of air the volume of water migrating through a membrane without flaws is negligible in a period under 8 hours. According to Bishop and Henkel, for longer periods the error is likely to be less than 0.02% of the specimen volume per day, which can be neglected for most tests on small specimens.

The rate of migration can be measured by enclosing a porous ceramic cylinder in a membrane and setting it up in a triaxial cell under the appropriate pressure. The permeability of latex rubber membrane material, as stated by Poulos (1964), is about 5×10^{-18} m/s, but the quality of membranes can be variable.

Migration of water can be substantially reduced by using two membranes, separated by a coating of silicone grease. The water permeability of the membrane itself is appreciably reduced if it is soaked in water for at least 24 hours before use. Water can also migrate along the interface between membrane and end caps. This form of leakage can be prevented by polishing the curved surfaces of the caps perfectly smooth and keeping them scrupulously clean. Before setting up, a thin smear of silicone grease should be applied, and at least two tight-fitting O-rings should be fitted at each end.

(9) Alternatives to use of membranes

Some of the disadvantages associated with the use of rubber membranes have been overcome by making special modifications to suit particular circumstances. Two of these are as follows.

(a) *Mercury surround.* A surround of mercury in a perspex cylinder within the cell is described by Bishop and Henkel (1962), Appendix 6(2), Fig. 134. The specimen is fitted with a membrane as usual, but air has negligible solubility in mercury. This arrangement was found to eliminate the error due to air diffusion in undrained tests of long duration carried out for research purposes.

(b) *Paraffin as cell fluid.* The use of paraffin as the cell fluid was reported by Iversen and Moum (1974). This method eliminated the need for membranes and the consequent risk of disturbance to very soft sensitive soils such as Norwegian quick clays. It was found that paraffin was immiscible with the pore water in clays under confining pressures up to about 1000 kPa. Special end caps had to be made, using fine porous stones with a very high air-entry value. This method could cause problems in soils containing features such as root holes or shells.

17.4.3 Side Drain Corrections

BARRELLING

A correction additional to the membrane correction is applied to the measured deviator stress to allow for the restraint imposed by conventional filter paper side drains, when fitted. This correction increases rapidly with strain up to about 2% strain, then remains fairly constant. For a 38 mm diameter specimen fitted with side drains of Whatman No. 54 filter paper, as described in Section 18.4.7, step 5, it is about 9.5 kPa. For specimens of other diameters fitted with drains of the same grade of filter paper in a similar manner, the correction is inversely proportional to the diameter. Corrections for the most usual specimen diameters, suitably rounded, are summarised as follows.

Specimen diameter	(mm)	38	50	70	100	150
Drain correction (additional to membrane correction)	(kPa)	10	7	5	3.5	2.5

These values are given in Table 2 of BS 1377:Part 8:1990, and apply to strains from 2% upwards. For strains from zero up to 2% the correction should increase linearly from zero with increasing strain. For example, for a 38 mm diameter specimen the correction would be 0.5 kPa for every 0.1% strain up to 2%, and 10 kPa thereafter. The total correction is deducted from the calculated deviator stress.

Spiral drains of the type described in Section 16.7.5 are claimed to be as effective as conventional drains for drainage, without appearing to affect the measured strength in either compression or extension (Gens, 1982).

SLIP PLANE

Tests to determine the correction for side drains under conditions of single-plane slip were carried out by La Rochelle (1967), Symons (1967), and Balkir and Marsh (1974), using 38 mm diameter 'specimens' of Perspex fitted with ball bearings on the sliding surface. There was some evidence of a small increase in resistance with increasing cell pressure.

The author suggests the use of the simplified drain correction curves, based on data from La Rochelle (1967), shown dashed in Fig. 17.17 (a). This is for drains on a 38 mm diameter specimen, and is related to the displacement (as percentage strain) measured from the start of slip. Drain corrections for specimens of any other diameter (D mm) can be obtained by multiplying the correction by $38/D$. If barrelling deformation to a strain of 2% or more precedes the start of movement along the slip surface, the drain correction for barrelling should be added. The total drain correction is additional to the membrane correction derived as explained in Section 17.4.2.

Slip plane corrections for both membrane and side drain can be avoided by not extending a compression test to a large strain when trying to measure the residual shear strength. A better approach, when the peak strength is already known or is not of importance, is to start with a specimen that has been 'sliced' along a suitable orientated surface. The residual strength will then be reached after a relatively small displacement for which only the area correction need be applied.

Corrections for slip plane failure are not given in BS 1377.

17.4.4   Volume Change Corrections

BACK PRESSURE LINE

The influence of the effects of the rubber membrane and side drains on volume change measurements in the back pressure line is discussed above in Sections 17.4.2 and 17.4.3. Provided that there is no leakage in the system, and that connecting leads are allowed to expand or contract before opening the drainage valve to admit a pressure change to the specimen, no other corrections should be necessary to these measurements.

The sign convention for movement of water in the back pressure line is summarised as follows.

Water draining out of specimen (consolidation): positive change (+)

Water moving into specimen (saturation and swelling): negative change (−)

CELL VOLUME CHANGE

Measurements of volume change on the cell pressure line require several corrections and are consequently less accurate than volume change measurements made in the back pressure system. Cell volume change measurements are not required in the test procedures specified in BS 1377:Part 8:1990. However, the movement of water into or out of a partially saturated soil from the back pressure system is not a true measure of the specimen volume change, and for some soils (e.g. expansive clays) independent measurements based on changes in the cell line are necessary.

Factors which affect the movement of water into or out of the cell, and which must therefore be taken into account in cell line volume change measurements, are as follows (see Fig. 17.20).

(1) Irregularities on specimen surface.

(2) Air trapped between specimen and rubber membrane, and side drains (if fitted).

(3) Expansion of cell due to pressure increase.

(4) Continued expansion of cell (creep) with time under constant pressure.

(5) Absorption of water into, and migration through, cell body.

(6) Air trapped at top of cell.

(7) Leakage of water: through piston bushing
                            through membrane and bindings
                            from connecting lines and valves.

(8) Voids in specimen.

(9) Change in specimen volume due to drainage of water (consolidation or swelling).

(10) Movement of piston.

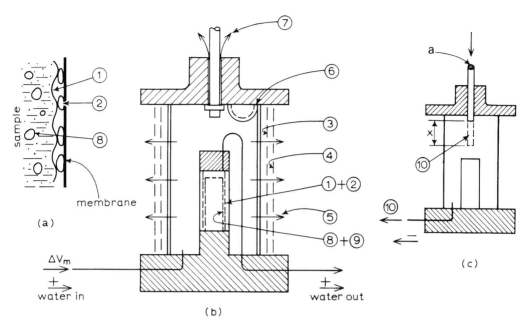

Fig. 17.20 *Representation of factors which influence cell volume change measurements:* (a) *components due to specimen,* (b) *effects during saturation and consolidation stages,* (c) *effect of piston penetration during compression*

These factors are represented diagrammatically in Fig. 17.20 (a) and (b) where they are numbered as above. The effect of each when the pressure of the water in the cell is increased will be considered, first assuming that the cell piston is secured at the upper limit of its travel so that it cannot move.

CORRECTIONS DURING SATURATION AND CONSOLIDATION

Movement of water *into* the cell is positive $(+)$ because this is observed when there is a positive change in volume (consolidation) of the specimen (see Fig. 17.20 (b)), and movement *out* of the cell is negative $(-)$.

(1) Increased cell pressure will cause the membrane to penetrate more tightly into surface irregularities, resulting in water moving *into* the cell $(+)$.

(2) Air trapped under the membrane will be compressed; water moves *into* the cell $(+)$.

(3) The cell body will expand, bringing more water *into* the cell $(+)$.

(4) Further expansion will continue with time due to creep; more water *into* the cell $(+)$.

(5) Absorption by and migration of water into the cell wall (if of Perspex or acrylic materials) is replaced by water moving *into* the cell $(+)$.

(6) Trapped air will be compressed; water moves *into* the cell $(+)$.

(7) Water leakage will be replaced by more water moving *into* the cell $(+)$.

(8) Voids in the specimen will tend to close, causing water to move *into* the cell $(+)$.

(9) Specimen decrease due to consolidation will cause water to move *into* the cell (+). If the specimen is fully saturated, this volume change should equal the volume draining out of the specimen into the back pressure system.

(10) No change if piston does not move.

All the changes numbered 1–9 are in the same direction, positive for an increase (positive change) in cell pressure. The measured cell volume change ΔV_m for a given pressure increment is the sum of all effects 1–9. The actual change in volume of the specimen, ΔV, is the sum of items 8 and 9, and is related to ΔV_m by the equation

$$\Delta V = \Delta V_m - [(1) + (2) + (3) + (4) + (5) + (6) + (7)] \qquad (17.16)$$

Changes 1, 2, 3, 6 and 8 occur almost instantaneously, or within a very short time. In the consolidation stage of a test, item 8 will usually take much longer and therefore the long-term effects 4, 5 and 7 should also be allowed for. The calibration of a triaxial cell, described in Section 17.3.1, takes into account items 3 and 6 as immediate effects, and items 4, 5 and 7 as time-dependent effects. These corrections will be more or less constant for a particular cell under a given pressure and at a certain temperature, but they should be checked from time to time.

   The effects of membrane penetration, and of the air trapped beneath it, items 1 and 2, are discussed in Section 17.4.2, part (3).

CORRECTIONS DURING COMPRESSION

Item 10 above must be considered during the compression stage because the piston moves into the cell as the axial deformation of the specimen increases (positive change). The effect is to push water *out of* the cell, giving a negative (−) volume change reading (Fig. 17.20 (c)). The volume ΔV_p displaced by the piston, having a cross-sectional area of a mm^2, during an axial deformation of y mm, is given by the equation

$$\Delta V_p = -\frac{a \cdot y}{1000} \ \text{cm}^3 \qquad (17.17)$$

The actual change in volume of the specimen, ΔV, during the compression stage, when the cell pressure remains constant, is given by the equation

$$\Delta V = \Delta V_m - [(4) + (5) + (7) + \Delta V_p]$$

i.e.   (17.18)

$$\Delta V = \Delta V_m - \left[(4) + (5) + (7) - \frac{a \cdot y}{1000} \right]$$

In a drained test the drainage component 9 is measured directly by the volume change indicator in the back pressure line, where movement *out* is positive. In an undrained test Equation (17.18) gives the volume change due to the voids component 8, which is zero in a fully saturated specimen.

17.4.5 Piston Friction

STANDARD BUSH

The correction to the measured axial force due to friction of the piston in the cell bushing can be allowed for by running the compression machine, with the cell under pressure, at the rate of displacement required for the test, but with the piston not in contact with the specimen top cap. If this is done immediately before starting a compression test, and the load ring dial gauge is set to read zero, no further correction will be necessary at that machine speed so long as the load remains truly axial.

If there are no lateral forces, the friction loss for a piston and bush in good condition should be small if oil is inserted at the top of the cell as a lubricant. Bishop and Henkel quoted errors of about 1–3% of the axial force, increasing with axial strain. Tests on 100 mm diameter specimens in a cell with a 19 mm diameter piston indicated that a correction of about 1% of the axial force should be deducted for every 5% strain.

When a specimen fails by slipping along a single plane, lateral forces are introduced which can appreciably increase the bush friction. Under these conditions Bishop and Henkel suggested that the error could rise to about 5% of the axial force. The correction needed would also depend on the axial strain, and a deduction of about 1% of the measured force for every 2% strain from the start of slip seems appropriate.

The nominal corrections suggested above are accurate enough for most practical purposes. If the frictional force is likely to exceed about 2% of the measured axial force it is better to measure the force with a device mounted inside the triaxial cell (see below).

ROTATING BUSH

Where the effect of piston friction is significant, such as with very soft soils and stiff soils that fail along a slip surface, a cell fitted with a rotating bush reduces the friction to a negligible amount, if oil is used in the top of the cell. Provided that the piston and bush are in good condition, the effect of lateral thrust caused by slip-plane failure can be neglected.

The rotating-bush cell is described in Section 16.2.3.

SUBMERSIBLE LOAD CELL

The effect of friction between piston and cell bush can be eliminated altogether if a submersible force measuring device can be mounted inside the triaxial cell. Electrical devices of this kind, known as submersible load transducers, are now often used in place of externally mounted rings in commercial testing, especially for use with automatic recording and data-processing systems. Details of force transducers are given in Section 16.6.4.

17.4.6 Pipeline Losses

For a permeability test carried out in a triaxial cell, the pipeline loss calibration determined as in Section 17.3.7 is applied as follows.

The pressures applied by the constant pressure systems to the specimen are denoted by p_1 and p_2, p_1 being the greater. The overall pressure difference $(p_1 - p_2)$ causes a steady flow of water through the specimen of q_m cm^3 per minute (Fig. 17.21 (a)). From the

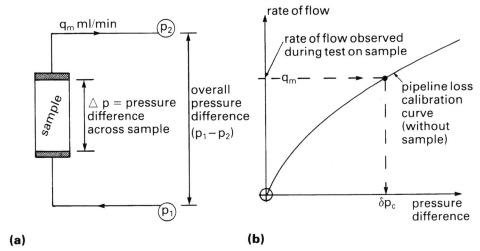

(a)   (b)

Fig. 17.21   *Application of pipeline loss calibration: (a) pressures and measured
flow, (b) derivation of pressure loss*

calibration curve represented in Fig. 17.21 (b) the pressure difference correction corresponding to q_m is read off, and is denoted by δp_c. This represents the total pressure loss in the connections to the specimen. It is deducted from the overall pressure difference to give the actual pressure difference across the specimen, Δp, i.e.

$$\Delta p = (p_1 - p_2) - \delta p_c \qquad (17.19)$$

The value of Δp is then used for calculating permeability from the tests described in Section 20.3, using the same cell and connecting pipelines.

17.4.7   Corrections for Routine Tests

The corrections described in the preceding sections that are necessary for general routine work are listed in Table 17.2. Other corrections would normally be used only where appropriate in special applications, or where very high accuracy is essential such as in research.

17.5   CORRECTIONS TO HYDRAULIC CELL TEST DATA

Details of corrections applicable to data from hydraulic cell (Rowe cell) tests are given in this section. The first is the diaphragm load correction, which should be applied to every loading pressure. The second is the drainage volume correction, applicable to each drainage stage of consolidation tests, and the third is the pipeline correction for permeability tests.

17.5.1   Diaphragm Load Correction

The calibration curves relating diaphragm pressure correction to true pressure on the specimen, obtained as described in Section 17.3.4, are used either for the application of a

Table 17.2. CORRECTIONS NECESSARY IN ROUTINE TESTS

Section	Type of correction	Comment
17.4.1	Area correction — barrelling — single plane slip	
17.4.2 (1) (2) (8)	Membrane correction — barrelling — single plane slip — water permeability	
17.4.3	Side drains — barrelling — single plane slip	no correction necessary with spiral drains
17.4.4	Volume change — back pressure line	pre-pressurising eliminates errors due to line expansion
	Cell volume — saturation — consolidation — compression	
17.4.5	Piston friction	no correction needed if external load ring is set to zero as piston is moved against cell pressure
17.4.6	Pipeline losses	permeability tests

specified stress to the specimen, or for the determination of the actual stress from a known pressure. Reference must be made to the particular graph that is appropriate to the test conditions, the diaphragm extension and whether the pressure is increasing or decreasing. Corrections at low pressures, especially in small cells, are likely to be unreliable.

APPLICATION OF A SPECIFIED VERTICAL STRESS

The vertical stress to be applied to the specimen is denoted by σ. From the appropriate calibration graph (shown diagrammatically in Fig. 17.22) the value of the pressure correction δp corresponding to σ (point A) is read off. The required diaphragm pressure p_d is given by the equation

$$p_d = \sigma + \delta p \qquad (17.20)$$

DETERMINATION OF ACTUAL STRESS

To determine the actual vertical stress on the specimen under a diaphragm pressure p_d, mark the value p_d on the horizontal axis of the appropriate calibration curve (point B in Fig. 17.22).

Draw the line through B perpendicular to the horizontal axis. Taking a point E at a convenient value on the pressure correction scale, measure horizontally a distance on the horizontal scale equal to that value to obtain the point D. Join BD, intersecting the calibration curve at C, and read off the value of δp for this point. The actual stress on the specimen is given by the equation

$$\sigma = p_d - \delta p \qquad (17.21)$$

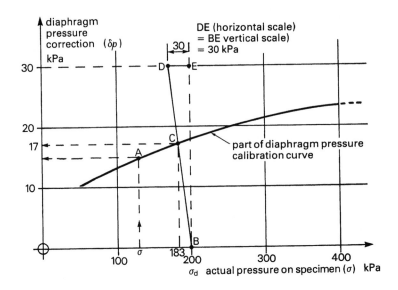

Fig. 17.22   *Use of Rowe cell diaphragm calibration curve*

In the example shown in Fig. 17.22, $p_d = 200$ kPa and the point E corresponds to 30 kPa; therefore DE = 30 kPa on the horizontal scale. Joining BD, the point C is found to be at $\delta p = 17$ kPa, hence the actual pressure on the specimen is equal to $200 - 17 = 183$ kPa. This is confirmed by projecting from C onto the horizontal axis.

17.5.2   Rim Drain Correction

The data obtained as described in Section 17.3.6 are used as a guide to the volume of surplus water to be drained off from behind the diaphragm before starting the drained stage of a consolidation test. The volume of water, and the time required to get rid of it, depend upon the effective pressure applied, and should be allowed for at every stage. The effect is greatest during the initial loading stages, and may become negligible at higher pressures. The spindle should be locked in place while increasing the diaphragm pressure to prevent water lifting the diaphragm clear of the specimen.

Provided that movement of water is not restricted by the folds of the diaphragm (e.g. by fitting a small drainage element as in Fig. 17.10 (b)) the rim drain valve F need be opened for only a few seconds just before opening the drainage valve D to start consolidation. This is not likely to have a noticeable effect on low-permeability soils.

17.5.3   Pipeline Losses

The correction for pipeline losses in the connections to the specimen during a permeability test in a Rowe cell is similar to that described in Section 17.4.6 for permeability tests in a triaxial cell. The calibration curve appropriate to the type of test (e.g. vertical flow; radial flow, inward or outward) must be used. With vertical flow, corrections are likely to be more significant than for triaxial specimens because of their lower ratio of height to diameter.

17.6 GENERAL LABORATORY PRACTICE

17.6.1 Scope

Some aspects of good laboratory practice and techniques were given in Volume 1, Section 1.3, and Volume 2, Section 8.1.4. This section makes reference to and supplements those recommendations, with reference to the equipment and procedures covered in this volume.

A safe working environment is of prime importance, but safety measures can be easily overlooked. All laboratory staff should be given instruction in basic safety precautions (Section 17.6.2).

The general laboratory environment is discussed briefly in Section 17.6.3. Errors that can occur in laboratory testing and analysis of results are reviewed in Section 17.6.4, and the importance of examining and considering test data and results is stressed in Section 17.6.5.

The importance of regular calibration and checking is discussed in Sections 17.2 and 17.3 above.

17.6.2 Safety

First aid equipment should be readily available at all times, and clearly marked. The staff of a large laboratory should include someone trained in first aid procedures, preferably holding a valid nationally recognised certificate.

General laboratory safety was discussed in Volume 1, Section 1.6. Additional aspects were included in Volume 2, Section 8.5. Particularly relevant to the tests described in this volume are the safe handling of compressed air (Section 8.5.3) and mercury (Section 8.5.4). Sulphur and slaked lime should be kept available for immediate treatment of areas where mercury is spilled, as described in Section 1.6.7.

Some additional safety precautions are as follows.

(1) One limb of a glass mercury manometer should always be open to atmosphere. Do not connect a source of pressure to both limbs in an attempt to use the manometer to measure differential pressure if the tube is not designed to withstand pressures greater than 100 kPa above atmospheric.

(2) The open limb of a mercury manometer should be connected to a catchpot to retain mercury in case of high pressure being inadvertently applied to the other limb.

(3) When using a vacuum, make sure that all vessels, tubing and connections will withstand the external atmospheric pressure.

(4) High pressures should be released by lowering the pressure at source, not by suddenly opening a valve or plug to atmosphere.

(5) Compressed air should not be used as a pressurising fluid in triaxial cells unless they have been designed and pressure-tested and certificated for that purpose.

17.6.3 Environment

Effective stress tests should be carried out at an even temperature, because temperature changes cause significant variations to pore pressure readings, whether observed manually or automatically. The ideal temperature is between about 20° and 25°C, but a higher temperature may have to be accepted in a hot climate. According to Parts 6 and 8 of BS 1377:1990, the ambient temperature should be controlled to within ±2 °C.

Test apparatus, including electronic instrumentation, should not be exposed to direct sunlight at any time of day. Apparatus should not be placed too close to sources of heat such as radiators, and should be protected from currents of air from ventilation or cooling systems, or from opening doors and windows.

Preparation of soil specimens should be carried out in a separate room with humidity control, so that the main testing area can be kept spotlessly clean and free from dust.

Samples should be stored in a temperature-controlled humidified room. Humidity can be maintained by continuously spraying fine jets of water on to non-absorbent walls, for instance of glazed tiles or copper sheets. The water if recirculated can be cooled or heated and treated with an anti-bacteria agent. A system of this kind was described by Bozozuk (1976). Racks should be of aluminium alloy, and shelves of plastics or aluminium. Ferrous components (except stainless steel) should not be used. Electric light fittings should be of weatherproof type, and the use of other electrical appliances should be prohibited.

The design of a laboratory for effective stress tests should incorporate the following features:

Floor surface drainage

Adequate lighting and ventilation

Sinks, water taps and drains

Adequate electric power sockets

Distribution lines for compressed air; vacuum; de-aired water

Durable bench top surfaces

Storage cupboards and drawers of various sizes

Desk space for writing, plotting, calculating.

Regular cleaning of the laboratory should be part of routine procedure. Areas of the laboratory where long term tests are being carried out, and where sensitive equipment (including computers) is used, should be clearly marked out as 'restricted areas' into which cleaning staff are not allowed entry. Cleaning within these areas should be done only by the laboratory staff responsible for the equipment concerned, as a regular duty.

17.6.4 Errors

Errors in scientific measurements may be due to human error, or may originate in the equipment in the procedures used. They can be divided into three categories:

(1) Mistakes

(2) Systematic errors

(3) Observational errors

The relevance of errors to the laboratory testing of soils, especially to the tests described in this volume, are discussed below. Ways of eliminating errors, or of minimising their effects, are outlined here or described in greater detail in other chapters as indicated.

(1) *Mistakes*

Mistakes, including gross errors, can be attributed to the following causes.

(a) Incorrect initial setting of a measuring device, e.g. a displacement dial gauge set with its stem not properly seated.

(b) Accidental movement or upsetting of a gauge.

(c) Mis-reading the scale of a dial gauge, or the digits of a display.

(d) Mis-counting the revolution counter of a dial gauge, or misreading the decimal point on a digital display (typical gross errors).

(e) Wrong recording of observed data.

(f) Wrong connection of electrical lines.

(g) Inadvertent omission of a stage of the test.

(h) Wrong plotting of points on a graph.

(j) Erroneous interpretation or reading-off from a graph.

(k) Arithmetical errors in calculation.

Mistakes are generally due to lack of attention. They can be detected by careful and systematic checking, and are usually eliminated by using correct techniques and following consistent procedures (see Volume 1, Section 1.3, and Volume 2, Section 8.1.4). Constant vigilance during a test should be practised (see Section 17.6.5).

The first course of action when results appear to be wrong should be to check the arithmetic. If there are no arithmetical mistakes, other possible sources of error should be investigated.

In effective stress triaxial tests, one of the common errors is not to run the compression stage slowly enough to allow pore pressure equalisation in undrained tests, or full drainage in drained tests. This leads to fictitious values of ϕ', which are too low for soils in which pore pressures increase with increasing deviator stress, and too high for dilatant soils. Another mistake is to curtail a test stage too soon, before pore pressures have fully dissipated or equalised. This practice leads to erroneous pore pressure measurements in subsequent stages.

The recommendations made in the appropriate sections regarding suitable rates of strain and end-of-stage criteria have been derived from experience and should always be followed. Short-circuiting of established procedures in an attempt to save time is, in the end, wasteful of both time and effort if the end results cannot be relied upon.

(2) *Systematic errors*

Systematic errors are those due to persistent causes, and are cumulative. The main sources of error are as follows.

(a) Small errors in the instruments used.

(b) Errors in technique due to personal habit.

(c) False data due to neglecting pre-test check routines or to not completing, or not taking sufficient care with, the preparation of equipment immediately before a test (e.g. de-airing procedures before pore pressure tests).

(d) Environmental effects—variations in temperature, pressure, humidity; vibrations or other disturbances.

(e) Influence of parts of the apparatus on observed readings (e.g. the rubber membrane in triaxial tests; clogged drainage discs).

(f) Wrong values derived from nominal measurements or data that have not been verified (often in the urge to save time).

These errors can be eliminated or minimised by good operator techniques (referred to above), by carefully following systematic pre-test procedures, and by proper calibration of instruments and test apparatus. They are not reduced by averaging a large number of results.

Nominal or 'standard' measurements should always be verified. For example the dimensions of a specimen cutting ring or tube can change with use due to wear, especially at a cutting edge. A test specimen may not be of the same diameter as the tube from which it was extruded. The thickness and stiffness of rubber membranes can vary considerably, and several sample membranes from each batch should be checked (Volume 2, Section 13.7.4).

Correct preparation and thorough checking are important for any apparatus, but are absolutely vital for the correct measurement of pore water pressures. Detailed procedures for equipment preparation are given in Section 16.8, and routine pre-test checks are described in Section 18.3. If these are not carried out properly, test results are likely to be meaningless.

Inadequate or wrong calibration is a common source of error. Environmental effects (of which those due to temperature variations are usually the most significant) should always be taken into account.

(3) *Observational errors*

Observational errors are those due to numerous small effects, some being beyond the control of the observer, and tend to be compensating. The usual causes are as follows.

(a) Small variations in reading a gauge (e.g. in interpolating between graduation markings).

(b) Small variations in technique (e.g. in adjusting a mercury null indicator for pore pressure readings).

(c) Random effects such as those due to electrical 'noise' in transducer readout or recording systems.

Errors of this kind depend on the perfection of the instrument or the skill of the observer. They cannot be eliminated entirely but are a matter of chance and are subject to the laws of probability. In general, positive and negative errors are equally probable, small errors are more frequent than large ones and very large errors seldom occur. The greater the number of readings taken, the smaller the error in the average result is likely to be.

17.6.5 Test Data and Results

The comments given in Volume 2, Section 8.1.4 under this heading are particularly relevant to the data obtained from effective stress tests. The technician should be constantly on the look-out for possible inaccuracies or erroneous data. During a long-duration test the question should be constantly asked — are the readings and results sensible for the test conditions and type of soil? It is better to terminate a dubious test and start again with another specimen than to complete a test that will give dubious or meaningless results.

Some typical questions which should be raised are:

Is the rate of drainage or absorption of water compatible with the estimated permeability characteristics of the soil?

Is the rate of consolidation likewise reasonable?

Are volume-change and pore pressure dissipation rates compatible?

Could very rapid pore pressure response be due to a leak in the membrane?

Has virtual termination of primary consolidation been reached?

Does the stress/strain relationship look reasonable?

Are the observed changes in pore water pressure, or specimen volume, following a reasonable pattern?

A correction applied to laboratory observations should be carefully scrutinised if it becomes of appreciable significance (e.g. more than 5% of the value to which it relates). The type and magnitude of any applied corrections should always be reported with the test results.

Derivation of permeability and consolidation parameters from triaxial tests in which filter paper side drains have been used should not be made. These drains may not be fully effective even for soils of very low permeability.

On completion of a set of triaxial tests the Mohr circles or stress paths should be critically examined before attempting to draw a failure envelope. If a good fit is obtained the question should be asked — is this envelope reasonable, or sensible, for this type of soil tested in this way? When a set of circles or stress paths does not give a clear-cut envelope the engineer should make a careful judgement taking into account the detailed behaviour of each specimen, or if necessary call for additional tests on similar material.

This questioning attitude is all the more necessary when electronic instruments are being used, because an electronic system introduces its own source of possible errors. For instance a wide scatter of observed readings may be due to mains voltage surges or other electrical 'noise' within the laboratory, or to a daily cycle of temperature fluctuations. A change in character of a graphical plot, such as a sudden flattening out, could be caused either by a mechanical defect or by a fault in the electrical measuring or read-out system.

The operator should always be fully aware of how the test is proceeding, and should be in overall control at every stage. Where automatic data acquisition or automatic control facilities are introduced the operator should continue making observations of the results from time to time throughout the test so as to ensure proper execution of the procedures, and to enable any necessary adjustments to be made. In effective stress testing it is essential to obtain an understanding of soil behaviour. Reliance should not be placed entirely on the end results.

REFERENCES

Balkir, T. and Marsh, A. D. (1974). 'Triaxial tests on soils: corrections for effect of membrane and filter drain'. *TRRL Supplementary Report 90 UC*. Transport and Road Research Laboratory, Crowthorne.

Blight, G. E. (1967). 'Observations on the shear testing of indurated fissured clays'. *Proc. Geotech. Conf.*, Oslo, Vol. 1, pp 97–102.

Bozozuk, M. (1976). 'Soil specimen preparation for laboratory testing'. *ASTM STP*, No. 599, p 113. Am. Soc. for Testing and Materials, Philadelphia, USA.

BS EN 10002, *Tensile Testing of Metallic Materials. Part 3: Calibration of Force Proving Instruments Used for the Verification of Uniaxial Testing Machines*. BSI, London.

BS EN 30012:1992, *Quality Assurance Requirements for Measuring Equipment*. BSI, London.

BS EN 45001, General criteria for the operation of testing laboratories. BSI, London.

Chandler, R. J. (1966). 'The measurement of residual strength in triaxial compression'. *Géotechnique*, 16:3:181–186.

Chandler, R. J. (1968). 'A note on the measurement of strength in the undrained triaxial compression test'. *Géotechnique*, 18:2:261.

European Standard EN 450001 (1989). *General Criteria for the Operation of Testing Laboratories*. CEN/CENELEC, Brussels.

Frydman, S., Zeitlen, J. G. and Alpan, I. (1973). 'The membrane effect in triaxial testing of granular soils'. *Journal of Testing and Evaluation*, 1:1:37.

Gens, A. (1982). 'Stress–strain and strength characteristics of a low plasticity clay'. PhD Thesis, Imperial College of Science and Technology, University of London.

Imperial Chemical Industries PLP, Petrochemicals and Plastics Division, Welwyn Garden City. Technical Note No. 265, 'Water vapour permeability of 'Perspex' '.

Iversen, K. and Moum, J. (1974). 'The paraffin method — triaxial testing without a rubber membrane'. Technical Note, *Géotechnique* 24:4:665.

Lade, P. V. and Hernandez, S. B. (1977). 'Membrane penetration effects in undrained tests'. *J. Geo. Eng. Div. ASCE*, Vol. 103, No. GT2, pp 109–125.

La Rochelle, P. (1967). 'Membrane, drain and area correction in triaxial test on soil samples failing along a single shear plane'. *Proc. 3rd Pan-Am. Conf. on Soil Mechanics & Foundation Engineering*, Caracas, Venezuela, Vol. 1, pp 273–292.

Molenkamp, F. and Luger, H. J. (1981). 'Modelling and minimization of membrane penetration effects in tests on granular soils'. *Géotechnique*, 31:4:471.

Newland, P. L. and Allely, B. H. (1959). 'Volume changes in drained triaxial tests on granular materials'. *Géotechnique*, 7:1:17.

Poulos, S. J. (1964). 'Report on control of leakage in the triaxial test'. *Harvard University Soil Mechanics Series*, No. 71, Cambridge, Mass., USA.

Roscoe, K. H., Schofield, A. N. and Thurairajah, A. (1963). 'An evaluation of test data for selecting a yield criterion for soils'. Laboratory shear testing of soils. *ASTM Special Technical Publication*, No. 361, pp 111–128. American Society for Testing and Materials, Philadelphia, USA.

Sandroni, S. S. (1977). 'The strength of London Clay in total and effective shear stress terms'. PhD Thesis, Imperial College of Science and Technology, University of London.

Shields, D. H. (1976). 'Consolidation tests'. Technical Note, *Géotechnique*, 26:1:209.

Sodha, V. (1974). 'The stability of embankment dam fills of plastic clay'. M Phil Thesis, Imperial College of Science and Technology, University of London.

Symons, I. F. (1967). Discussion, *Proc. Geotech. Conf.*, Oslo, Vol. 2, pp 175–177.

Symons, I. F. and Cross, M. R. (1968). 'The determination of the shear-strength parameters along natural slip surfaces encountered during Sevenoaks by-pass investigations'. Report LR 139. Transport and Road Research Laboratory, Crowthorne, Berks.

Wesley, L. D. (1975). 'Influence of stress path and anisotropy on the behaviour of a soft alluvial clay'. PhD Thesis, Imperial College of Science and Technology, University of London.

Chapter 18

Routine effective stress triaxial tests

18.1 INTRODUCTION

18.1.1 Scope

This chapter deals with the two most common routine triaxial compression tests for the measurement of the effective shear strength of soils. The full designation of these procedures and the abbreviations used here are:

(1) Consolidated undrained triaxial compression test with measurement of pore water pressure (the 'CU test').

(2) Consolidated-drained triaxial compression test with measurement of volume change (the 'CD test').

These tests are covered in BS 1377: Part 8 : 1990, Shear strength tests (effective stress). A similar test for undrained triaxial compression is described in ASTM Standards, Designation D 4767-88. The procedures given in this chapter are in accordance with those given in BS 1377, but requirements of the ASTM standard are also referred to where they differ significantly from the BS. The BS procedures were derived from those established by Bishop and Henkel (1962), which had been accepted as standard practice in many countries. These routine tests also provide the basis for other types of effective stress test.

Reference is made to Chapter 15 for relevant theoretical background, to Chapter 16 for descriptions of the equipment used, and to Chapter 17 for calibration of equipment and corrections to test data.

18.1.2 Presentation of Test Procedures

The CU and CD triaxial tests have much in common, and the differences arise only during the actual compression stage and in the associated analysis and presentation of results. The equipment required is summarised in Section 18.2.

Test procedures are dealt with in the following stages:

(1) Pre-test checks on apparatus   (Section 18.3)

(2) Prepare test specimen   } (Section 18.4)
(3) Set up specimen

(4) Saturate   }
(5) Consolidate   } (Section 18.6)
(6) Apply compression to failure }

(7) Analyse data

(8) Prepare graphical and tabulated data   } (Section 18.7)

(9) Report results

Stages 1 to 5 are identical for the two types of test. Stage 6 is covered in Section 18.6.3 for the CU test, and in Section 18.6.4 for the CD test. Stages 7, 8 and 9 are described in Section 18.7, with separate descriptions for the two types of compression test in Sections 18.7.3 and 18.7.4. Some general comments on test procedures are given in Section 18.5.

Details of corrections which are applied to the test data are given in Chapter 17, Section 17.4, and some examples of their use are included in this chapter.

The detailed test procedures given in Sections 18.4, 18.6 and 18.7 relate to a test on a single specimen. To derive the effective shear strength parameters for the appropriate failure criterion it is necessary to test a set of specimens, all from the same material, under different effective confining pressures, and to draw the envelope to the resulting set of Mohr circles of stress paths. Three specimens normally constitute a set, which is convenient for taking three 38 mm diameter specimens from one horizon in a 100 mm diameter sample. The analysis of data from a complete set is described at the end of Sections 18.7.3 (for the CU test) and 18.7.4 (for the CD test).

Test procedures include saturation of the specimen by the application of back pressure, before consolidation, as standard practice. Although widely accepted in Britain, this practice may not always be necessary or desirable. The use of back pressure is discussed in Section 15.6.

18.1.3 Equipment Described in Test Procedures

In the description of test procedures emphasis is placed on the use of conventional equipment requiring manual operation, observation and recording, except for the measurement of pore water pressures, for the following reasons.

(1) The author believes that one can learn the principles of the test, and obtain a feel for the procedures, more readily by this means.

(2) Manual operation maintains the operator's awareness of the progress of the test, enabling discretion to be applied where necessary and any errors to be detected at an early stage.

(3) Electronic systems are not infallible and knowledge of the use of manual procedures enables tests to be continued in the event of their malfunction for any reason.

(4) Automatic systems demand special care in the checking of results.

Measurement of pore water pressure is the exception because the use of pore pressure transducers is now universally accepted, and is therefore described here as standard practice. However, the now obsolete manual mercury null indicator device does have other uses which are referred to in Section 16.5.6.

Many laboratories are now making increasing use of electronic methods of measuring pressures, loads and displacements, therefore the use of electronic devices is also described. Both types of equipment are listed in Section 18.2. The BS and the ASTM Standards allow for the use of either conventional or electronic measuring instruments.

Routine pre-test checks on the equipment to ensure that pressure systems are performing satisfactorily, and other preparatory measures, are given in Section 18.3.

These checks are an essential part of the test procedure, and are mandatory in both British and ASTM Standards.

18.2 TEST EQUIPMENT

18.2.1 Equipment for Manual Operation

Items required for carrying out effective stress shear strength triaxial tests using conventional equipment for manual operation and recording, in accordance with BS 1377 : Part 8 : 1990, are listed below. Detailed descriptions and requirements are given elsewhere, as indicated.

(1) Triaxial cell, with valves, connections and fittings, as described in Section 16.2.2.

(2) Two independently controlled constant pressure systems, capacity up to 1000 kPa (see Section 16.3 and Volume 2, Section 8.2.4). Compressed-air/water systems with air pressure regulator valves and butyl rubber bladder air/water interchange cells (Section 16.3.2) are referred to here because they are commonly used in commercial laboratories, but pressure systems using mercury pots, motorised oil/water systems or dead-weight pressure cells are also suitable. Each constant pressure system should preferably include a pressure gauge of 'test' grade (see (3) below) which is permanently connected into the system. Alternatively, one pressure gauge can be shared between two systems if it can be isolated from each in turn by a suitable arrangement of valves.

(3) Two calibrated pressure gauges of 'test' grade (see Section 16.5.2 and Volume 2, Section 8.2.1). Readability to 5 kPa is required, and this is easily obtained from typical gauges of 1000 or 1700 kPa capacity which have graduation markings at intervals of not more than 10 kPa.

For accurate measurement of pressures below 50 or 100 kPa a mercury manometer (Section 16.3.5 and Volume 2, Section 8.3.5), or a calibrated pressure transducer of the appropriate resolution, should be used.

If a high degree of precision is justifiable, the level of the pressure gauge, manometer or transducer relative to a known datum level should be taken into account when determining the pressure acting on or in the test specimen. The datum would normally be the mid-height of the specimen, therefore it is convenient to mount gauges at about that level to avoid the need to apply corrections in routine tests.

(4) Calibrated pore water pressure transducer, range 0–1000 kPa, mounted on a de-airing block (Section 16.6.3).

(5) Rotary hand pump, screw controlled (the control cylinder), for flushing and checking the pressure systems (Section 16.3.5).

(6) Calibrated volume-change indicator, paraffin twin-tube burette type (Section 16.5.6), connected into the back pressure line between the pressure gauge and the cell. For some tests it is desirable to include an additional volume-change indicator on the cell pressure line.

(7) Glass burette, open to atmosphere, with graduations at intervals of 0.2 ml. A thin layer of coloured paraffin on the exposed water surface will prevent loss of water by evaporation, and make for easier reading.

(8) Tubing, usually of nylon and 65 mm outside diameter, 4 mm internal diameter, for

connecting the components of the pressure systems to the triaxial cell. The expansion coefficient due to internal pressure should not exceed 0.001 ml per metre length for every 1 kPa increase in pressure.

(9) Nylon tubing for compressed air connections, suitable for working pressures up to 1700 kPa.

(10) Calibrated timer, readable to 1 second.

(11) Plentiful supply of freshly de-aired tap water (not de-ionised) from a direct supply system or overhead reservoir (Section 16.7.6).

(12) Wash-bottle containing de-aired water.

(13) Triaxial compression load frame with multispeed drive, minimum capacity to 10 kN for 38 mm diameter specimens; 50 kN for 100 mm diameter specimen (Section 16.4.1).

(14) Calibrated force-measuring device (load ring) of capacity and sensitivity appropriate to the test specimen, with calibration chart. The usual range of suitable load ring capacities is from 2 to 50 kN (Section 16.5.3 and Volume 2, Sections 8.2.1. and 8.3.3). Guidance for the selection of a load ring of appropriate capacity is given in Section 18.3.6 (3).

(15) Calibrated dial gauge, 25 or 50 mm travel (depending on specimen size) reading to 0.01 mm. The dial gauge is mounted on a bracket fixed to the load ring near its seating on the cell piston, so that it can measure the axial deformation of the specimen (Section 16.5.4 and Volume 2, Sections 8.2.1 and 8.3.2).

(16) Two porous drainage discs of the same diameter as the test specimen (Section 16.7.3):

    'coarse' grade for sandy and silty soils;

    'fine' grade for clay soils.

(17) Rubber membranes (Section 16.7.4 and Volume 2, Section 13.7.4).

(18) Four O-ring sealing rings (Volume 2, Section 13.7.4).

(19) Suction membrane stretcher of the appropriate size, with rubber tube and pinch clip (Volume 2, Fig. 13.59).

(20) Split-tube stretcher for O-rings (Fig. 16.44).

(21) Filter paper side drains (Section 16.7.5). Side drains should be used only with specimens of very low permeability soil for which the testing time would be unacceptably prolonged.

(22) Split former, for preparing the test specimen (Volume 2, Section 9.1.2).

(23) Tube of silicone grease or petroleum jelly.

(24) Large sponge, wiping cloths, tissues.

(25) Thin rubber gloves.

(26) Small corrosion-resistant metal or plastics tray.

(27) Clinometer or protractor.

(28) Connection to a vacuum line, or electric vacuum pump, with vacuum tubing and water trap (Volume 1, Section 1.2.5(3)). Alternatively, connection can be made to a filter pump on a water tap.

The arrangement of the triaxial cell in the load frame is shown in Fig. 18.1. A typical general layout of the whole system is shown in Fig. 18.2.

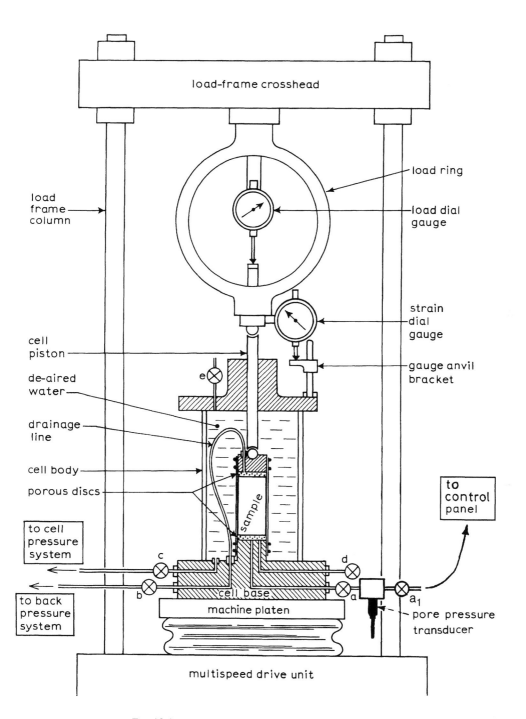

load-frame crosshead

load ring

load dial gauge

load frame column

strain dial gauge

gauge anvil bracket

cell piston

de-aired water

drainage line

cell body

porous discs

sample

to control panel

to cell pressure system

to back pressure system

cell base

machine platen

multispeed drive unit

pore pressure transducer

Fig. 18.1 *Arrangement of triaxial cell in load frame*

182

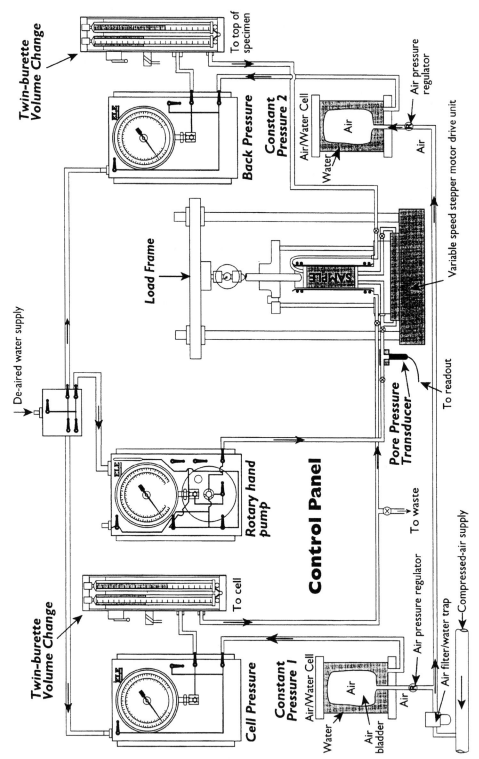

Fig. 18.2 General layout of triaxial apparatus and ancillary equipment using air/water pressure systems

The equipment specified in ASTM D 4767 is similar in principle to the above, but differs in some details as follows.

(a) Direct contact between compressed air and water is permitted in air/water pressure systems.

(b) Use of de-aired water is not mandatory, but its use is stated to be advantageous.

(c) The rigidity of the pore water pressure measuring system is specified in terms of the volume of the test specimen by the condition

$$\frac{\Delta V/V}{\Delta u} < 3.2 \times 10^{-6} \text{ m}^2/\text{kN}$$

in which, ΔV is the change in volume of the pore water pressure measurement system (mm^3), V is the total volume of the specimen (mm^3) and Δu is the change in pore pressure (kPa). (Note that if V is expressed in ml or cm^3, the above limiting value becomes $3.2 \times 10^{-3} \text{ m}^2/\text{kN}$).

(d) For the porous drainage discs, a minimum permeability of approximately 10^{-6} m/s is specified, i.e. about the same as for fine sand.

(e) The filter paper for side drains should have a permeability of not less than 10^{-7} m/s under a normal pressure of 550 kPa.

(f) A leakage check for rubber membranes (as described in Section 16.7.4) is required.

(g) The rate of loading of the compression machine should not deviate by more than ±1% of the selected value.

(h) Vibrations produced by the machine should be small enough not to produce visible ripples on a glass of water placed on the loading platform.

(i) There are minor differences in the mass and dimensional tolerances of the top cap, membranes and O-rings.

18.2.2   Equipment for Electronic Instrumentation

If electronic instrumentation is used the following items are required, some of which replace certain items listed above, as indicated in brackets. A mixed system is possible, in which some measurements are made electronically while others continue to be taken manually.

(1) Two electric pressure transducers in de-airing blocks, one connected to the cell pressure line and one to the back pressure line. It is advantageous to incorporate pressure gauges into the pressure lines as well, so as to provide continuous visual indication of the pressures.

(2) Electric displacement transducer, with suitable mounting bracket, for measuring axial deformation. (Replaces strain dial gauge.)

(3) Electric load cell, either submersible type fitted in triaxial cell chamber, or load ring with transducer or strain gauge load cell mounted externally. (Replaces load ring with dial gauge.)

(4) Volume-change indicators with electric displacement transducers, suitably calibrated, to display and record movement of water in the back pressure line and (if required) in the cell pressure line. (Replaces twin-tube volume change burettes.)

Fig. 18.3   Typical triaxial test installation for effective stress tests (courtesy Sage Engineering Ltd)

(5) Power supply/digital readout unit for each transducer; or an integrated unit for all transducers used.

Readings from electronic measuring devices are sensitive to temperature variations. The laboratory temperature should be maintained constant to within $\pm 2\,°C$ day and night.

Data logging, computer control and automatic data processing systems are referred to in Section 16.6.8.

18.2.3   Layout of Apparatus

GENERAL LAYOUT

The arrangement of the triaxial cell and load frame is shown in Fig. 18.1. The general layout of a complete system, using compressed air, is indicated in Fig. 18.2. In the upper half is shown conventional measuring and control apparatus which is positioned adjacent to the load frame. The lower half includes two regulated air/water pressure systems. Valve designations in both diagrams are the same as those used in Chapter 16. A typical installation is shown in Fig. 18.3. Components are usually connected as shown in Fig. 18.4.

PRESSURE CONNECTIONS TO CELL

The three essential connections to the triaxial cell (Fig. 18.1 and 18.2) are:

the cell pressure connection to the cell chamber at valve c;

the back pressure connection to the top of the specimen at valve b, also referred to as the drainage line;

the pore pressure connection to the base of the specimen at valve a.

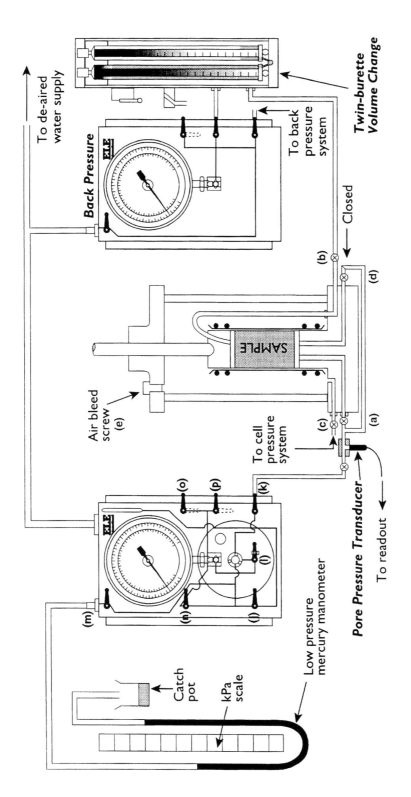

Fig. 18.4   *Connections to triaxial cell for effective stress tests*

In some triaxial cells an additional connection from the base pedestal terminates at valve d (Fig. 18.1). This connection can be used for flushing, and for base drainage when necessary.

Valves a, b, c and d should be fitted to the base of the cell itself. This allows all flexible pipework connections to be on the side of these valves remote from the cell.

PORE PRESSURE MEASUREMENT

The arrangement shown in Figs 18.1 and 18.4, with pore pressure measurement at the specimen base and drainage from the top, is normally used. This ensures that the volume of water enclosed between the specimen and pore pressure device is as small as practicable, and is rigidly confined.

PRESSURE TRANSDUCERS

When pressure transducers are used they should be fitted close to the cell but on the remote sides of valves a, b, c, as indicated in Fig. 18.5. The block housing the pore pressure transducer is attached to valve a which is fixed to the cell base, so that the transducer is as close to the specimen as possible. The valve a_1 is connected to a control panel for flushing and de-airing before a test.

18.2.4   Use of Volume Change Indicator

For most tests only one volume-change indicator is required. This is connected into the back pressue line for measuring the flow of water into or out of the specimen.

However if it is necessary to determine volume changes of a specimen that is not fully saturated, an additional volume-change indicator is required in the cell pressure line. This procedure is not covered in BS 1377, and is outlined separately in Section 18.6.6. The by-pass is used when large volumes of water are flushed through the cell system during de-airing operations.

18.3   PRE-TEST CHECKS   (BS 1377 : Part 8 : 1990 : 3.5)

18.3.1   Scope

The checks described in this section are an essential part of the triaxial test procedures, and must never be overlooked. Effective stress tests are lengthy, often running into several days, and the relatively short time spent in making sure the apparatus is in good order is amply justified. A small fault such as a slightly leaky joint can invalidate the results from a costly test.

The pore pressure system (Section 18.3.2) demands the most careful attention. Routine checks on pressure systems, and their preparation immediately prior to use, are covered in Sections 18.3.3 to 18.3.5. Some miscellaneous checks are dealt with in Section 18.3.6.

The checking procedures given below refer to the use of a manual pressure control cylinder, which is a very useful aid for rapid pressurising and flushing. If a control cylinder is not available the pressure systems themselves can be used for flushing and pre-test checks.

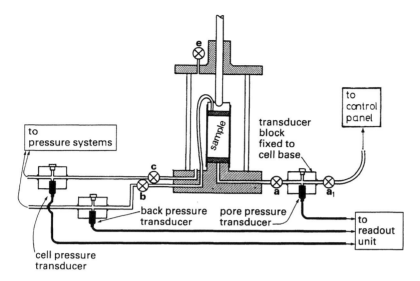

Fig. 18.5   *Arrangement of pressure transducers for measuring cell pressure, back pressure and pore pressure*

A mercury null indicator, which is now obsolescent for pore pressure measurement, can provide a useful and versatile means of fault-finding if it has been correctly prepared.

Checking procedures for the triaxial test pressure systems are specified in BS 1377 : Part 8 : 1990 in Clause 3.5. Checks fall into two categories, as follows.

(1) 'Complete' checks — to be carried out

(a) when an item of new equipment is introduced into a system;

(b) if part of a system has been removed, stripped down, overhauled or repaired;

(c) at intervals of not more than 3 months.

(2) 'Routine' checks — to be carried out before starting each test.

The valve designations quoted below relate to Figs. 18.2, 18.4 and 18.5. All valves initially are closed.

18.3.2   Checking the Transducer Pore Pressure System

NEED FOR CHECKING

The pore pressure system must be very carefully checked before use, because it is essential to eliminate all traces of air and the slightest leak must not be tolerated. The checking sequence described below provides a routine procedure for tracing faults and should be followed through systematically. Any defects brought to light should be remedied at once. The checking routine should not be skimped.

A plentiful supply of freshly de-aired water (Section 18.3.6(1)) should be readily available for flushing, either direct from the de-airing apparatus or from an elevated reservoir. The line from the source of de-aired water should include a control cylinder, and is connected to valve a_1. A suitable arrangement is shown in Fig. 18.4.

A means of detecting movement of water due to leaks that is more positive than visual inspection is to connect a sensitive volume-change indicator between the source of flushing water and valve a_1, but it must incorporate a by-pass. Alternatively a mercury null indicator could be used, and this would provide even greater sensitivity.

COMPLETE CHECK

(1) Open valves a and a_1 (Figs. 18.1 and 18.4). Using the control cylinder, pass freshly de-aired water through the transducer mounting block and cell base and out through the base pedestal port. Continue until no entrapped air or bubbles are seen to emerge. This is to ensure that the system is filled with de-aired water. Close valve a_1.

(2) Fit the cell body on the cell base, taking care not to pinch the top cap drainage line, and secure it in place.

(3) Open the cell air bleed e and open valve a_1 to allow de-aired water from the flushing system to fill the cell.

(4) Remove the bleed plug in the pore pressure transducer mounting block and close valves a and a_1.

(5) Inject a solution of soft soap into the bleed plug hole. If a hypodermic needle is used, take care to keep it well clear of the transducer diaphragm.

(6) Open valve a to allow water from the cell to flow out of the bleed hole, then open valve a_1 as well so that water also flows from the de-aired supply.

(7) Replace the bleed plug in the transducer mounting block while water continues to emerge, to avoid trapping air, and make it watertight. Refill the cell until water emerges from the air bleed, then close the bleed plug e.

(8) Open valve d briefly to allow about 500 ml of de-aired water to pass through the pedestal to waste, to ensure that any further air, or water containing air, in the mounting block is removed.

(9) Raise the pressure in the system to 700 kPa, and again allow about 500 ml of water to pass out through valve d.

(10) Leave the system pressurised overnight, or for at least 12 hours. Record the reading of the volume-change indicator, or the level of the mercury thread in the null indicator, if fitted in the system, after allowing for expansion of the connecting lines.

(11) After this period, record the reading of the indicator, if fitted, and check the system carefully for leaks, especially at joints and connections. A change in the reading of the indicator will provide immediate confirmation of leakage. Any leaks that are detected should be dealt with immediately and rectified, and the above check repeated.

(12) When it is confirmed that the system is leak-free, close valve a_1 and release the pressure in the cell by opening valve c. Drain the water from the cell through valve c with the air bleed e open.

(13) Remove the cell body. Seal the pore pressure measurement port on the base pedestal with a watertight plug without entrapping air. A rubber plug or eraser, held fast with a

G-clamp, has been found satisfactory for this; alternatively a purpose-made plug could be turned from a piece of nylon.

(14) Open valve a_1 and apply the maximum available pressure (consistent with the limitations of the pressure system and transducer) to the base pedestal.

(15) Close valve a_1 and record the reading of the pore pressure transducer. Leave for a minimum of 6 hours.

(16) If after this period the pressure reading remains constant the pore pressure connections can be assumed to be free of air and free of leaks.

(17) A decrease in the pressure reading indicates that there is a defect in the system — either a leak, or air was present and has been forced into solution. The defect must be rectified, and stages (1)–(15) repeated, until it is confirmed that the system is free of entrapped air and leaks.

ROUTINE CHECK

Immediately before setting up each test specimen, follow steps (1)–(12) above to ensure that the base pedestal is flushed clear of any remaining soil particles and that the system is still free of leaks and entrapped air.

Remove the cell body, and keep the base pedestal covered with de-aired water until ready for setting up the next test specimen. This is to ensure that no air enters the base ports. A small 'pond' of water can be maintained by fitting a cut-down rubber membrane around the pedestal, secured with O-rings, as shown in Fig. 16.48.

18.3.3   Checking the Back Pressure System

NEED FOR DE-AIRING

The back pressure system, including the flexible connection from the cell base to the specimen top cap, must be made free of air and obstructions, not only when first setting up the apparatus but also before starting a new test. If the specimen previously tested was not fully saturated initially, air which dissolved in the water under pressure would have drained into the back pressure system to re-appear as bubbles when the pressure was reduced to atmospheric.

COMPLETE CHECK OF BACK PRESSURE SYSTEM

(1) Ensure that the back pressure system is fully charged with a supply of freshly de-aired water.

(2) Open valve b and flush freshly de-aired water from the back pressure system, through the volume-change indicator and the specimen drainage line and out through the port in the top cap. Apply enough pressure to maintain a reasonable rate of flow. During this operation the volume-change indicator should be worked to the limits of its travel at least twice. If the top cap is immersed in a beaker of water any air emerging will become apparent (see Fig. 18.6). Replenish the water in the system with freshly de-aired water as necessary.

(3) Continue until the absence of emerging bubbles indicates that the system is

substantially free of air. If bubbles persist, the volume-change indicator might need de-airing, as described below.

(4) Insert a suitable watertight plug into the port in the top cap while submerged, and hold the plug fast with a G-clamp.

(5) Increase the pressure in the back pressure system with valve b open. Observe the volume-change indicator and record the reading when steady, i.e. after allowing for initial expansion of the connecting lines.

(6) Leave the system under pressure overnight, or for at least 12 hours, then record the reading of the volume-change indicator again.

(7) Take the difference between the two readings and deduct the volume change due to further expansion of the tubing. If this corrected difference does not exceed 0.1 ml the system can be considered to be leak-free and air-free, and ready for a rest.

(8) If the corrected difference exceeds 0.1 ml, the system needs to be investigated and the leaks rectified until the above requirement is achieved.

VOLUME CHANGE INDICATOR

The following action might be necessary to remove air from the volume-change indicator. The manufacturer's instructions should also be followed carefully.

(1) Air in the burettes can be released by momentarily opening bleed screw D or E (Fig. 18.6) while under a small pressure. The valves on the indicator manifold are opened to allow flow through the burettes.

(2) Air trapped in the valves, manifold or connecting pipework, should if possible be flushed out through the line at y.

(3) If this is not possible, pressurise the system until the bubbles have been forced into solution, then flush out either at y, or into the burette, so that the bubbles re-form on reducing the pressure and can then be released at bleed screw D or E.

(4) If there is an appreciable amount of air which cannot be eliminated in this way it may be necessary to drain the burettes and re-fill them as described in Section 16.8.3.

ROUTINE CHECK

(1) Immediately before setting up each test specimen, follow steps (1)–(3) of the complete check to ensure that the drainage line is free of air and obstructions. Close valve b.

(2) Increase the pressure in the back pressure line to 750 kPa, and after 5 minutes record the reading of the volume-change indicator.

(3) Leave the system under pressure, and determine the change in indicated volume, as in steps (6)–(8) above. Carry out remedial measures if found necessary, then recheck.

18.3.4 Checking the Cell Pressure System

A complete check of the cell pressure system should be made by applying the maximum test pressure to the cell, filled with water. It should be verified that the applied pressure can

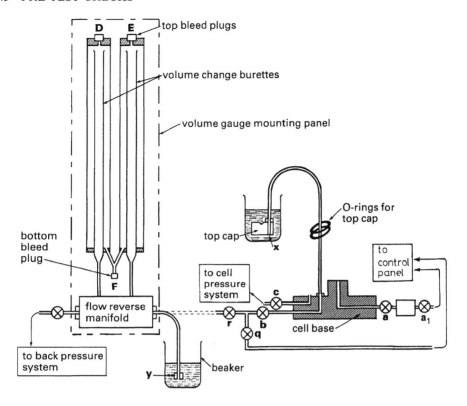

Fig. 18.6   *Arrangement for flushing drainage line*

be maintained constant to within ± 0.5% of the indicated reading over a period of time at least equal to the duration of a typical test.

If a volume-change gauge is connected into the cell pressure line, the system and the gauge should be made air-free and checked for leaks in the same way as outlined above for the back pressure system. Removal of the last trace of air is not critical if a volume gauge is not present.

Nevertheless, the water used in the cell should be de-aired if samples are to be tested in the saturated condition, to avoid migration of air through the membrane (see Section 17.4.2(7)).

18.3.5   Checks on Constant Pressure Sources

COMPRESSED AIR SYSTEM

Before starting a test the air bladder on the cell pressure line should be almost deflated, since it will be used to push water into the cell as the cell pressure is increased. The bladder on the back pressure line should be about half inflated, to allow first for driving water into the specimen and then to accept water draining out of the specimen during consolidation.

The air regulator valve on each pressure line should be checked by rapidly increasing then decreasing the line pressure. A delay in response may be due to dirt or moisture in the diaphragm orifice.

OIL-WATER SYSTEM

The oil-water interchange cylinder of each system should be charged with enough oil to allow for displacement of water similar to that described above for the two pressure lines.

PRESSURE GAUGES

Whichever type of constant pressure system is used it is essential that pressure gauges have been accurately calibrated and that calibration data are readily available (see Volume 2, Section 8.4.4). Accurate determination of small pressure differences between two gauge readings is necessary at some stages of a test. A pair of gauges used in this way should be cross-checked against each other, or a differential pressure gauge or transducer can be used (Section 16.5.2).

18.3.6   Preparation and Checking of Ancillary Items

(1) *De-aired water*

A plentiful supply of freshly de-aired water should be available. A simple vacuum de-airing and storage system is described in Volume 2, Section 10.6.2(1) and a more convenient apparatus is referred to in Section 16.7.6. The water should be stored in an air-tight container until required for use.

Clean tap water should be used, not distilled or demineralised (de-ionised) water. The latter, when air-free, can be surprisingly corrosive, especially towards materials used for seals and diaphragms.

(2) *Porous discs*

Porous discs should be inspected to ensure that they are clean and that water drains through them freely (see Section 16.7.3). Discs that are clogged by soil particles should be rejected.

Porous discs should be boiled for 10 minutes in distilled water to remove air, and then kept under de-aired water until required (Section 16.7.3). Filter paper should not be interposed between a porous disc and the soil specimen.

(3) *Load ring*

The load ring selected should be appropriate to the likely strength of the specimen to be tested, with enough sensitivity to obtain good resolution at small loads. It may be desirable to use rings of different capacity for specimens prepared from the same soil sample but tested under different effective confining pressures.

A guide to the selection of a suitable load ring is given in Fig. 18.7. The theoretical axial force P at peak deviator stress in drained compression tests for a given value of ϕ' is derived from the equation derived in Section 15.5.7:

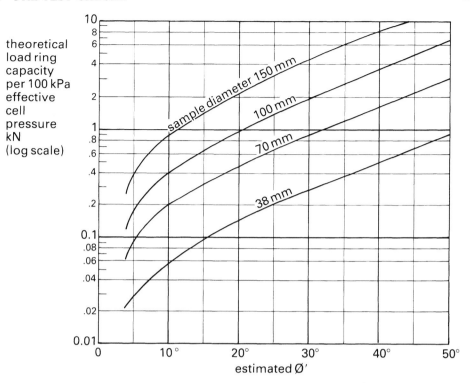

Fig. 18.7   *Curves for estimating required capacity of load ring*

$$P = \frac{2.4 \sin \phi'}{1 - \sin \phi'} \frac{A_0}{1000} \sigma_3' \ \text{N} \qquad (15.44)$$

in which A_0 is the nominal area of specimen initially (mm^2), ϕ' is the estimated value for the soil (if in doubt, err on the high side), and σ_3' is the effective cell pressure for the test (kPa). The constant 2.4 includes a factor of 1.2 as an allowance for specimen barrelling, equivalent to about 17% strain.

The curves in Fig. 18.7 give the theoretical maximum force in terms of 'kN per 100 kPa of effective cell pressure' for a range of values of ϕ', and four specimen diameters. The load ring selected should have a capacity at least 50% in excess of that indicated, so that the maximum force remains well within that capacity. For an undrained test on a dilatant solid in which pore pressure decreases (A_f negative) the maximum force would be very much greater than indicated by Equation (15.44) and Fig. 18.7, but if pore pressure increases (A_f positive) it will be less.

The load ring also has to sustain the upward force on the cell piston from the cell pressure (σ_3 kPa), equal to $a\sigma_3/10^6$ kN where a is the piston area (mm^2). This additional force is small; for a cell pressure of 1000 kPa it amounts to less than 0.3 kN on a piston of 19 mm diameter, and about 0.8 kN on 31.75 mm diameter.

As an example of the application of the curves in Fig. 18.7, a 100 mm diameter/specimen is to be tested at an effective cell pressure of 600 kPa and the estimated ϕ' is 30°. The graph gives a failure load of 1.9 kN for 100 kPa cell pressure, therefore the force required at failure for $\sigma_3' = 600\,\text{kPa}$ will be $1.9 \times 6 = 11.4\,\text{kN}$. A $14\,\text{kN}$ ring might be adequate, but a $20\,\text{kN}$ ring would allow for a greater margin of uncertainty.

The load ring must be securely fixed to the crosshead of the load frame. The stem of the dial gauge should be in contact with the anvil on the ring, and the anvil locking screw should be tight. The dial gauge should be set to read zero, or a convenient initial reading, and reset to zero when the cell pressure for the compression stage has been applied.

The care and use of load rings, and their calibration, is covered in Volume 2, Sections 8.3.3 and 8.4.3. The calibration chart for the selected ring should be available for reference.

(4) *Strain dial gauge*

The gauge need not be fitted into place until just before the compression stage, but it should be checked for free movement throughout its range. The necessary fixings and support brackets should be complete and ready for assembly. For $38\,\text{mm}$ diameter specimens a gauge with $10\,\text{mm}$ travel, reading to $0.002\,\text{mm}$ or $0.01\,\text{mm}$, is suitable, but for larger specimens a travel of $50\,\text{mm}$ will be needed unless only small strains are to be measured.

(5) *Electronic instrumentation*

Electronic measuring devices should be already calibrated, and installed in accordance with the supplier's instructions. The momentary application of a small pressure, load or displacement will indicate whether the instrument and readout are responding. It is particularly important to ascertain the direction of the response (i.e. $+$ or $-$) for displacement and volume-change transducers.

The power supply/readout unit should be switched on several hours beforehand (preferably overnight) to allow for warming and 'soaking' at a steady working temperature. The laboratory area where these units and transducers are used should be temperature controlled to within $2\,^\circ\text{C}$.

Further observations on the use of electronic equipment are discussed in Section 16.6.10.

(6) *Triaxial cell*

A cell that is in constant use needs little attention other than to ensure that the sealing ring is clean and in sound condition, and that the mating recesses in both cell body and base are free from dirt and soil particles. The piston should be wiped dry with a clean cloth and checked for free movement in its bushing.

Cell calibration data for pressure change and for 'creep' under constant pressure should be available (Section 17.3.1).

(7) *Rubber membranes*

A new rubber membrane should be used for every test. Membranes are low-cost items and it is false economy to re-use membranes for these types of test. Replacing a specimen spoilt by a leaking membrane can be very costly. The membrane should be carefully inspected for flaws or pinholes by stretching it in two directions while holding it against a bright light. An air check for leaks is outlined in Section 16.7.4. A faulty membrane should be cut up and discarded immediately.

Before use membranes should be soaked in water for 24 hours.

(8) *Rubber O-rings*

Rings should be inspected before use and stretched a little to ensure that they are free from necking or weak points. They should be clean, dry and free from grease.

(9) *Filter paper drains*

Filter paper side drains (Section 16.7.5) should be fully saturated by immersing in de-aired water before use. Allow surplus water to drain off before fitting on to the specimen.

(10) *Timer*

A calibrated seconds timer (fully wound) and wall clock, or a calibrated wall clock with seconds hand, should be conveniently visible.

(11) *Test forms*

The following test forms and graph paper should be to hand at the start of either a CU or a CD test:

Specimen data sheet (Fig. 18.14, Section 18.6.1)

Saturation sheet (Fig. 18.15, Section 18.6.1)

Consolidation sheet (Fig. 18.18, Section 18.6.2)

Compression test sheet (several sheets are usually needed) (Fig. 18.20, Sections 18.6.3 and 18.6.4)

Graph paper, 1 mm and 10 mm squares

Semilogarithmic paper, 5 cycles × 1 mm

For a set of specimens, a summary sheet for tabular data and graphs is also needed (Fig. 18.22, section 18.7.3 for CU tests; Fig. 18.24, Section 18.7.4 for CD tests).

18.4 PREPARATION OF TEST SPECIMENS   (BS 1377:Part 8:1990:4)

18.4.1. General Principles

SELECTION OF SAMPLES

The selection of samples for shear strength testing should be under the control of the engineer who is familiar with both the site from which they were taken and the type of analysis for which the test results are required. The portion of available samples selected for preparing test specimens should be properly representative of the soil profile. This is particularly important for the types of test described here, which are time-consuming and therefore expensive.

When a sample is removed from its container or sampling tube it should be carefully inspected to ascertain its condition and suitability for test. The soil description should be recorded. Any unusual or detrimental features should be reported immediately. If these result in the sample being not truly representative an alternative sample should be used instead, and the fact reported.

SIZE OF SPECIMEN

The test specimen should be large enough to represent adequately the material whose properties are to be determined. For instance, a specimen of stony soil should be of a diameter at least 4–5 times that of the largest particle. Suggested specimen sizes related to the largest particle size present are given in Volume 2, Table 13.4. Specimens from fissured soils should be at least 100 mm diameter. Orientation of fissures and other discontinuities should be taken into account. The in-situ shear strength of fissured soil may be little over 50% of that measured in the laboratory on a small specimen (Bishop and Little, 1967).

SAMPLE DISTURBANCE

Test specimens should be prepared with the minimum of disturbance. Precautions include avoiding changes of strain (i.e. linear distortion) as much as possible, and elimination of changes in moisture content whether due to drying out or absorbing free water. The engineer should recognise disturbance, and should judge the extent to which disturbance is likely to affect the data obtained from tests. Some general principles are as follows.

Stress/strain characteristics are particularly susceptible to error by the slightest disturbance, especially in normally consolidated clays.

Pore pressure response to stress is also affected by disturbance.

Measurements of undrained (immediate) strength require rather greater disturbance to produce significant errors.

Overconsolidated soils are generally less affected by disturbance than normally consolidated soils, but this can lead to the effects of unsatisfactory procedures being obscured unless comparisons are made with field measurements.

Effective shear strength parameters c', ϕ' are least affected by disturbance.

Undisturbed samples should be tested with the minimum of delay after recovery. This factor can have an important bearing on a testing programme.

Test specimens should be set up and confined as soon as possible after extrusion.

Better specimens are generally obtained from piston samplers than from open-drive samplers.

SAMPLE HANDLING

Notes on the care of samples are given in Volume 1, Section 1.4. Equipment used for preparing test specimens is described in Volume 2, Section 9.1.2 and some general principles are outlined in section 9.1.3.

Detailed procedures for specimen preparation given in Chapter 9 are referred to below, together with additional comments where appropriate.

18.4.2 Undisturbed Samples

Preparation of cylindrical specimens for triaxial tests from undisturbed samples is described in Chapter 9 under the following headings:

Single specimen from 38 mm diameter tube:	Section 9.2.3
Set of 3 specimens from 100 mm diameter sampling tube:	Section 9.2.4
Single 100 mm diameter specimen from U-100 tube:	Section 9.2.5
Small specimen from block sample:	Section 9.3.2
Large specimen from block sample:	Section 9.3.3
Use of soil lathe:	Section 9.4.1

Some procedures for setting up triaxial test specimens are given in Chapter 13, and these are referred to and amplified, where appropriate, in Section 18.4.7.

Three 38 mm diameter specimens are often taken as a set at the same horizon from a 100 mm diameter undisturbed sample. Small specimens of this kind are satisfactory in intact homogenous fine-grained soil. If the soil is uniform with depth the effect of disturbance due to specimen preparation can be reduced by taking three successive specimens from the centre-line of the sample tube, instead of side by side.

Specimens should be prepared in a humidified atmosphere (relative humidity about 50%) to prevent drying out, ideally in a humidified sample trimming room. Failing that, a localised humid zone can be improvised on the work-bench by setting up a 'tent' of dampened cloths, supported by burette stands. Thin rubber gloves should be worn when handling the soil. Material not being used immediately should be wrapped in thin 'cling-film' or a polythene bag to prevent moisture loss.

Visible structural features in the soil, and other discontinuities, should be carefully recorded. Particularly significant are zones of potential weakness on which failure by slip might occur. In fissured or laminated soils, small specimens should not be used but the whole sample should be tested.

For the routine tests described in this chapter, specimens should be of a length : diameter ratio of at least 2:1, but not greater than about 2.5:1. Shorter specimens can be used when lubricated end caps are fitted. This is a special procedure which is beyond the scope of this book.

Specimen ends must be trimmed flat, and square with the sample axis to within $\frac{1}{2}°$. The effect of 'smear' from cutting or trimming can be offset to some extent by lightly brushing the ends with a brass-bristle brush. The specimen should be weighed to an accuracy within 0.1% (e.g. a 38 mm diameter specimen should be weighed to the nearest 0.1 g). Length and diameter should be measured with vernier calipers to 0.1 mm, avoiding disturbance and moisture loss. These tolerances enable the specimen density to be calculated to an accuracy well within the $\pm 1\%$ called for in BS 1377, even for 38 mm diameter specimens.

The immediate trimmings from the specimen should be retained (without loss of moisture) and used for determination of moisture content. Some material should be set aside for measurement of the particle density (required for voids ratio calculations), and for relevant classification tests such as Atterberg limits and particle size distribution.

The initial specimen data referred to above are entered on a test form of the kind shown in Fig. 14.14 (Section 18.6.1).

18.4.3 Remoulded Specimens

Effective stress tests on soil specimens that have been remoulded by 'kneading', in the manner described in Volume 2, Section 13.5.3, are carried out less often commercially than in research work. Important requirements are to obtain uniformity throughout the specimen and to avoid trapping air.

Preparation of remoulded soil by consolidating from slurry, using a large Rowe consolidation cell, is described in Chapter 22. A large mass of uniform homogeneous soil can be prepared in this way. Tube samples can easily be taken from the material so formed.

18.4.4 Reconstituted Compacted Specimens

Preparation of compacted cylindrical specimens is described in Volume 2, Chapter 9, as follows:

Compaction criteria:	Section 9.5.1
Specimens 38 mm diameter:	Section 9.5.4
Use of compaction mould:	Section 9.5.5
Large specimens (100 mm and 150 mm diameter)	Section 9.5.6

Standard compaction procedures are given in Volume 1, Chapter 6. The degree of compaction to be applied to the soil should be related to the in-situ compaction control requirements, or other relevant field condition.

A specimen compacted into a split mould lined with a rubber membrane should be fitted with a second membrane before testing, in case of damage to the first during compaction. Specimens containing angular particles should be enclosed in at least two thick membranes as a safeguard against puncturing. The outside of a membrane already fitted should be coated with silicone grease before placing the next. Specimens may be formed on the triaxial cell base pedestal if the split mould and membrane are first fitted and clamped around the pedestal.

The moisture content of the soil to be compacted should be measured, but a more reliable determination of the moisture content of the test specimen can usually be obtained from wet and dry weighings made after testing, together with the initial specimen mass.

The length and diameter of the specimen should be carefully measured after removal of the split mould, in case it has slumped.

Compacted specimens should be sealed and left to mature for at least 24 hours after compaction, to allow equalisation of pore pressure set up during compaction. Specimens of clay of low permeability should be left for several days.

Small compacted specimens (e.g. 38 mm diameter) are best prepared by first compacting the soil into a compaction mould, if enough soil is available, and then extruding into tubes after maturing, as explained in Section 9.5.5. Specimens of stiff soils are better hand-trimmed.

18.4.5 Saturated Cohesionless Specimens

A saturated sand specimen is prepared on the triaxial pedestal as described in Volume 2, Section 13.6.9, and illustrated in Figs. 13.53, 13.54 and 13.55. The procedure provides support to maintain the shape of the specimen until an effective stress large enough to

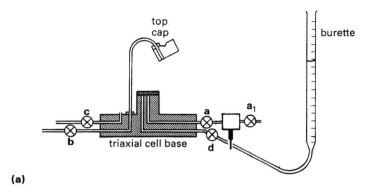

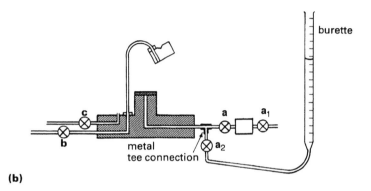

Fig. 18.8   *Connections to burette for setting up saturated cohesionless specimens:*
            (a) cell with two base pedestal connections, (b) cell with one connection

make it self-supporting can be applied. The effective stress is provided by inducing a small negative pore pressure when the burette connected to the base of the sample is lowered (Fig. 13.55).

Placing the specimen dry and subsequently flooding is not advisable because it is difficult to ensure complete saturation afterwards. Use of vacuum or a soluble gas to displace the air can induce unknown effective stresses, and flooding after setting up dry can modify the structure and the effective strength.

When using a triaxial base with two connecting ports to the base pedestal, one outlet can be connected to the control panel, and the other to the burette (valve d in Fig. 18.8(a)). If there is only one port, a brass or copper tee connection and an additional valve (a_2 in Fig. 18.8 (b)) can be fitted to the cell base. The volume enclosed between valves a, a_2 and the cell must be as small as possible and the joints and valves must be completely free of leaks, and de-aired.

Before placing the top loading cap in position, a porous disc, slightly smaller than the specimen diameter, is placed on top of the specimen. Otherwise the procedures for forming and measuring the specimen are the same as those given in the second part of Section 13.6.9.

18.4.6   Dry Cohesionless Specimens

When it is necessary to set up a dry specimen of sand or other granular material, it is formed on the triaxial pedestal as described in Volume 2, Section 13.6.9, Fig. 13.56. Dry porous stones of a coarse grade should be used. Suction is applied to the base of the specimen until the cell is assembled, filled with water and a pressure applied. Water can then be admitted through the base pedestal, and allowed to percolate upwards through the specimen, displacing most of the air to atmosphere through the top drainage line. The cell pressure and subsequent effective stress must be high enough to prevent collapse of the specimen.

18.4.7   Setting Up

GENERAL PRACTICE

The following procedures relate to undisturbed, remoulded or compacted specimens prepared as described in Sections 18.4.2 to 18.4.4. Specimens formed in-situ on the triaxial pedestal (Sections 18.4.5. and 18.4.6) are already set up ready for test.

(1) In order to improve the seal when the membrane is fitted, rub a smear of silicone grease or rubber grease on to the clean curved cylindrical surfaces of the base pedestal and top cap. But avoid contaminating the porous disc and the filter paper drains (if used) with grease. Then cover the cleaned base pedestal with a film of de-aired water from a wash-bottle.

(2) Take a de-aired saturated porous disc from under water, where it has been kept, and slide it on to the pedestal without trapping any air. Remove excess surface water from the disc with a quick wipe with a finger, but ensure that it remains saturated.

(3) Place the prepared specimen on the porous disc without delay, and so as not to trap any air.

(4) Place a second saturated porous disc, with excess surface water removed, on the top end of the specimen.

(5) If side drains are to be used, place the saturated filter paper drain (formed as described in Section 16.7.5) around the specimen, as shown in Fig. 18.9 (a) for conventional drains, or Fig. 18.9(b) for spiral filter paper strips. The saturated drain should first be held up for a few seconds to allow surplus water to drip off. Press the filter paper lightly against the side of the sample with the fingers, avoiding ruckling and trapping air. Suction from the soil will normally hold the drain in place. The top of the filter paper should overlap the upper porous disc in order to make a drainage connection. Normal practice is for the ends of the filter paper to overlap both the upper and the lower porous discs. However, the author suggests that for undrained tests a gap should be left between the lower edge of the filter paper and the bottom disc, as shown in Figs. 18.9 and 18.10 for a 38 mm diameter specimen. This helps to prevent a 'short circuit' effect between the back pressure system and the pore pressure measuring system. In the UU test procedure described by Hight (Section 19.3.2) a gap of at least 10 mm is required for a 100 mm diameter specimen. The filter paper drain may be fitted to the specimen before mounting it on the pedestal if it is more convenient to do so.

(6) Fit two rubber O-rings over the membrane stretcher. Take a rubber membrane from the container in which it has been soaking, allow surplus water to drain off, and fit it inside

*Fig. 18.9   Filter paper side drains fitted to triaxial specimen: (a) (left) vertical
drains, (b) (right) spiral drains (courtesy LTG Laboratories)*

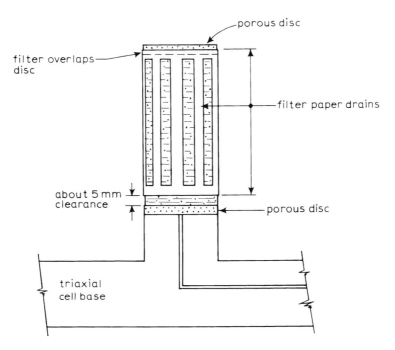

*Fig. 18.10   Vertical drains on 38 mm diameter specimen avoiding overlap of lower
porous disc*

the membrane stretcher as explained in Volume 2, Section 13.6.3, step 6, and shown in Fig. 13.32 (a) and (b).

(7) Place the membrane over the specimen while applying suction to the rubber tube (Fig. 13.32 (c)), then release the suction so that the membrane clings to the specimen at the correct height (Fig. 13.32 (d)).

(8) Roll the lower end of the membrane over the base pedestal and seal it in place with both O-rings.

(9) Unroll the upper end of the membrane and remove the stretcher.

With experience it is possible to implement steps (8) and (9) in one operation by suddenly blowing into the tube connected to the membrane stretcher. This causes the membrane to jump into position at both ends simultaneously. This expedient is useful with a sensitive clay or silt sample, because it reduces the risk of disturbance by hand manipulation.

(10) Remove as much air as possible from between the specimen and membrane by stroking upwards with the fingers. A length of fine plastics-coated wire, with the wire itself removed, serves as a small tube which can be inserted under the membrane to expedite the removal of air pockets as it is withdrawn. Do not insert any water between the specimen and the membrane.

(11) For a free-draining soil, and if a second connection to the base pedestal is available as in Fig. 18.8 (a) or (b), air may be displaced by allowing water from the burette to flow upwards through the specimen. This procedure must be carried out with care, and only if saturation in this way does not disturb the soil structure.

(12) Slip two O-rings over and past the top loading cap so that they encircle the drainage lead to the cell base (Fig. 18.6).

(13) Open valve b for an instant to allow a little water to moisten the top cap, before fitting it on to the porous disc covering the specimen, but avoid excess free water. Do not entrap air between the cap and disc or between the cap and membrane. Turn the top cap to a position where the drainage connection will not interfere with fitting the cell body.

(14) Seal the membrane on to the top cap with the two O-rings, without disturbing the specimen. When using only the fingers it is difficult to avoid disturbance or damage to the specimen. This operation is made easier by using the split steel tube stretcher shown in Fig. 16.44 (Section 16.7.4). The stretcher is held in position while the rings are rolled off it and on to the top cap and it can then be taken from around the drainage connection (Fig. 18.11).

Fig. 18.11   Use of split tube O-ring stretcher

(15) If water has been used to displace air upwards from the specimen, lower the burette connected to the base pedestal to give a negative pressure of 100 mm or so of water, and open valve a_2 (Fig. 18.8) for a few minutes. This should remove excess water and seat the specimen firmly on the pedestal. Close the valve and return the burette to its normal position.

(16) Check that the specimen axis is vertical and that the specimen and top cap are properly seated. Place the ball-bearing in the cap recess if needed. Ensure that the surface of the cell base is clean, and that the sealing ring is clean and properly fitted.

(17) Carefully lower the cell body into position over the specimen with the cell piston raised to its maximum extent. Take care not to knock against the specimen or the drainage connection. Position the cell correctly, and progressively tighten the clamping screws, or nuts on the tie-rods (see also Volume 2, Section 13.6.3, step 8).

(18) Allow the piston to fall slowly into contact with the ball or dome on the top cap. A good fit confirms proper alignment. If there is any eccentricity of the top cap, remove the cell body and correct the alignment.

(19) Retract the piston to its maximum extent if it is fitted with an internal collar as a positive stop. Otherwise the outer end of the piston should be brought up against a bearing surface rigid enough not to be deflected by the upward force due to subsequent cell pressure. If cell volume change measurements are to be made the piston should not bear against a load ring, which would deflect when cell pressure is applied. A sufficiently strong bracket on the cell, or the crosshead of the load frame, would be suitable, but the strain dial gauge datum bracket should not be used unless it is designed for this purpose.

(20) Fill the cell steadily with water (de-aired tap water for a saturated soil) through valve c from the supply system or reservoir, with the air bleed e (Fig. 18.4) open. Close valve c when water begins to emerge from e. Do not allow turbulence during filling.

(21) If the test is likely to take several days, and in any case if the cell piston bushing is not watertight, introduce a layer of castor oil through the plug at e (or through a separate plug if fitted) before the cell is completely filled. More water can then be admitted to the cell to expel the remaining air.

(22) Keep the bleed e open until the cell is about to be pressurised, to maintain the cell at atmospheric pressure. The specimen is now ready for starting the saturation stage of a routine test (Section 18.6.1), or one of the tests outlined in Chapter 19 and 20.

SETTING UP SENSITIVE SOILS

Sensitive undistorted soils could become distorted while the membrane and O-rings are being fitted as described in steps (7) to (14) above. An alternative method is to follow the procedure upside-down, as follows.

Clamp the inverted top cap in a burette stand, and place the porous disc and specimen on the cap in the same way as in steps (1) to (3). Fit side drains (if needed), and place the membrane and O-rings as in steps (6) to (8). Roll back the membrane carefully to expose the present upper end of the specimen. Release the top cap, invert the specimen and top cap assembly and place the specimen on the porous disc on the base pedestal. Roll the membrane into place and secure it with two O-rings previously looped over the drainage lead, using the split-ring stretcher.

18.5   COMMENTS ON TRIAXIAL TEST PROCEDURES

18.5.1   Outline of Test Stages

SUMMARY

A routine effective stress triaxial compression test consists of three stages:

Saturation: Section 18.6.1

Consolidation: Section 18.6.2

Compression: Section 18.6.3 (CU test)
                           Section 18.6.4 (CD) test)

The first two stages are the same for both the CU and the CD test.

SATURATION

The term 'saturation' as a stage of the test refers to ways by which the pore water pressure in the specimen is increased so that air as a separate phase in the void spaces is eliminated. This enables reliable readings of pore pressure changes to be obtained during subsequent testing stages for the determination of the effective stresses. The pore water pressure is increased in a controlled manner, usually by the application of back pressure, so that air in the voids is forced into solution. The process is discussed in Section 15.6, and is represented diagrammatically in Fig. 18.12 (a). Procedures are described in Section 18.6.1.

CONSOLIDATION

In the consolidation stage the specimen is consolidated isotropically under a confining pressure by allowing water to drain out into the back pressure system (as represented in Fig. 18.12 (b)) so that the pore water pressure gradually falls until it virtually equals the back pressure. Drainage of water results in a decrease in volume and an increase in the effective stress, which after consolidation is equal to the difference between the confining pressure and the mean pore pressure remaining in the specimen. A pore pressure dissipation of 100% would be reached if the pore pressure eventually falls enough to equal the back pressure. This is not usually achieved, and in practice 95% dissipation of pore pressure is normally acceptable for termination of consolidation.

COMPRESSION

In the compression stage the axial force is gradually increased while the total confining pressure remains constant, until failure occurs due to the maximum available shear strength of the specimen being overcome. Compression can be extended to reach an 'ultimate' condition.

   This process may also be referred to as the 'loading stage', or the 'shearing stage', or 'shearing to failure'. Drainage conditions differ for the two types of test. In the CU test, no drainage is permitted, but measurements are made of pore water pressure changes

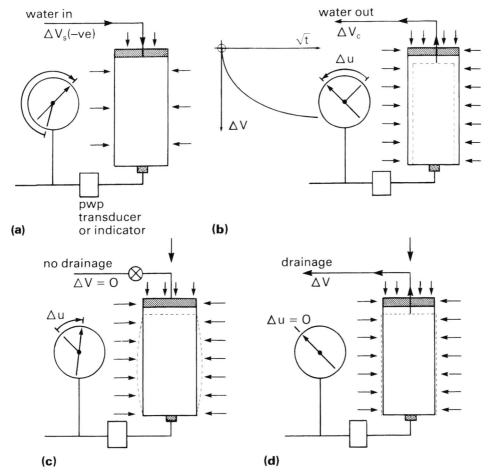

Fig. 18.12   *Representation of stages in effective stress triaxial tests:* (a) *saturation,*
(b) *consolidation,* (c) *undrained compression,* (d) *drained compression*

(Fig. 18.12 (c)); in the CD test, full drainage is allowed in order to prevent pore pressure change (Fig. 18.12 (d)), and the resulting volume changes are measured.

Usually three identical specimens are prepared from one soil sample for a set of tests, each specimen being isotropically consolidated to a different effective stress for the compression stage. The treatment of a set of results is described in Section 18.7.3 (CU test) and 18.7.4 (CD test).

18.5.2   Saturation

Saturation of the specimen by the application of back pressure has become an established procedure for routine effective strength tests. However, work at Imperial College indicates that the achievement of full saturation is not always necessary and, in some cases, not desirable.

In presenting the test procedures (Section 18.6), five methods of increasing the initial pore water pressure are given, including the procedure described in BS 1377 to achieve virtually 100% saturation by raising the confining pressure and back pressure in alternate increments. An automatic version of this method, and other procedures which are simpler or exercise a greater degree of control on the effective stress, are also described. Selection of the appropriate method is discussed in Section 15.6.6.

18.5.3 Test Conditions

STRESSES

Before carrying out the compression test the effective stress in the specimen must be brought to the value required for the test. This is done by adjusting the cell pressure and back pressure so that their difference is equal to that value, and then allowing the resulting pore pressure in the specimen to dissipate by the process of consolidation (Section 18.6.2). The effective stress selected should be related to the in-situ conditions, either initially or resulting from the imposition of the proposed loading, as appropriate.

When a set of three specimens is to be tested for deriving shear strength parameters from the Mohr envelope, it is often appropriate to test one specimen at or about the in-situ effective stress, one at about twice that value and one at about three or four times that value. An effective confining pressure of less than the mean in-situ effective stress would leave a specimen of normally consolidated clay in a state of overconsolidation, which could give a misleading failure envelope (Section 15.5.4). The effective pressures should be selected by the geotechnical engineer dealing with the project, so that they are relevant to the particular application; they cannot be standardised.

ENGINEER'S SPECIFICATION

The following list summarises the test conditions that should be specified by the engineer at the time the tests are requested.

(a) Type of test (e.g. CU or CD).

(b) Size of test specimens.

(c) Number of specimens to be tested as a set.

(d) Drainage conditions.

(e) Method of saturation, including cell pressure increments and differential pressure if appropriate; or whether saturation should be omitted.

(f) Effective cell confining pressures.

(g) Criterion for failure (usually one of the criteria defined in Section 15.4.2).

(h) Method of correction for the effect of the membrane, and of side drains if used.

ENVIRONMENTAL REQUIREMENTS

To comply with BS 1377:1990, the temperature of the laboratory in which these tests are carried out should be maintained constant to within $\pm 2\,°C$. The actual mean temperature level is not specified; the important factor is to confine the temperature variations to within the specified limits. Apparatus should be protected from direct sunlight, from local sources of heat and from draughts.

18.5.4 Duration of Test Stages

SATURATION

The time required for saturation depends upon the type of soil and size of specimen, as well as the initial degree of saturation, and is discussed in Section 15.6.4. Automatic saturation (Section 18.6.1(5)) enables the process to continue unattended, but the rate of pressure increase must be appropriate for the soil. Incremental checks are necessary to ascertain the B value from time to time.

The saturation stage should not be excessively prolonged. In some soils there is the possibility of 'undrained creep' occurring if the specimen is left under pressure with the drainage valve closed for a long period. This can cause the pore pressure to rise beyond the increase due to the B value effect, even under isotropic stress conditions (Arulanandan, Shen and Young, 1971). The increase can be significant over a period of 24 hours, and may continue over longer periods due to effects such as particle reorientation, similar to those causing secondary compression in consolidation. This is one reason for avoiding the saturation process if it is not necessary. If a specimen must be left unattended for any length of time (e.g. overnight), it would be better to leave it during a 'water in' stage, with the back pressure line valve open, than during a 'build up' stage when the valve is closed.

The use of side drains with soils of low permeability can reduce the time required for saturation, but pore pressure response readings should be interpreted with care (see Section 18.5.5).

CONSOLIDATION

Permeability controls the time required for the sample to consolidate. Consolidation must be allowed to continue until at least 95% of the excess pore pressure has dissipated. This may take from less than an hour to many days, depending upon the type of soil and size of specimen. Use of side drains appreciably shortens the consolidation time for soils of low permeability.

If after consolidation the specimen is left with the drainage valve closed the pore pressure may increase again over a period of several hours until it reaches a steady value. This is another manifestation of undrained creep, probably caused by secondary compression effects (Arulanandan, Shen and Young, 1971; Sangrey, Pollard and Egan, 1978). It can be avoided by proceeding with the compression stage as soon as it is clear that consolidation is complete, according to the above criterion.

COMPRESSION

Data from the consolidation stage are used to derive a suitable rate of strain for the compression test. For small specimens of soils of low permeability, shearing to failure usually requires a day or two, but for large specimens the time to failure may run into weeks.

The permissible rate of deformation that can be applied depends upon the following factors:

Type of test (undrained or drained)

Type of soil (drainage characteristics, stiffness)

Size of specimen

Whether or not side drains are fitted

Intervals of strain at which significant readings are to be taken.

These are allowed for in routine calculations. Some comments on the influence of the type of test are as follows.

In a drained test the rate of strain must be slow enough to allow water to drain out of the specimen so that no appreciable pore pressure can build up. However, a small increase of pore pressure within the specimen is necessary in order to push the drainage water out; the usual requirement is that 95% dissipation of excess pore pressure should have taken place when failure occurs (Bishop and Henkel, 1962). This criterion enables a suitable rate of strain to be calculated, if the strain at failure can be assumed. Details are given in Section 15.5.6, and the method is illustrated in Section 18.6.2.

An undrained test on non-sensitive plastic clays can be run at a faster rate of strain than a drained test on similar material because no movement of water through and out of the specimen is involved. However, the rate of strain must be slow enough to permit equalisation of pore pressure within the specimen. Pore pressure is measured at the base, but it is the value within the middle third that most affects the measured shear strength. For this type of soil the rate of strain for an undrained test from which only the strength parameters at failure (c', ϕ') are significant, can be nine or more times faster than for a drained test. If intermediate values of stresses and pore pressure are significant the test should be run more slowly. This applies to tests in which values of the principal effective stress ratio σ'_1/σ'_3 are to be derived, and to undrained stress path tests. The test period should never be less than about two hours, for practical reasons, even if a faster rate is calculated. Further details are given in Section 18.6.2.

For soils that fail in a brittle manner (e.g. stiff fissured clays), and for sensitive soils, the rate of strain in an undrained test should be the same as that calculated for a drained test.

Research at Imperial College has shown that in tests of very long duration the measured strength at failure decreases with increasing time to failure, by about 5% for every log cycle increase of time (Wesley, 1975).

Use of a piezometer probe to measure pore pressure at the mid-height of a large specimen avoids having to run the test as slowly as indicated above (Hight, 1982). This significantly reduces the time to failure, and therefore the cost of tests. Use of lubricated end platens can also reduce the testing time by providing better equalisation of pore pressures with the sample. These procedures are beyond the scope of this book.

OVERNIGHT RUNNING

Slow tests of a duration longer than one working day pose the problem of what to do when leaving the laboratory unattended overnight. Once shearing has started it should not be stopped until completion; on the other hand it is essential that critical readings such as those leading up to 'peak' deviator stress should not be missed. The usual compromise is to reduce the rate of displacement to a speed that is slow enough to ensure that critical readings will not be reached while the test is unattended. This may result in a smaller increase in deviator stress overnight than would be expected from the measured axial displacement (i.e. a decrease in stiffness; see Fig. 18.13 (a)), but resumption of the normal rate of displacement brings the readings back on to the original curve, with a negligible effect on the measured peak strength (Sandroni, 1977). However, if the machine is stopped, relaxation of the load ring reduces the deviator stress, and this, together with

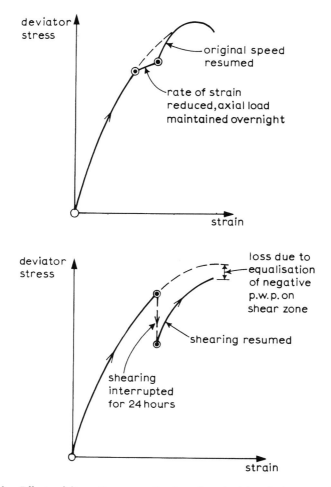

Fig. 18.13   *Effects of discontinuous application of strain:* (*a*) *reducing rate of strain overnight,* (*b*) *stopping the test for 24 hours* (*after Sandroni, 1977*)

creep effects in the specimen can cause changes in the pore water pressure. These may not be cancelled out when compression is resumed, resulting in loss of strength (Fig. 18.13 (b)), giving incorrect values of c' and ϕ'.

The period during which tests are observed can be extended to 10 or 12 hours per day, without the need for overtime working by staggering the staff working hours. This could become standard practice when effectivess stress tests are carried out regularly.

If an automatic data-logging system is used there is no problem because the machine can be kept running normally and data will accumulate for attention next morning. It is necessary to ensure that the limit of travel will not be reached while unattended (if an automatic limit switch is not fitted), and to check that there is adequate capacity in the recording medium. Prominent signs reading DO NOT SWITCH OFF should be displayed.

EFFECTS OF ACCELERATING TESTS

Attempts to accelerate the completion of each stage of a test can lead to erroneous results. The most likely effects are summarised as follows.

Where full saturation is required, incomplete saturation leaves air pockets in the specimen, giving unreliable pore pressure readings.

Early curtailment of consolidation leaves undissipated pore pressure in the specimen, resulting in consolidation continuing during the compression stage.

The effect on the measured shear strength of running compression tests too quickly was investigated (among others) by Lumb (1968). In general the result is to give false values of effective strength parameters, in which the value of c' is too high and ϕ' is too low.

18.5.5   Effect of Side Drains

The use of filter paper drains (Section 16.7.5) fitted to the side of the specimen can lead to some saving in time in testing soils of low permeability. The permeability of the filter paper itself is fairly low, of the same order as that of silt, and therefore drains are effective only if the soil permeability is less than about 10^{-8} m/s. Drains help to equalise pore pressures as well as providing a short radial drainage path.

If side drains are contemplated the following limitations should be borne in mind.

(1) The main benefit of side drains is obtained in drained tests on soils of low permeability. Since the permeability is lower when a soil is partially saturated than when fully saturated, drains are relatively more effective before saturation than after.

(2)   During saturation the 'short-circuit' effect due to the drains making contact with the porous discs at each end may give a false indication that the build-up of pore pressure is higher than it really is, and this must be allowed for (Section 18.6.1(1), step 7).

(3) The presence of drains must be allowed for in the calculation of time to failure from the consolidation data (Sections 18.6.3 and 18.6.4), bearing in mind the permeability qualification mentioned above.

(4) Corrections may need to be applied to allow for the contribution of drains to the measured shear strength of the specimen, unless spiral drains (Section 16.7.5) are used.

(5) The value of c_{vi} calculated from the consolidation data is known to be grossly in error when side drains are used.

18.6   TEST PROCEDURES

18.6.1   Saturation (BS 1377 : Part 8 : 1990 : 5 and ASTM D 4767)

INITIAL ASSUMPTIONS

It is assumed that the specimen has been prepared and set up on the triaxial cell pedestal by one of the procedures described in Section 18.4, and that the cell body has just been fitted and filled with water (Section 18.4.7, step 22). All valves are closed except the air bleed e on top of the cell. The initial dimensions of the specimen, and other relevant data, have been recorded (Fig. 18.14).

Triaxial Specimen

Location Stourford Bridge	Loc. No. 58861
Operator A. F. W	Sample No. 14/8 B
Test Type (CU.) CD. C–QU. DISS. PERM.	Date started 12.9.84
Specimen type UNDISTURBED. ~~REMOULDED~~. ~~RECOMPACTED~~.	Nominal diameter 38 mm

Specimen preparation
Extruded from U100 Sampler into 3 no. 38 mm dia. tubes

Soil type and description
Firm brown silty clay

MEASUREMENTS (mm)			INITIAL CONDITIONS	
Initial length	76.2	76.1	Area $A_o = \frac{\pi}{4} \times D_o^2$	1134 mm^2
	76.3	76.3		
	76.2	76.0	Volume $V_o = \frac{L_o \times A_o}{1000}$	86.4 cm^3
Average L_o mm		76.2		
Initial diameter	38.1	38.1	Density $= \frac{W_o}{V_o}$	2.06 Mg/m^3
	38.0	37.9		
	38.0	38.0	Moisture content	18.1 %
Average D_o mm		38.0	Dry density $= \frac{W_d}{V_o}$	1.74 Mg/m^3

WEIGHINGS	Trimmings		Specimen initially	Specimen after test	
Container no.	(12)		(17)	(9)	
Specimen + cont.	73.03		194.14	186.68	
Container	16.27		16.05	15.31	
Mass of specimen	–	W_o	178.09	171.37	W_f
Dry spec. + container	64.32		–	166.11	
Dry mass of specimen	48.05		150.80	150.80	W_d
Moisture	8.71		27.29	20.57	M_f
Moisture content	18.13		18.10	13.64	

LOCATION OF SPECIMEN

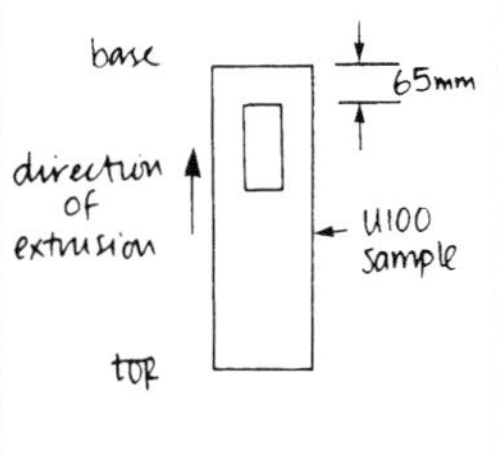

Fig. 18.14 Initial specimen data for effective stress triaxial test

Pore pressure is measured by means of a pressure transducer. The use of a manual pore pressure mercury null indicator for this purpose is now obsolete and is not included here. Reference is made to Fig. 18.4.

MEASUREMENT OF INITIAL PORE WATER PRESSURE

Immediately after setting up the specimen, and while the air bleed (valve e) remains open, carefully open valve a and observe and record the pore water pressure. It might or might not represent the pore pressure throughout the specimen, which could take up to 24 hours to equalise. Even if this time is allowed, the measured pore pressure does not necessarily reflect the in-situ value, but it provides a datum value for the test.

Record the initial pore pressure, and the initial readings of the volume-change indicators on the pressure lines, on the saturation stage test form as shown in Fig. 18.15. Ensure that the top of the piston is rigidly restrained, and close the air bleed e.

SATURATION PROCEDURES OUTLINED

For the reasons outlined in Section 15.6.1, normal practice is to achieve saturation by raising the pore pressure in the specimen before proceeding with an effective stress test. Five ways of doing this are described below, namely:

Triaxial Saturation

Location	Stourford Bridge						Test type		Loc. No.	58861	
Operator	A.F.W						CU		Sample No.	14/8	B
Membranes thickness	1 no. 0.2 mm	~~With~~ Without	side drains				Cell No. 6		Specimen dia	38 mm	
							Panel No. 2		Length	76 mm	
Remarks	Cell volume change not measured								Date started	12.9.84	

line	Cell pressure kN/m²	Back pressure kN/m²	Pore pressure kN/m²	pwp difference kN/m²	B value	Back pressure volume change before	after	diff	Cell volume change LHS/RHS before	after	diff	+ consol cm³	− exp cm³
0	0	0	−3										
1	50	—	5.5	8.5	0.17								
2	50	40	39			89.5	89.2	0.3					
3	100	—	57	18	0.36								
4	100	90	88			88.9	88.4	0.5					
5	200	—	153	65	0.65								
6	200	190	190			88.0	87.3	0.7					
7	300	—	276	86	0.86								
8	300	290	290			86.9	86.5	0.4					
9	400	—	385	95	0.95								
10	400	390	390			86.1	85.9	0.2					
11	500	—	488	98	(0.98)								
	Saturation Stage Completed							2.1	total				
	600		586	98	0.98				for consolidation				

Fig. 18.15   *Triaxial saturation data sheet*

(1) Application of pressure increments to achieve full saturation.

(2) One-stage elevation of back pressure.

(3) Use of 'initial effective stress'.

(4) Saturation at constant water content.

(5) Automatic saturation.

Methods (1) and (4) are given in Clause 5 of BS 1377 : Part 8. A method similar to method (1), but with application of cell pressure and back pressure increments simultaneously, is given in ASTM D 4767, which also includes a saturation method using a partial vacuum.

Method (1) is the procedure widely used in Britain for commercial testing, and is described in detail. This is not always the most suitable method, but it has little effect on measured shear strengths from routine CU and CD tests. Method (2) is that given by Bishop and Henkel (1962). Methods (3) and (4) have been used at Imperial College (Hight, 1980). Method (4) is necessary when the possibility of swelling, or when cyclic changes in effective stress, would significantly affect the specimen behaviour, for example when the in-situ stress is to be reimposed. Method (5) is similar in principle to (1).

At the outset of each procedure, all valves are assumed to be closed. Valve a_1 remains closed throughout.

(1) *Saturation by increments of back pressure*

The suitability of this method is discussed in Section 15.6.6. Before starting, two factors must be decided:

(a) the cell pressure increments to apply;

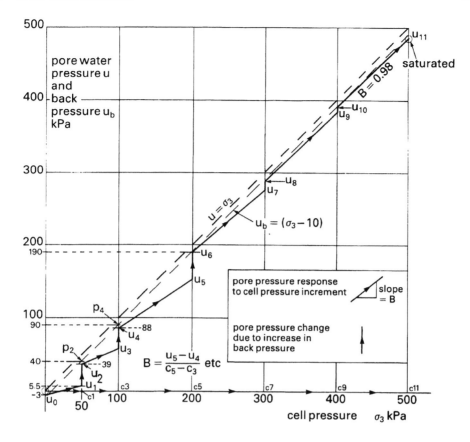

Fig. 18.16 *Plot of pore pressure, back pressure and cell pressure changes during saturation stage*

(b) the differential between applied back pressure and cell pressure, which controls the lowest effective stress to which the specimen will be subjected.

Regarding (a), to comply with BS 1377 the cell pressure increments should not exceed 50 kPa, or the effective stress to which the specimen is to be consolidated for the test, whichever is the less, unless the engineer specifies otherwise. An acceptable sequence that has been found to be suitable for many soils is to apply cell pressure increments of 50 kPa until the value B reaches 0.8, and 100 kPa thereafter, provided that the effective consolidation pressure is greater than 100 kPa.

Regarding (b), the BS states that the differential pressure should not be greater than the effective consolidation pressure, or 20 kPa, whichever is less, and should not be less than 5 kPa. For soils not susceptible to swelling at this level of effective stress, a differential pressure of 10 kPa is appropriate. For soils that are susceptible to swelling, the differential pressure should be sufficient to prevent swelling (see Section 15.6.6).

The procedure described below represents typical practice in which the first two cell pressure increments are of 50 kPa and the subsequent increments 100 kPa, and a differential of 10 kPa is maintained after each pressure increment. This is the same procedure as that given in BS 1377 except in one respect, which is indicated below.

It is assumed that the initial measured pore pressure is a small negative value, represented by u_0 on the graphical plot used to illustrate this procedure (Fig. 18.16), and recorded on line 0 in Fig. 18.15.

(1) Increase the pressure in the cell line to the first desired value (e.g. 50 kPa), as soon as possible after setting up the specimen.

(2) Open valve c to admit the pressure into the cell, represented by c_1 in Fig. 18.16. The cell pressure increase causes an increase of pore pressure in the specimen, which should be observed until an equilibrium value (u_1) is reached, and then recorded (line 1 in Fig. 18.15). If the pore pressure continues to increase and the increment exceeds the cell pressure increment, this indicates a leakage of cell water into the specimen and the test should be abandoned and restarted on a replacement specimen.

If the pore pressure has not reached a steady value after about 10 minutes, plot a graph of pore pressure against time so that the steady-state value can be inferred from the eventual flattening of the curve.

If the pore pressure starts to decrease, even after an initial increase, prevent it from dropping to zero by proceeding to step (7) without waiting for equilibrium.

(3) Calculate the initial value of the pore pressure coefficient B from the equation

$$B = \frac{\Delta u}{\Delta \sigma_3} = \frac{u_1 - u_0}{\Delta \sigma_3}$$

In this example,

$$B = \frac{5.5 - (-3)}{50 - 0} = \frac{8.5}{50} = 0.17$$

Enter the B value in line 1 (Fig. 18.15).

(4) Keep valve c open and valve b closed. Increase the pressure in the back pressure line to a value equal to the cell pressure less the selected differential pressure. In this case the back pressure would be increased to $(50 - 10) = 40$ kPa (p_2 in Fig. 18.16). Wait about 5 or 10 minutes until the back pressure volume-change indicator reading reaches a steady value, to ensure that the connecting lines have fully expanded, then record the reading in the 'before' column under 'back pressure volume change' (Fig. 18.15, line 2).

(5) Open valve b to admit the back pressure into the specimen and observe the increase in pore pressure. Also keep under observation the flow of water into the specimen through the back pressure volume-change indicator. Allow the pore pressure to build up until it virtually equals the applied back pressure; this might take several hours for a clay soil.

If side drains are fitted, the pore pressure reading may increase rapidly due to the 'short circuit' effect of the drains. But the pore pressure may not have equalised throughout the specimen, and the pressure must be maintained long enough to ensure equalisation. To check, close valve b and observe the pore water pressure for a time. If it decreases it has not equalised and the back pressure should be re-applied by opening valve b.

(6) When the pore pressure is virtually equal to the back pressure, or when pore pressure and volume changes have ceased, record the pore pressure u_2, and the back pressure volume-change indicator reading (under 'after') (Fig. 18.15, line 3). In this example the final pore pressure does not quite reach the pressure p_2. Close valve b.

(7) Increase the cell pressure by 50 kPa (e.g. in this case to 100 kPa). Observe the resulting

pore pressure as in step (2) and record the steady value (u_3) (line 3 in Fig. 18.15). Calculate the value of the pore pressure coefficient B as in step (3).

(8) Increase the back pressure as described in step (4), again giving a differential pressure of 10 kPa, and follow the procedures given in steps (4)–(6).

(9) Repeat step (7), but with a cell pressure increment of 100 kPa.

(10) Repeat steps (8) and (9) as many times as necessary until the B value calculated as in step (3) reaches 0.95, or a value appropriate to the specimen (see Section 15.6.5). For stiff clays in which it is not possible to achieve a B value of 0.95, a value of 0.90 is acceptable, according to BS 1377, if it remains unchanged after three successive increments of pressure as described above.

Readings are shown in lines 4 to 11 in Fig. 18.15, where pore pressure responses to cell pressure increments are recorded on the odd-numbered lines, and back pressure increments on the even-numbered lines. They are represented graphically in Fig. 18.16 from u_4 onwards.

(11) When the coefficient B reaches 0.95 (or the appropriate value from Section 15.6.5), further increments of back pressure and cell pressure are not necessary. Saturation is terminated by closing valves c and a (i.e. all valves are closed). Alternatively, an additional back pressure increment may be applied to the specimen to allow for any uncertainty, in which case steps 4 to 6 are followed, and then the valves are closed. The specimen is then ready for consolidation (Section 18.6.2).

(12) Plot the value of B obtained from each pressure increment against cell pressure, or pore pressure at the start of the increment, as shown in Fig. 18.17. Report the final value to two significant figures.

(13) The volume of water taken up into the specimen during saturation can be calculated, if required, by totalling the differences between the 'before' and 'after' volume-change indicator readings at each back pressure increment stage. These differences are entered and totalled in column 8 of Fig. 18.15.

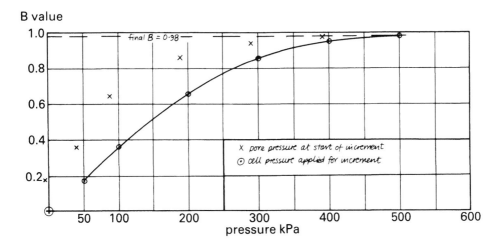

Fig. 18.17 *Pore pressure coefficient B related to pore pressure and cell pressure during saturation*

In BS 1377 : Part 8 : Clause 5.3.2, steps (3)–(6) above are omitted for the first cell pressure increment, and the second increment is applied without introducing a back pressure.

If it is required to record the cell volume changes during saturation (e.g. to evaluate specimen volume changes), a volume change indicator can be included in the cell pressure line. Readings are taken in the same way as those on the back pressure line, i.e. making allowance for expansion of the lines, and the differences between the 'before' and 'after' readings at each stage are totalled. The expansion of the cell itself is obtained from the cell calibration data (Section 17.3.1), and is deducted from that total (see Section 18.6.6).

(2) *One-stage elevation of back pressure*

The simplest way of applying back pressure is to increase the pore pressure enough to dissolve air in the specimen voids and proceed immediately with the consolidation stage. The back pressure needed depends on the initial degree of saturation (Table 15.5, Section 15.6.3). The cell pressure required is equal to the back pressure plus a suitable differential pressure (which could be up to the in-situ mean effective stress).

The required cell pressure is applied to the specimen as described in steps (1) and (2) above. Meanwhile, the pressure in the back pressure system is raised to the selected value keeping valve b (Fig. 18.4) closed, and the volume change indicator is observed, as in step (4). The back pressure is admitted to the specimen by opening valve b, and when the pore pressure is steady the volume-change indicator reading indicates the volume of water taken up by the specimen. Valve b is closed, and pressures are adjusted for the consolidation stage as described in Section 18.6.2.

(3) *Use of 'initial effective stress'*

As soon as possible after the specimen is sealed in place on the triaxial pedestal, a cell pressure is applied and pore pressure changes are recorded. The cell pressure is held constant with valve a open and the back pressure line valve b (Fig. 18.4) kept closed. The cell pressure should be at least equal to the in-situ total vertical stress, or greater if necessary, to bring the measured pore pressure up to 50 kPa or more. Up to 48 hours may be required to reach pore pressure equilibrium.

The difference between the applied cell pressure and the measured pore pressure, when stabilised, is referred to as the 'initial effective stress'. It may not necessarily be the same as the in-situ effective stress. Accurate measurement of the difference between the two pressures is essential, and the use of a differential pressure gauge or transducer might be desirable.

Saturation is continued by the application of alternate back pressure and cell pressure increments, as described in method (1) above. Pressures may be applied either by using a small differential (5 or 10 kPa), or by using a minimum differential pressure equal to the initial effective stress. The latter is considered to be preferable because it tends to reduce the scatter of measured strengths, but it takes longer to complete.

(4) *Saturation at constant water content*

This procedure is given in Clause 5.4 of BS 1377 : Part 8, and is suitable for low-plasticity clays. No water is allowed to enter or leave the specimen, therefore the drainage line valve b (Fig. 18.4) remains closed.

The cell pressure is raised in increments of typically 50 or 100 kPa. Enough time must be allowed for the pore pressure to reach a steady value after each increment. The steady state might not be easy to define, but by plotting a graph of pore pressure change against time or square-root time the flattening out of the curve gives a reasonable indication. The value of B is calculated from the eventual pore pressure response after each increment of cell pressure. According to BS 1377, the specimen is considered to be saturated when one of the conditions given in step (10) of method (1) is achieved.

This method requires a longer time to reach saturation than using a back pressure. The necessary applied confining pressure depends on the initial degree of saturation (Table 15.5, Section 15.6.3).

At the end of this procedure the specimen is subjected to a high effective stress. The effective stress can be reduced if the pore pressure is raised by introducing a suitable back pressure, but the specimen is then likely to swell. Unless high effective stresses are appropriate this method should not be applied to soils that are susceptible to swelling.

(5) *Automatic saturation*

Automatic control of cell pressure and back pressure enables the two pressures to be increased continuously, avoiding the cyclic changes in effective stress that are applied to the specimen by stage increases (method 1), which can be detrimental to partially saturated soils. This method is possible where compressed air is available for operating constant pressure systems. The two air pressure regulator valves controlling the cell pressure and back pressure systems are fitted with a variable-speed electric motor drive which operates the two valves simultaneously.

After setting up the specimen and fitting the triaxial cell, the initial value of B is obtained by the manual method given in method (1) above. A back pressure of 10 kPa less than the cell pressure is then introduced to the specimen, until equilibrium is established. The automatic system can then take over.

The time period required for saturation is estimated, on the basis of experience. This might typically range from about 2 days to 10 days, depending on the type of soil, initial degree of saturation, size of specimen and whether side drains are fitted. The final saturation back pressure is assumed, from which the average rate of increase of pressure is calculated. The controller of the motor driving the air regulator valves is set so that cell pressure and back pressure both increase steadily at that rate, maintaining the initial differential of 10 kPa. Pressure and volume-change indicator readings are recorded at suitable intervals. The pressure differential should be checked once or twice daily, and the cell pressure adjusted if necessary to maintain a constant differential.

At any stage the automatic process can be interrupted in order to make a manual check of the B value. Before resuming automatic operation the cell pressure is reduced to the value immediately before interruption. When the pre-set pressures are reached the B value should be checked manually. Saturation is complete if an appropriate value is indicated. The recorded data are summarised and plotted as in method (1).

When side drains are fitted, saturation should be checked by closing the back pressure valve b; if the measured pressure falls, the whole of the specimen is not fully saturated.

This method can be developed further if electronic measuring devices are used and connected to a data acquisition system with automatic pressure control (see Section 16.6.9). The pore pressure response to small cell pressure increments can be used in a feedback loop to adjust the cell and back pressures in turn automatically as soon as a steady condition is reached. The process would stop when a predetermined B value is achieved. No assumptions need to be made beforehand regarding the rate of pressure increase.

18.6.2   Consolidation (BS 1377 : Part 8 : 1990 : 6 and ASTM D 4767)

ADJUSTMENT OF EFFECTIVE STRESS

The effective stress to which the specimen is subjected at the end of the saturation stage is usually much lower than the effective confining pressure required for the compression test. The effective stress is increased either by raising the cell pressure, or by reducing the back pressure, or by a combination of both, and then allowing the resulting excess pore pressure to dissipate against an appropriate back presure, i.e. the specimen consolidates. Normally the cell pressure should be raised, but if the pressure needed is greater than the maximum working pressure available, some reduction of back pressure will be necessary in addition to a cell pressure increase. This is acceptable provided that the back pressure is not reduced to a level below that of the pore pressure at the end of the final step in the saturation stage, or 300 kPa, whichever is greater. For compacted specimens with a high initial air content the back pressure should not be less than 400 kPa.

If the measured pore pressure at the end of the saturation stage is denoted by $\dot{u}_s$, and the effective stress to be used for compression by σ'_3, the cell pressure required (σ_3) is calculated from the equation

$$\sigma_3 = \sigma'_3 + u_s \tag{18.1}$$

If σ_3 exceeds the maximum working pressure $\sigma_{3\,\text{max}}$, the back pressure should be set to a value u_b such that

$$u_b = \sigma_{3\,\text{max}} - \sigma'_3 \tag{18.2}$$

and it might then be unavoidable for u_b to be less than u_s. The cell pressure is increased to $\sigma_{3\,\text{max}}$. It is usually more convenient to set the back pressure u_b to an exact multiple of 100 kPa such that the cell pressure is rather less than $\sigma_{3\,\text{max}}$.

If the required change in effective confining pressure is very large it might be desirable to consolidate the specimen in two or more stages, especially if side drains are used.

CONSOLIDATION PROCEDURE

(1) Close valves a, b and c (Fig. 18.4). Increase the pressure in the cell line and adjust the back pressure (u_b) if necessary, to give a difference equal to the required effective stress, as described above. Record the back pressure volume-change indicator reading.

(2) Open valve c to admit the pressure into the cell, and open valve a to observe the consequent rise of pore pressure. Record the pore pressure reading (u_i) when steady. Calculate the new B value from the changes of pore pressure and cell pressure; this may not be quite as high as the final value in the saturation stage if it was terminated at step 4 because of the increased effective stress.

The difference between the final steady pore pressure and the back pressure (as yet isolated from the specimen) is the excess pore pressure ($u_i - u_b$) to be dissipated during consolidation, as indicated graphically in Fig. 18.19.

(3) Set the timer to zero, and record pore pressure and volume-change indicator readings on the consolidation test sheet (Fig. 18.18 (a)) as the zero values.

(4) Start the consolidation stage by opening the drainage valve b and at the same time start the clock.

(5) Record pore water pressure and back pressure volume-change indicator readings at time intervals similar to those used in the oedometer consolidation test (Volume 2, Section 14.5.5(14), Table 14.11). The intervals should normally be those to give a regular spacing of points on a square-root time scale, but more frequent intervals might be necessary for soils which compress very rapidly. In this case accurate readings might be difficult to obtain in the early stages of consolidation. Typical readings are shown in Fig. 18.18 (a).

(6) Plot graphs of specimen volume change against square-root time (Fig. 18.18 (b)) and pore pressure dissipation (%) against log time (Fig. 18.18 (c)). The consolidation stage can be terminated when at least 95% dissipation is reached. The method of calculating pore pressure dissipation is given by Equation (15.28), Section 15.5.5, and is shown graphically in Fig. 18.19.

(7) Terminate the consolidation stage by closing valve b. Leave valve c open. Record the end of stage pore pressure measured at the specimen base, denoted by u_c. The effective confining pressure σ'_3 is now given by the effective stress equation

$$\sigma'_3 = \sigma_3 - \bar{u} \tag{15.1}$$

where $\bar{u}$ is the mean pore pressure within the specimen. The pore pressure at the top is equal to the back pressure, u_b. If the pore pressure distribution within the specimen is assumed to be parabolic,

$$\bar{u} = \tfrac{2}{3}u_c + \tfrac{1}{3}u_b \tag{15.26}$$

However, if the measured pore pressure u_c is within a few kPa of the back pressure, for practical purposes $\bar{u}$ can be assumed equal to the arithmetical mean pressure $\tfrac{1}{2}(u_c + u_b)$.

(8) Record the final reading of the volume-change indicator and calculate the total change in volume (ΔV_c) during consolidation. The consolidated specimen is now ready for the compression stage.

DERIVATION OF TIME TO FAILURE

(9) The graph of volume change against square-root time is used to derive the time intercept t_{100} by the following method.

The initial portion to the graph, as far as about half the total volume change, can usually be represented by a straight line. Extend this line to intersect the horizontal line representing 100% consolidation at point X (Fig. 15.20, Section 15.5.5). Read off the value $\sqrt{t_{100}}$ from the horizontal scale, and multiply it by itself to obtain t_{100} (minutes).

(10) Multiply t_{100} by the appropriate factor to determine the 'significant testing time' for the compression stage, t_f (minutes). The time t_f could be either (a) the time to failure, or (b) the time between successive sets of readings taken at the selected intervals of strain. The factor depends upon whether or not side drains are fitted, and the type of test (drained or undrained). For specimens of $L:D$ ratio of about 2:1 the factors are independent of the size of the specimen, and are summarised in Table 18.1.

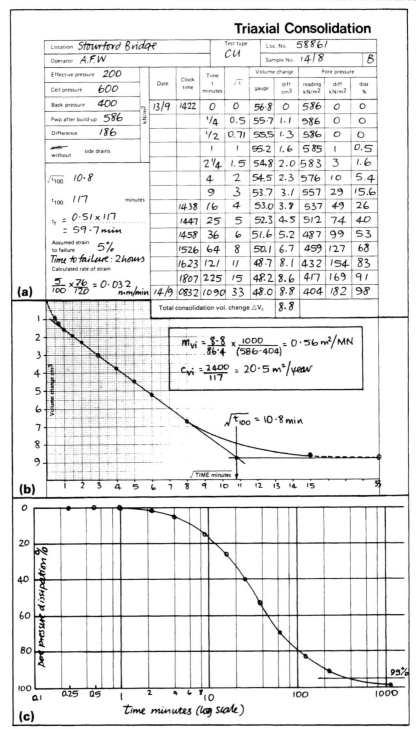

Fig. 18.18   *Triaxial consolidation stage data sheet:* (a) *test data and readings,* (b) *volume change against square-root time,* (c) *pore pressure dissipation against log time*

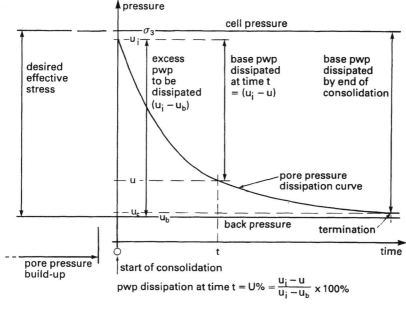

pwp dissipation at time t = U% = $\dfrac{u_i - u}{u_i - u_b}$ × 100%

at termination of consolidation, $\sigma_3' = \sigma_3 - \tfrac{1}{2}(u_c + u_b)$

Fig. 18.19   *Representation of pore pressure dissipation*

Table 18.1. FACTORS FOR CALCULATING TIME TO FAILURE

Time of test (any diameter) (2:1 ratio)	*NO side drains*	*WITH side drains*
Undrained (CU)*	$0.53 \times t_{100}$	$1.8 \times t_{100}$
Drained (CD)	$8.5 \times t_{100}$	$14 \times t_{100}$

*For plastic deformation of non-sensitive soils only.
(Based on Table 1 of BS 1377: Part 8: 1990).

With other values of the $L:D$ ratio, for drained tests Equation (15.36) is used together with the data from Table 15.4 (Section 15.5.5). For undrained tests on non-sensitive plastic clays this calculated value of t_f is divided by 16 if side drains are not used, or by 8 if drains are used, bearing in mind the practical lower limit of time to failure referred to in Section 18.6.3.

(11) The calculated time to failure is applied to a CU test as described in Section 18.6.3, and to a CD test as described in Section 18.6.4. But it is important to appreciate that for soils that fail in a brittle manner (e.g. stiff fissured clays), and for sensitive soils, the factors given for drained tests should be used for undrained tests as well.

18.6.3   Undrained Compression (CU Test) (BS 1377 : Part 8 : 1990 : 7 and ASTM D 4767)

TEST CONDITIONS

The test procedure described below relates to a specimen in the triaxial cell which has been saturated and brought to the required state of effective stress by consolidation, as described in Sections 18.6.1 and 18.6.2. In this test no change of water content of the specimen is allowed while it is being sheared at a constant rate of axial deformation with the cell confining pressure remaining constant. Valve b (Fig. 18.4) is kept closed to prevent movement of water out of or into the specimen. If the specimen is fully saturated the undrained condition means that there is no change in its volume during compression. However, the increasing shear stresses resulting from the imposed axial load give rise to changes in pore water pressure, and therefore the effective stress changes are not equal to the changes in total stress. Compression must be applied slowly enough to allow pore pressure changes to equalise throughout the specimen by the time failure is reached. Pore pressure is measured at the base, and if the test is run too quickly the measurements will not be a true representation of the conditions within the middle of the specimen where the shear stresses have their most significant effect.

TIME TO FAILURE

In routine tests for which the only significant pore pressure reading is that at failure, the value of t_f calculated from t_{100} (Section 18.6.2, steps 9 and 10) relates to the time from the start of compression to failure. Using the factors from Table 18.1, t_f is obtained as follows:

Without side drains: $t_f = 0.53 \times t_{100}$ minutes.

With side drains: $t_f = 1.8 \times t_{100}$ minutes.

However for brittle or sensitive soils, the value of t_f should be not less than that obtained as in Section 18.6.4 for a drained test.

The term 'failure' usually relates to the maximum (or 'peak') deviator stress which the specimen can withstand, but any of the criteria discussed in Section 15.4.2 could apply. The strain ε_f at which failure is likely to occur must be estimated, and this depends on the type and condition of the soil. The data summarised in Table 18.2 are provided as a general guide, but the estimate should preferably be based on experience. If in doubt a low estimate of failure strain will err on the safe side.

If the calculated value of t_f is less than 120 minutes (2 hours), the actual time to failure should not be less than 2 hours.

In tests where it is necessary to know the effective stresses at intermediate values of strain before failure is reached (such as when a stress path is to be derived), the strain ε_f used in the calculation given below is then the required strain increment between each set of significant readings. This will result in a very much slower rate of strain than when failure is the only significant criterion.

If failure is considered to occur when the principal stress ratio σ'_1/σ'_3 reaches its maximum value, the strain at which this is likely should be substituted for ε_f in the calculation of rate of displacement. In overconsolidated and compacted clays maximum principal stress ratio usually occurs before maximum deviator stress, and a strain of about two-thirds of the estimated strain at 'peak' would be a reasonable assumption.

Table 18.2. SUGGESTED FAILURE STRAINS IN TRIAXIAL TESTS

Soil type	Typical ranges of strain at failure $\varepsilon_f\%$ (maximum deviator stress)	
	CU test	CD test
Undisturbed clay		
normally consolidated	15–20	15–20
overconsolidated	20 +	4–15
Remoulded clay	20–30	20–25
Brittle soils	1–5	1–5
Compacted 'boulder clay':		
dry of o.m.c.	3–10	4–6
wet of o.m.c.	15–20	6–10
Compacted sandy silt	8–15	10–15
Saturated sand:		
dense	25 +	5–7
loose	12–18	15–20

If in doubt, assume a failure strain less than the tabulated values.

RATE OF STRAIN AND DISPLACEMENT

The maximum rate of strain to be applied is equal to $\varepsilon_f/t_f\%$ per minute.

The axial compression of the specimen corresponding to a strain of $\varepsilon_f\%$ is $(\varepsilon_f\%/100) \times L$ mm, where L is the specimen length in mm. The rate of axial displacement should therefore not exceed

$$\frac{\varepsilon_f L}{100 t_f} \text{ mm/minute}$$

If the 'critical' or 'ultimate' condition (Section 15.4.3) is to be reached, application of strain is continued at the same rate beyond the 'peak' point.

Speed control settings on the compression machine to give various rates of displacement (mm/minute) are provided by the manufacturer. The setting required is that which gives the calculated rate of displacement, or the next slower speed.

COMPRESSION TEST PROCEDURE

Procedures marked * are identical for the CD test (Section 18.6.4), where they are referred to but not repeated.

Adjustments for Shearing

(1) Referring to Fig. 18.4, valve b must remain closed. Valve c remains open to the cell pressure line and valve a is open to enable pore pressure readings to be observed.

(2)* Set the speed controller on the compression machine to give the required rate of displacement. Check that the reversing switch is set correctly for upward movement of the machine platen, in accordance with the manufacturer's instructions.

(3)* While the cell piston is still well clear of the top loading cap, switch on the motor. As the piston is pushed into the cell adjust the dial gauge on the load ring to read zero. This will compensate for the combined effects of cell pressure and piston friction on the load ring reading. If the machine speed for the test is very slow a faster speed may be used for this adjustment, without significantly altering the frictional resistance. If adjustment of the load ring dial gauge is not possible, record its reading during this operation.

(4)* Switch off the motor, engage the hand drive gear and continue winding up the platen by hand until the piston makes contact with the top cap. Make sure that the conical taper seats correctly on the ball or hemispherical surface. A seating load, which should be as small as possible, will be indicated on the load ring dial gauge.

(5)* Secure the strain dial gauge or transducer vertically in position, and adjust the bracket on which the stem rests to set the dial to zero, or to a convenient initial reading. Ensure that the dial gauge has enough travel, and that clearances are large enough, to permit a strain movement of at least 25% of the specimen length.

(6)* Re-engage the machine speed setting required for the test. Check that contact is maintained between specimen top cap and piston; if necessary make small adjustments by switching on the motor momentarily. Set the timer to zero.

(7) Observe the following and record them as the initial readings for the compression stage:

   Clock time and date

   Displacement gauge

   Load ring dial gauge

   Pore water pressure

   Cell pressure (check)

A suitable test form is shown in Fig. 18.20, and the above readings are entered as the first line of data. The columns relating to specimen volume change (back pressure line) are not normally used.

Loading to failure

(8) Switch on the motor to start the test, and at the same instant start the clock.

(9) Taking readings at regular intervals of axial displacement, usually corresponding to strain intervals of 0.2% strain up to 1% strain, and 0.5% strain thereafter. At least 20 sets of readings should be obtained up to failure so that stress–strain and other curves can be clearly defined. When it is evident that a critical condition (e.g. peak deviator stress, or maximum pore pressure) is approached, or if readings are changing rapidly, readings should be observed more frequently. Each set of data comprises readings of the gauges listed in step 7.

   Data from a typical test are shown in Fig. 18.20.

   Brittle or hard specimens are likely to fail suddenly at very small strains, perhaps less than 1%. For this type of material, sets of readings should be taken at regular intervals of

TRIAXIAL SHEARING

Loc. No. 58861	Location *Stourford Bridge*		Sample No. 14/8 B
Test Type *CU*	Proving Ring No. 1153-21	~~With~~ Without — Side Drains	Effective Cell Pressure *initially* 196 kN/m² Cell Pressure 600 kN/m²
Rate of Strain 2.4 % per hour	Operator *A.F.W.*	Membranes 1 x 0.2mm	Back Pressure 404 kN/m²
Consolidated Length 73.6 mm	Consolidated Area 1057 mm²		Consolidated Volume 77.8 cm³

DATE	TIME	STRAIN DIAL DIV	DIAL DIV	ε %	LOAD DIAL DIV	RING CALIB N/DIV	LOAD N	U kN/m²	U DIFF	V/C GAUGE	DIFF	$\Delta V/Vc$ %	STRESS kN/m²	MEMBR + DRAINS kN/m²	$\sigma_1-\sigma_3$ kN/m²	σ_1 kN/m²	σ_1' kN/m²	σ_3' kN/m²	σ_1'/σ_3'
14.9.84	10.20	100	0	0	0		0	404	0				0		0	600	196	196	1
	10.29	110	10	0.13	8.5	1.461	12	407	3				11.7	0	12	612	205	193	1.06
		120	20	0.27	23	1.460	34	414	10				31.7	0	32	632	218	186	1.17
	10.41	140	40	0.54	63	1.457	92	428	24				86.4	0	86	686	258	172	1.50
		160	60	0.82	80	1.455	116	442	38				109.2	0	109	709	267	158	1.69
	10.54	180	80	1.09	91		132	452	48				123.9	0.1	124	724	272	148	1.84
		200	100	1.36	99		144	460	56				134.4	0.2	134	734	274	140	1.96
	1109	220	120	1.63	107		156	468	64				144.8	0.2	145	745	277	132	2.10
		240	140	1.90	114	1.453	166	474	70				153.7	0.3	153	753	279	126	2.21
	1122	260	160	2.17	119		173	480	76				160.0	0.3	160	760	280	120	2.33
		280	180	2.45	124		180	483	79				166.3	0.3	166	766	283	117	2.42
	1136	300	200	2.72	130		189	487	83				173.8	0.4	173	773	286	113	2.53
		320	220	2.98	134		195	490	86				178.7	0.4	178	778	288	110	2.62
	1151	340	240	3.26	140	1.452	203	494	90				186.2	0.5	186	786	292	106	2.75
		360	260	3.53	144		209	496	92				190.8	0.5	190	790	294	104	2.83

From Fig. 18.21 (a), $\Delta V_c = 8.8$ cm³. $V_c = 86.4 - 8.8 = 77.6$ cm³

$\therefore \varepsilon_v = \dfrac{8.8}{86.4} \times 100 = 10.2\%$, $\frac{1}{3}\varepsilon_v = 3.4\%$, $\frac{2}{3}\varepsilon_v = 6.8\%$

$\therefore L_c = 76.2\left(1 - \dfrac{3.4}{100}\right) = 73.6$ mm. $A_c = 1134\left(1 - \dfrac{6.8}{100}\right) = 1057$ mm²

Fig. 18.20   *Data sheet for consolidated-undrained (CU) triaxial compression test with pore pressure measurement*

axial force, rather than of strain, in order to obtain enough readings to define the curves leading up to failure. Otherwise the stress/strain curve may have only one or two points before the 'peak'. A set of readings should also be taken at the peak stress, or as close to it as possible, after which the load is likely to fall dramatically.

(10) As compression continues, calculate each line of data as described in Section 18.7.3 and make a rough plot of the deviator stress/strain curve as the test proceeds. If that is not practicable, a fair indication of the trend of the curve can be obtained by plotting the load ring dial readings against strain (%) on an area-correction graph sheet (Volume 2, Fig. 13.17 (a)), and that at least should be kept up with the readings.

(11) Allow the test to continue until the relevant criterion of failure has been clearly identified, as indicated by one of the following:

(1) Maximum deviator stress

(2) Maximum principal stress ratio

(3) Defined limit of strain

(4) Shear stress and pore pressure remaining constant with increasing strain.

If none of the above is evident, terminate the test when an axial strain of 20% is reached (15% strain in ASTM D 4767).

(12) If an attempt is to be made to measure the 'ultimate' shear strength, allow the test to

continue to 20% strain if practicable. However, the test should be terminated earlier if the specimen becomes so distorted that further readings would have little meaning.

Completion of test

(13)* When the displacement reaches the intended limit, and the final readings have been taken, switch off the motor, close valve a, and unload the specimen, either by switching the motor to reverse (allowing it to stop first), or by winding down the machine platen by hand.

(14)* Reduce the cell pressure to zero.

(15)* Open the bleed plug e (Fig. 18.4) and empty the cell by draining the water to the waste line or into a bucket from valve c.

(16)* When the cell is empty, progressively slacken the cell body securing screws or clamps, pull up the piston and carefully ease the cell body off the base. Take care not to knock against the specimen or drainage lead when removing the cell body. Mop up any water remaining on the cell base with a sponge. (An alternative dismantling procedure is outlined in Section 18.6.5.)

(17)* Roll the O-rings upwards off the top cap and roll down the membrane to expose the specimen.

(18)* Make sketches of the specimen as viewed from two directions at right angles to illustrate the mode of failure. If a slip surface is visible measure its inclination to the horizontal. Record any other features observed.

(19)* Carefully take the specimen off the base pedestal and remove the drainage discs, and filter paper drains if used. Weigh the specimen immediately together with any soil which may have adhered to the discs or drains (mass m_f grams).

It is sometimes easier to weight the specimen with the discs (and filters) which are then weighed separately to obtain the sample mass by difference. Comparison with the known mass of the discs when fully saturated gives an indication of the mass of water which has migrated into the specimen after removing the confining stresses.

(20)* A small specimen may be placed in a moisture container and dried overnight in an oven, then weighed dry. If there is a failure slip surface the soil immediately adjacent to it should be first cut out and the moisture content measured separately. The moisture content of a large specimen should be determined either by breaking down the whole before placing in the oven, or by taking three or more representative portions of convenient size from recorded locations within the specimen.

The dry mass of the whole specimen is needed for check calculations of the initial soil properties.

(21)* If the soil fabric is considered to be significant, cut the sample longitudinally about half-way through, then break it open. Allow one half to air-dry, and sketch and photograph it after 18–24 hours by which time the fabric structure is usually seen more clearly. The other half can be used for moisture content measurements as described above.

(22)* Remove the rubber O-rings and the membrane from the cell pedestal and top cap. Wash and dry the O-rings; discard the membrane. Wash the porous discs and boil them in distilled water before re-use.

Clean the cell pedestal and top cap, and flush de-aired water through the connections to remove any soil particles. Leave the pore pressure system under a small pressure.

(23) Calculations, plotting and analysis of test data are described in Section 18.7.1 to 18.7.3.

18.6.4 Drained Compression (CD Test) (BS 1377: Part 8: 1990: 8)

TEST CONDITIONS

The test procedure described below relates to a specimen in the triaxial cell which has been saturated and brought to the required state of effective stress by consolidation, as described in Sections 18.6.1 and 18.6.2. In this test water is allowed to drain out of or into the specimen while it is being sheared at a constant rate of axial deformation with the cell confining pressure remaining constant, so that no excess pore pressure (positive or negative) develops. Valve b (Fig. 18.4) is kept open to permit drainage into the back pressure system, or to an open burette if back pressure is not applied. The pore pressure remains substantially constant but the volume of the specimen changes due to the movement of water.

As shear stresses develop a contracting soil expels water into the drainage line, and a dilating soil draws in water from it. A dilatant soil may show some contraction at small strains. The volume of pore water draining out of or into the specimen is measured by means of the volume-change indicator or burette in the back pressure line, and is assumed to be equal to the change in volume of the specimen.

Compression must be applied slowly enough to ensure that pore pressure changes due to shearing are negligible, so that the virtually fully drained condition (95% pore pressure dissipation) is applicable by the time failure is reached. The required rate of strain is usually very much slower than for an undrained test on a similar specimen under similar conditions.

TIME TO FAILURE

The time to failure, t_f, based on the above criterion is calculated from t_{100}, which is derived as shown in Section 18.6.2, step (9). The factors relating to a drained test given in Table 18.1 are used, as follows:

Without side drains: $t_f = 8.5 \times t_{100}$ minutes.

With side drains: $t_f = 14 \times t_{100}$ minutes.

The term 'failure' usually relates to the maximum (or 'peak') deviator stress which the specimen can withstand, but any of the criteria discussed in Section 15.4.2 could apply.

The strain at failure (ε_f%) must be estimated, and this depends on the type and condition of the soil. The data summarised in Table 18.2 are provided as a general guide, but the estimate should preferably be based on experience. If in doubt a low estimate of failure strain will err on the safe side.

If the calculated value of t_f is less than 120 minutes (2 hours) the actual time to failure should not be less than 2 hours.

In a drained test the usual significant criterion for determining rate of strain is the peak deviator stress.

RATE OF STRAIN AND DISPLACEMENT

The maximum rate of axial displacement to apply in a drained test is equal to

$$\frac{\varepsilon_f\% L}{100 t_f} \text{ mm/minute}$$

where L (mm) is the length of the specimen.

The speed control setting on the compression machine that gives this rate of displacement, or the next slower setting, is selected for the test.

COMPRESSION TEST PROCEDURE

Adjustments for shearing

(1) Referring to Fig. 18.4, open valve b to connect the specimen to the back pressure system, and keep it open throughout the test. Valve c remains open to the cell pressure line. Valve a remains open so that the pore pressure can be checked throughout the test.

(2) to (6) See steps (2) to (6) for the CU test, Section 18.6.3.

(7) Observe the following and record them as the initial readings for the compression stage:

Date and clock time

Displacement gauge

Load dial gauge

Pore water pressure

Back pressure volume-change indicator

Cell pressure (check)

Back pressure (check)

The above readings are entered as the first line of data in the test form shown in Fig. 18.21. The columns relating to principal effective stresses are not normally required for the CD test.

Loading to failure

(8) Switch on the motor to start the test, and at the same instant start the clock.

(9) Take readings at regular intervals of axial displacement, usually corresponding to strain intervals of 0.2% strain up to 1% strain, and 0.5% strain thereafter (see Section 18.6.3, step (9)). Readings should be taken more frequently when peak deviator stress is approached.

Each set of data consists of readings of the gauges listed in step (7).

Pore water pressure is observed as a safeguard against running the test too fast; if there is any significant change (whether increase or decrease) the rate of strain should be reduced by 50% or more until the excess (positive or negative) pressure has dissipated. A small excess pressure is necessary to motivate the drainage process, and a change in measured pore pressure of up to about 4% of the effective confining pressure is usually acceptable.

Data from a typical test are shown in Fig. 18.21.

TRIAXIAL SHEARING

Loc. No. 62703		Location *Downside Dam*				Sample No. *C 21/3 A*			
Test Type *CD*		Proving Ring No. *1382-11*		With/~~Without~~ Side Drains		Effective Cell Pressure 200 kN/m²; Cell Pressure 540 kN/m²; Back Pressure 340 kN/m²			
Rate of Strain 0.2 % per hour		Operator *H.M.R*		Membranes *1 x 0.2 mm*					
Consolidated Length 74.8 mm		Consolidated Area 1098 mm²				Consolidated Volume 82.1 cm³			

TIME		STRAIN		LOAD			PORE PRESSURE		SAMPLE VOLUME			DEVIATOR STRESS			STRESS RATIO			
DATE	TIME	DIAL DIV	ε %	DIAL DIV	RING CALIB N/DIV	LOAD N	U kN/m²	U DIFF	V/C GAUGE	DIFF	$\frac{\Delta V}{Vc}$%	STRESS kN/m²	MEMBR + DRAINS kN/m²	$\sigma_1 - \sigma_3$ kN/m²	σ_1 kN/m²	σ_1' kN/m²	σ_3' kN/m²	$\frac{\sigma_1'}{\sigma_3'}$
16.7.85	0918	0	0	0		0	340	0	17.0	0	0	0	0	0				
	1106	25	0.33	27	2.271	61	340	0	16.85	0.15	0.18	55.5	0	56				
	1249	50	0.67	38	2.266	86	341	+1	16.7	0.3	0.37	78.1	0	78				
	1427	75	1.00	49	2.263	111	342	+2	16.5	0.5	0.61	100.5	0.1	100				
	1602	100	1.34	58	2.261	131			16.4	0.6	0.73	118.7	0.2	118				
17.7.85	1532	450	6.02	137	2.260	310	344	+4	15.7	1.3	1.58	269.6	10.3	259				
	1941	500	6.68	142		321			15.6	1.4	1.71	277.5	10.3	267				
	2258	550	7.35	145		328			15.6	1.4	1.71	281.3	10.4	271				
18.7.85	0204	600	8.02	145.5		329			15.65	1.35	1.64	280.1	10.5	270				
	0617	650	8.69	146		330	343	+3	15.7	1.3	1.58	278.8	10.6	268				

Fig. 18.21   *Data sheet for consolidated-drained (CD) triaxial compression test*

(10) As compression continues, calculate each line of data as described in Section 18.7.4, step (10), and make a rough plot as the test proceeds.

(11) Allow the test to continue until the relevant criterion of failure has been clearly identified, as indicated by one of the following:

(1) Maximum deviator stress

(2) Defined limit of strain

(3) Shear stress and volume-change indicator remaining constant with increasing strain.

If none of the above is evident, terminate the test when an axial strain of 20% is reached.

(12) If an attempt is to be made to measure the 'ultimate' shear strength, allow the test to continue to 20% strain if practicable. However, the test should be terminated earlier if the specimen becomes so distorted that further readings would have little meaning.

(13) to (22) When the final readings have been taken, switch off the motor, close valve b, unload the specimen, dismantle the apparatus and remove and measure the specimen, as described for the CU test (Section 18.6.3, steps (13) to (22)). Reduce the back pressure to zero.

Alternatively, follow the rapid procedure outlined in Section 18.6.5.

(23) Calculations, plotting and analysis of test data are described in Section 18.7.4.

18.6.5   Dismantling after Test

NORMAL PROCEDURE

The usual procedure for dismantling the specimen after test, described in Section 18.6.3, inevitably leads to a non-uniform distribution of moisture content within the specimen. This is because water is sucked in from the porous discs on removal of the stresses. When a final moisture content measurement is important the specimen should be cut into at least three, or preferably five, equal discs and the moisture content of each measured separately. The moisture content of the central portion represents the significant value for correlation with strength parameters. Dismantling and cutting should be done as quickly as possible so as not to allow time for redistribution of moisture.

RAPID PROCEDURE

The procedure outlined below for minimising the unequal distribution of moisture content was described by Olson (1964). The objective was to separate the specimen from the porous discs as soon as possible after removal of the axial load.

(1) As soon as the test is completed, switch off the motor and immediately close valves a and b.

(2) Disconnect the pore pressure and back pressure lines.

(3) Lock the cell piston in place so that the axial load is maintained when the machine platen is wound down to remove the external load.

(4) Close valve c and disconnect the pressure line. Also remove any other attachments to the cell.

(5) With the cell pressure and axial load maintained on the specimen, transfer the cell to a sink or a suitably large receptacle.

(6) Open valve c to release the cell pressure and quickly loosen the cell body clamping bolts, allowing the cell body to be lifted up and the contained water to splash into the sink.

(7) Separate the specimen from the porous discs as quickly as possible, and proceed with the weighings and other operations as previously described.

If the above operations are carried out rapidly the time during which the specimen is in contact with the porous discs under zero pressure should not be more than about 15 seconds, which should appreciably reduce the amount of water absorbed from the discs.

This procedure can have no effect on the non-uniform water distribution due to migration of pore water during shear induced by frictional restraint at the specimen ends. Use of lubricated ends makes for greater uniformity within the specimen.

18.6.6 Measurement of Cell Volume Changes

SCOPE

It is sometimes desirable to make measurements to ascertain the actual volume change of the test specimen before it becomes saturated. In the unsaturated condition the change in specimen volume due to a change in applied stress is not necessarily equal to the volume of water moving into or out of it. Measurements of volume change might be important for soils with a swelling potential, where they can be used to enable the swelling of a specimen to be kept within defined limits.

An indication of the changes in volume of a specimen in the triaxial cell can be obtained by including a volume-change indicator in the cell pressure line and recording the volume of water moving into or out of the cell. For these measurements to be of value, the cell and its connections must be free from leaks, and the cell must be carefully calibrated to determine its volume change relationships with pressure and with time (see Sections 17.3.1

and 17.3.2). Because of the nature and magnitude of these corrections, volume-change measurements on small specimens are difficult to determine with accuracy from cell line measurements.

SATURATION

During the saturation stage, when the pressure in the cell line is being raised in increments, allowance must be made for the expansion of the connecting lines in the same way as for back pressure line volume change measurements (Section 18.6.1(6)–(8)). The difference between the 'before' and 'after' readings gives the volume of water moving into or out of the cell due to that pressure increment. This volume depends on both the expansion of the cell and the change in volume of the specimen. Water *into* the cell is a *positive* volume change because it represents a specimen volume decrease (see Section 17.4.4). Water *out of* the cell is a *negative* change. This sign convention must be strictly followed. These readings are used as outlined in Section 18.7.1.

CONSOLIDATION AND COMPRESSION

Changes in volume measured in the cell line during the consolidation and compression stages are irrelevant in routine tests on saturated specimens. These measurements might be required in tests on unsaturated soils, which are beyond the scope of this volume.

18.7 ANALYSIS OF DATA

Calculations necessary for the analysis of data from CU and CD triaxial tests, and the associated graphical plotting, are described in this section. Sections 18.7.1 and 18.7.2 are applicable to both kinds of test.

18.7.1 Specimen Details

INITIAL CONDITIONS

Symbols used to illustrate the calculations are summarised in Table 18.3. The following values are calculated:

Initial area of cross-section of specimen

$$A_0 = \frac{\pi D_0^2}{4} \text{ mm}^2 \tag{18.3}$$

Initial volume

$$V_0 = \frac{A_0 L_0}{1000} \text{ cm}^3 \tag{18.4}$$

Initial density

$$\rho = \frac{m_0}{V_0} \text{ Mg/m}^3 \tag{18.5}$$

The initial moisture content may be calculated in two ways, either from trimmings, or from the final dry mass of the specimen (since dry mass remains constant throughout), the latter by using the following equation:

Table 18.3. SYMBOLS FOR SPECIMEN DATA CALCULATIONS

Measurement	initial	saturated	consolidated	after test	unit
			Condition		
Specimen length*	L_0		L_c		mm
diameter*	D_0				mm
area	A_0		A_c		mm^2
volume	V_0	V_s	V_c		cm^3
mass*	m_0	m_s	m_c	m_f	g
dry mass	. . .	. . .	. . .	m_D	g
density	ρ				Mg/m^3
dry density	ρ_D		ρ_{D_c}		Mg/m^3
moisture content**	w_0		w_c	w_f	%
Particle density**	ρ_s	. . .	. . .	. . .	Mg/m^3
Voids ratio	e_0	e_s	e_c	e_f	—
Degree of saturation	S_0			S_f	%

*Specimen measurements.
**Measured on trimmings.
. . .Constant throughout the test.

$$w_0 = \frac{m_0 - m_D}{m_D} \times 100\% \tag{18.6}$$

The value of ρ_s may be measured or assumed, but it should be stated which is applicable. When the initial moisture content $w_0\%$ and the particle density ρ_s are known, the following can be calculated:

Initial dry density

$$\rho_D = \frac{100}{100 + w_0} \rho_D \text{ Mg/m}^3 \tag{18.7}$$

Initial voids ratio

$$e_0 = \frac{\rho_s}{\rho_D} - 1$$

Initial degree of saturation

$$S_0 = \frac{w_0 \rho_s}{e_0} \%$$

AFTER SATURATION

Symbols are summarised in Table 18.4.

The total volume of water entering the specimen during the saturation stage, ΔV_1, is obtained by adding the differences between the 'before' and 'after' readings of the volume-change indicator on the back pressure line, as shown in Fig. 18.15. Since this is measured

Table 18.4. SYMBOLS FOR CHANGES DURING SATURATION, CONSOLIDATION AND COMPRESSION

Measurement	Saturation stage	Cell pressure adjustment	Consolidation stage	Compression stage	Unit
Volume change on back pressure line	ΔV_1	–	ΔV_4	ΔV_6	ml
Volume change on cell pressure line	ΔV_2	ΔV_3	ΔV_5	ΔV_7	ml
Cell volume correction	δV_2	δV_3	δV_5	δV_7	ml
Actual specimen volume change	ΔV_s	ΔV_a	ΔV_c	ΔV_d	cm^3
Total specimen volume change to start of compression			ΔV		cm^3
Volumetric strain			ε_v		%
Sample volume: start	V_0	V_s	V_a	V_c	cm^3
end	V_s	V_a	V_c	V_f	cm^3

in millilitres or cubic centimetres it is equal to the increase in mass of the specimen in grams.

Therefore

$$m_s = m_0 + \Delta V_1 \text{ grams} \qquad (18.8)$$

The actual change in volume of the specimen, ΔV_s, is not equal to the measured volume change ΔV_1 because most of the additional water has taken the place of air in the voids by forcing it into solution under pressure. In most routine tests it is assumed that no volume change takes place during saturation for soils that are not susceptible to swelling.

If the cell pressure line includes a volume-change indicator, the overall cell volume change during saturation (ΔV_2) is obtained by totalling the differences between the 'before' and 'after' readings referred to in Section 18.6.6, taking into account whether the changes are positive or negative. Allowance must then be made for the expansion of the cell, as indicated in Section 17.3.1 (δV_2). No correction is required for piston movement if the piston is fully restrained. The volume change of the specimen during saturation (ΔV_s) is then given by the equation

$$\Delta V_s = \Delta V_2 - \delta V_2 \qquad (18.9)$$

The correction δV_2 is obtained from the cell volume calibration (Section 17.3.1) and is correction (3) in Section 17.4.4. If any of the other corrections (1) and (8) are likely to be significant they should first be added to correction (3).

If the specimen ends up fully saturated, the difference between the volume of water which has entered the specimen and the initial volume of voids should be equal to the change in volume of the specimen.

AFTER CONSOLIDATION

The volume of water draining out of the specimen into the back pressure line during
consolidation, and measured by the volume-change indicator on that line (ΔV_4), is equal to
the change in volume of the specimen, since it is fully saturated; i.e. $\Delta V_c = \Delta V_4$.

The volumetric strain at the end of consolidation is calculated from the equation

$$\varepsilon_{vc} = \frac{\Delta V_c}{V_0} \times 100\% \qquad (18.10)$$

The volume V_c, length L_c and area A_c of the specimen after consolidation are calculated
as follows. These equations are based on elastic theory for a small volume change,
assuming a Poisson ratio of 0.5 (Case and Chilver, 1959).

$$V_c = V_0 - \Delta V_c \qquad (18.11)$$

$$L_c = L_0 \left(1 - \frac{1}{3} \frac{\varepsilon_{vc}}{100} \right) \text{ mm} \qquad (18.12)$$

$$A_c = A_0 \left(1 - \frac{2}{3} \frac{\varepsilon_{vc}}{100} \right) \text{ mm}^2 \qquad (18.13)$$

as a check,

$$V_c = \frac{A_c L_c}{1000} = V_0 - \Delta V \qquad (18.14)$$

The consolidated length L_c is used for calculating axial strain, and the area A_c for
calculating stress, in the compression stage.

DURING COMPRESSION

In an undrained compression test on a saturated specimen it can be assumed that no
change in volume occurs.

In a drained test the change in volume of the specimen is equal to the volume change
readings taken from the volume-change indicator on the back pressure line. In accordance
with the usual sign convention, water draining out of the specimen (volume decrease)
represents a positive change; water moving into the specimen represents a negative change.

18.7.2 Consolidation Characteristics

COEFFICIENT OF CONSOLIDATION

A value for the coefficient of consolidation, c_{vi}, for isotropic consolidation of the specimen
can be calculated from t_{100} obtained from the square-root time/volume change curve in the
consolidation stage (Section 18.6.2, Fig. 18.18 (b)), by applying the general equation

$$c_{vi} = \frac{\pi D_c^2}{\lambda t_{100}} \qquad (15.29)$$

In this equation λ is a constant depending on the drainage conditions (Section 15.5.5), and
D_c is the specimen diameter after consolidation, calculated from the equation

Table 18.5. CALCULATION OF c_{vi} FROM t_{100}

$$c_{vi} = \frac{N}{t_{100}} \ \text{m}^2/\text{year (approximations)}$$

Nominal initial specimen diameter (mm)	values of N			
	no side drains		with side drains	
	$r = 2$	$r = 1$	$r = 2$	$r = 1$
38	2400	9500	30	83
50	4100	16500	52	140
100	17000	66000	210	570
150	37000	150000	460	1300

t_{100} is in minutes
$r = L/D$
'With side drains' applies to drainage from radial boundary and one end.

$$D_c = \sqrt{\frac{4A_c}{\pi}} \qquad (18.15)$$

Values of λ are summarised in Table 15.4.

Relationships between c_{vi} (m²/year) and t_{100} (minutes) for typical specimen sizes with $L:D$ ratios of 2:1 and 1:1, and for the two most usual drainage conditions (from one end only, and from one end and radial boundary) are summarised in Table 18.5. In this table the factor N replaces $\pi D^2/\lambda$ in Equation (15.29), and the values of N relate to typical nominal specimen diameters.

The calculated value of c_{vi} is reported to two significant figures.

The value of c_{vi} derived in this way should be used with caution. Calculated values of c_{vi} should be used only for estimating the rate of strain in the triaxial test and not in consolidation or permeability calculations. When side drains are used the value of c_{vi} derived in this way can be grossly in error.

COEFFICIENT OF VOLUME COMPRESSIBILITY

The coefficient of volume compressibility, m_{vi}, for isotropic consolidation can be calculated from the equation given in Section 15.5.5:

$$m_{vi} = \frac{\Delta V_c}{V_0} \times \frac{1000}{\Delta \sigma'} \ \text{m}^2/\text{MN} \qquad (15.31)$$

The change in effective stress, $\Delta \sigma'$, is equal to the change of pore water pressure $(u_i - u_c)$ in the specimen from start to end of the consolidation stage, since the cell confining pressure remains constant. The volume change ΔV_c is equal to the volume of water draining out of the specimen during the consolidation stage, and when divided by the specimen volume at the start of the stage gives the volumetric strain due to consolidation.

Table 18.6. SYMBOLS FOR SHEAR STRENGTH CALCULATIONS

Measurement	Symbol	Unit
Load ring calibration	C_r	N/division
Load ring reading at time t	R	divisions
Axial displacement at time t	ΔL	mm
Axial strain at time t	ε	%
Area of cross-section at time t	A	mm^2
Axial force at time t	P	N
Applied deviator stress (uncorrected)	$(\sigma_1 - \sigma_3)_m$	kPa
Correction for rubber membrane	σ_{mb}	kPa
Correction for side drain	σ_{dr}	kPa
Pore pressure	u	kPa
Corrected deviator stress	$(\sigma_1 - \sigma_3)$	kPa
Cell pressure	σ_3	kPa
Pore pressure at start of shearing	u_0	kPa
Volume change due to shear	ΔV	cm^3
Volumetric strain due to shear	ε_v	%

A typical calculation of m_{vi} is included in Fig. 18.18. The value is reported to two significant figures.

COEFFICIENT OF PERMEABILITY

A value for the coefficient of permeability k can be calculated from the equation given in Section 15.5.5:

$$k = c_{vi} m_{vi} \times 0.31 \times 10^{-9} \, \text{m/s} \tag{15.32}$$

in which the units are as stated above.

Its accuracy depends on the reliability of the value of c_{vi}. The result will not be realistic if its value is that of a silt or greater (i.e. more than 10^{-8} m/s). If side drains are used the value of k should not be quoted.

18.7.3 Shear Strength — CU Test

CALCULATED AXIAL STRESS

Symbols used to illustrate the calculations for shear strength are summarised in Table 18.6.

Each line of data on the test sheet (Fig. 18.20) is calculated as follows:

Axial strain $\varepsilon\% = \dfrac{\Delta L}{L_c} \times 100\%$

Axial force $\quad P = C_r \times R$ newtons

The load ring calibration C_r might not be constant, but can vary with ring deflection (see Volume 2, Section 8.4.3).

The area of cross-section (A) of the specimen, allowing for barrelling deformation (Volume 2, Section 13.3.7) is equal to

$$\frac{100 A_c}{100 - \varepsilon\%} \text{ mm}^2$$

The axial stress in the specimen (the measured deviator stress) induced by the force P is equal to P/A newtons per square millimetre, i.e. $(P/A) \times 1000$ kPa. Hence from Equation (13.7) (Volume 2, Section 13.3.7)

$$(\sigma_1 - \sigma_3)_m = \frac{P}{A_c} \times 1000 \times \frac{100 - \varepsilon\%}{100} \text{ kPa}$$

$$= \frac{P}{A_c} \times 10(100 - \varepsilon\%) \text{ kPa} \qquad (18.16)$$

If deformation occurs by failure on a definite surface ('single-plane slip'), the area correction has to be modified as described in Section 17.4.1.

MEMBRANE AND DRAIN CORRECTIONS

The calculated deviator stress must now be corrected to allow for the effects of the rubber membrane, and side drains if fitted. These corrections are discussed in Section 17.4.2. and 17.4.3, and are deducted from $(\sigma_1 - \sigma_3)_m$, i.e.

$$(\sigma_1 - \sigma_3) = (\sigma_1 - \sigma_3)_m - \sigma_{mb} - \sigma_{dr}$$

The membrane correction for barrelling, σ_{mb}, for the appropriate strain is obtained from the graph in Fig. 17.16 (Section 17.4.2), which applies to a 38 mm diameter specimen in a 0.2 mm thick membrane. For any other specimen diameter (D mm) and membrane thickness (t mm) this value is multiplied by

$$\frac{38}{D} \times \frac{t}{0.2}$$

to obtain the membrane correction to use.

The drain correction, σ_{dr} for several specimen diameters is given in Section 18.7.4.

CORRECTED STRESSES

The fully corrected deviator stress is used for the calculation of the principal stresses, and other parameters, from each set of recorded data as follows. The symbols are those given in Table 18.6.

Minor effective principle stress $\sigma_3' = \sigma_3 - u$

Major effective principal stress $\sigma'_1 = (\sigma_1 - \sigma_3) + \sigma'_3$

Principal stress ratio $= \sigma'_1 / \sigma'_3$

Pore pressure coefficient $\bar{A}(u - u_0)/(\sigma_1 - \sigma_3)$

In a fully saturated soil, $\bar{A} = A$

The stress path parameters, s' and t', in terms of effective stress, may be calculated from the equations derived in Section 21.1.4(2):

$$s' = \frac{\sigma'_1 + \sigma'_3}{2} \tag{21.3}$$

$$t = \frac{\sigma'_1 - \sigma'_3}{2} \tag{21.2}$$

The above calculations are tabulated in Fig. 18.20.

GRAPHICAL PLOTS

The calculated data are used for plotting the following graphs.

Deviator stress against strain.

Pore pressure against strain, with the value at the start of compression emphasised to show the datum to which pore pressure changes are related.

Principal effective stress ratio against strain, starting from unity at zero strain.

Pore pressure coefficient $\bar{A}$ against strain.

Stress path plot of t against s'.

FAILURE CRITERIA

From the curve of deviator stress against strain the maximum value (at the 'peak' point or 'failure' condition) is located, and denoted by $(\sigma_1 - \sigma_3)_f$. The corresponding values of strain (ε_f) and pore pressure (u_f) are also read off, as shown in Fig. 18.22. The curve might indicate that the maximum value lies between two sets of readings, in which case a sensible interpolation should be made. Values of σ'_3, σ'_1, σ'_1/σ'_3 and $\bar{A}$ (denoted by $\bar{A}_f$) corresponding to the 'failure' condition are calculated as described above.

Similar sets of values corresponding to the point of maximum principal stress ratio, or the 'critical state' condition (constant deviator stress and pore pressure) are read off from the curves and calculated, if either of these alternative criteria for failure is appropriate.

STRESS PATHS

Stress path plots of $t = \frac{1}{2}(\sigma_1 - \sigma_3)$ against $s' = \frac{1}{2}(\sigma'_1 + \sigma'_3)$, or q against p' (Section 21.3), are also drawn. These plots convey more information than Mohr circles on soil behaviour during the test.

MOHR CIRCLES

The sets of values of σ'_1 and σ'_3 obtained as described above enable Mohr circles of effective stress to be drawn to represent the relevant criterion for failure.

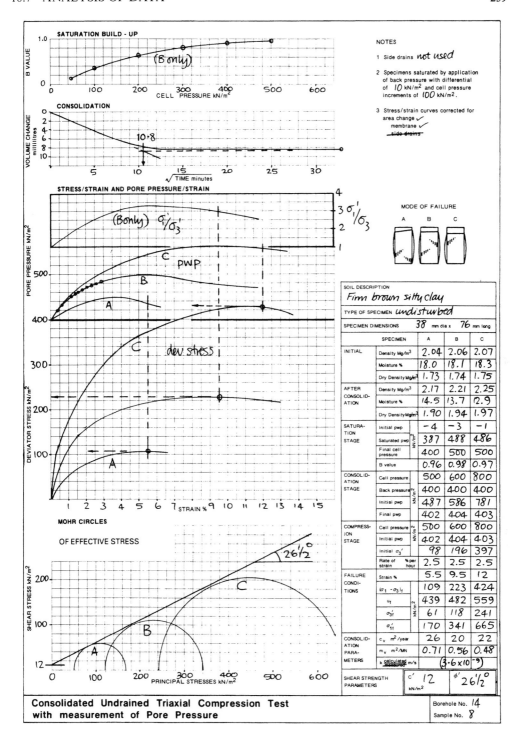

Fig. 18.22 *Graphical data from a set of consolidated-undrained (CU) triaxial compression tests*

Mohr circles of total stress are not relevant to the derivation of a strength envelope. They are not required by BS 1377 but they are in ASTM D 4767.

SET OF TESTS

Results from a set of three specimens taken from one sample can be presented on one test report sheet by grouping the graphs together as shown in Fig. 18.22.

The stress path plots of t against s' for the set of specimens are drawn on one sheet (Fig. 18.23). The point on each plot corresponding to the failure criterion of maximum deviator stress is marked, and the line of best fit is drawn through the set of points. The angle of slope of the line to the horizontal axis is denoted by θ, and its intercept with the t (vertical) axis is denoted by t_0.

The shear strength parameters (c', (ϕ') are calculated from the following relationships derived in Section 21.2.3:

$$\sin \phi' = \tan \theta \tag{21.8}$$

$$c' = \frac{t_0}{\cos \phi'} \tag{21.9}$$

Alternatively three Mohr circles from the set, representing the appropriate failure criterion, can be plotted as shown in Fig. 18.22, and the envelope of 'best fit' is drawn to them. The slope of the envelope gives the effective angle of shear resistance, ϕ', and the intercept with the vertical axis gives the effective cohesion, c'.

18.7.4   Shear Strength — CD Test

CALCULATIONS AND CORRECTIONS

The symbols used to illustrate shear strength calculations are the same as for the CU test (Table 18.6, Section 18.7.3).

Axial strain ($\varepsilon\%$) and axial force (P newtons) are calculated as for the CU test. In calculating the deviator stress (($\sigma_3 - \sigma_3)_m$ kPa) the change in volume of the specimen due to drainage is an additional factor to be taken into account, as explained in Section 17.4.1.

The volumetric strain is equal to $\varepsilon_v = (\Delta V/V_c) \times 100\%$ and the deviator stress is calculated from the equation

$$(\sigma_1 - \sigma_3)_m = \frac{P}{A_c} \times \frac{100 - \varepsilon}{100 - \varepsilon_{vs}} \times 1000 \text{ kPa} \tag{18.17}$$

which is derived from Equation (17.7).

If there is no significant change in pore pressure there is little to be gained from calculating the principal stress ratio because when plotted it will give a curve identical in shape to the stress/strain curve.

MEMBRANE AND DRAIN CORRECTIONS

The calculated deviator stress is corrected to make allowance for the effects of the membrane, as described in Section 18.7.3 for the CU test.

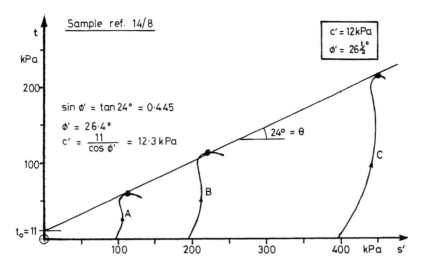

Fig. 18.23   *Stress path plots from the data in Fig.* 18.22

If vertical side drains are fitted the correction to apply for barrelling is constant for strains exceeding 2%. The correction depends on the specimen diameter (see Section 17.4.3) and is summarised as follows.

Specimen diameter (mm)	38	50	70	100	150
Drain correction σ_{dr} (kPa)	10	7	5	3.5	2.5

Up to 2% strain it can be assumed that the correction increases linearly to the values stated above, as explained in Section 17.4.3.

No correction is applied if spiral drains are used.

For single-plane slip the corrections for area change, membrane and side drains described in Sections 17.4.1 to 17.4.3 may be applied.

Details of the corrections applied for area change, membrane and side drains, and their magnitudes, should always be clearly shown in the test report. This is particularly important for single-plane slip corrections which can be of appreciable magnitude.

GRAPHICAL PLOTS

The following graphs are plotted.

Deviator stress against axial strain.

Volumetric strain against axial strain.

Pore pressure may also be plotted against strain if it changes significantly.

FAILURE

Read off the peak deviator stress $(\sigma_1 - \sigma_3)_f$, and the corresponding axial strain (ε_f) from the deviator stress/strain curve, as for the CU test. Also read off the volumetric strain and the actual pore pressure (u_f) coresponding to ε_f (see Fig. 18.24). Corresponding values for the condition of constant deviator stress and constant volume are obtained if relevant.

Calculate the stresses corresponding to failure in the same way as for the CU test (Section 18.7.3).

MOHR CIRCLES

The values of σ'_{3f}, σ'_{1f} are used for plotting stress paths or Mohr circles of effective stress at failure under drained conditions. If several specimens are tested from one sample, they are plotted together as a set and the failure envelope is drawn. The drained shear strength parameters c_d, ϕ_d are determined in the same way as for the CU test (see Fig. 18.26). For most practical purposes the drained parameters can be considered as being identical to the undrained parameters c', ϕ', and the latter symbols are normally used.

18.7.5 Reporting Results

The following data are reported, together with the relevant graphical plots as indicated. The data are similar for both CU and CD tests except where separately identified. Items marked * are additional to those listed in Clauses 7.6 and 8.6 of BS 1377.

GENERAL

Sample identification, reference number and location

Type of sample

Soil description

Type of specimens and initial dimensions

Location and orientation of test specimens within the original sample

Method of preparation of each specimen

Moisture content determined from trimmings*

Date test started

For each specimen:

INITIAL CONDITIONS

Moisture content

Bulk density, dry density

Voids ratio and degree of saturation*

Thickness of membrane*

Whether side drains were fitted, and if so what type

SATURATION STAGE

Saturation procedure

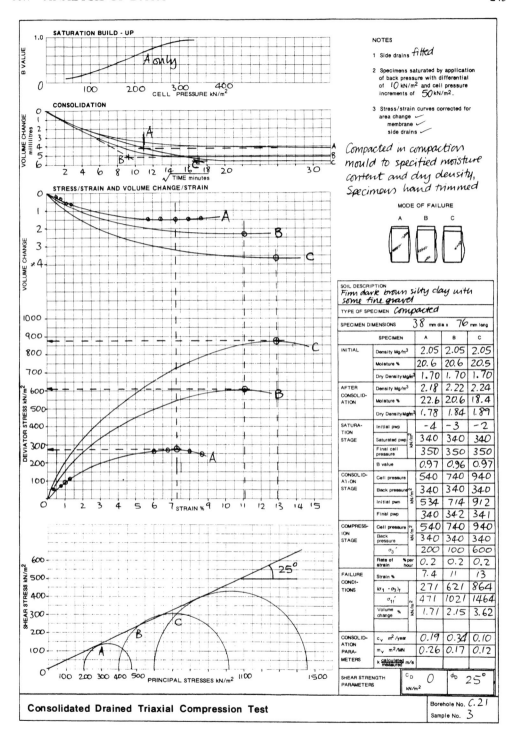

Fig. 18.24 *Graphical data from a set of consolidated-drained (CD) triaxial compression tests*

Cell pressure increments $\Big\}$ if by incremental pressures
Differential pressure

Pore pressure and cell pressure after saturation

Value of B achieved

Degree of saturation reached

Graphical plot of B value against pore pressure (or cell pressure)*

CONSOLIDATION STAGE

Cell pressure

Back pressure

Initial pore pressure

Final pore pressure

Percentage pore pressure dissipated

Coefficient of consolidation c_{vi}*

Coefficient of volume compressibility m_{vi}*

Calculated time to failure

Graphical plot of volume change against square-root time

COMPRESSION STAGE (CU TEST)

Cell pressure

Initial pore pressure

Initial effective confining pressure

Rate of strain applied

Failure criterion used

Data at failure: axial strain

                     deviator stress

                     pore pressure

                     effective major and minor principal stresses

                     effective principal stress ratio

Sketch of specimen at failure, illustrating mode of failure

Details and magnitude of membrane, drain and other corrections applied

Actual time to reach failure*

Final density and moisture content

Graphical plots: deviator stress

                   effective principal stress ratio $\Big\}$ against strain

                   pore pressure

Mohr circle of effective stress representing failure

Stress path for effective stresses with s' plotted against t

Statement that the test was carried out in accordance with Clauses 4, 5, 6 and 7 of BS 1377 : Part 8 : 1990.

COMPRESSION STAGE (CD) TEST

Cell pressure

Back pressure

Initial pore pressure

Effective confining pressure

Rate of strain applied

Failure criterion used

Data at failure: axial strain

deviator stress

volumetric strain

pore pressure

effective major and minor principal stresses

effective principal stress ratio

Sketch of specimen at failure, illustrating mode of failure

Details and magnitude of membrane, drain and other corrections applied

Actual time to reach failure*

Final density and moisture content

Graphical plots: deviator stress

volume change   } against strain

pore pressure changes (if significant)

Mohr circle of effective stress

Statement that the test was carried out in accordance with Clauses 4, 5, 6 and 8 of BS 1377 : Part 8 : 1990.

FOR A SET OF CU SPECIMENS

Numerical data as listed above, grouped together as a set.

Graphical plots as stated above, including stress path plots, grouped on common axes.

Envelope to points representing failure on stress path plots, indicating derivation of shear strength parameters c', ϕ'.

Set of Mohr circles corresponding to failure, with failure envelope indicating slope (ϕ') and intercept (c').

FOR A SET OF CD SPECIMENS

Numerical data as listed above, grouped together as a set.

Graphical plots as stated above, including stress path plots, grouped on common axes.

Envelope to points representing failure on stress path plots, indicating derivation of shear strength parameters c', ϕ'.

Set of Mohr circles corresponding to failure, with failure envelope indicating slope (ϕ') and intercept (c').

REFERENCES

ASTM Designation D 4767, *Standard Test Method for Consolidated-Undrained Triaxial Compression Test on Cohesive Soils*. American Society for Testing and Materials, Philadelphia, USA.

Arulanandan, K., Shen, C. K. and Young, R. B. (1971). 'Undrained creep behaviour of a coastal organic silty clay'. *Géotechnique*, 21:4:359.

Bishop, A. W. and Little, A. L. (1967). 'The influence of the size and orientation of the sample on the apparent strength of the London Clay at Maldon, Essex'. *Proc. Geotech. Conf.*, Oslo, Vol. 1.

Blight, G. E. (1963). 'The effect of non-uniform pore pressures on laboratory measurements of the shear strength of soils'. Symposium on Laboratory Shear Testing of Soils, pp 173–184. *ASTM Special Technical Publication* No. 361. American Society for Testing and Materials, Philadelphia, USA.

BS 1377: Parts 1 to 8: 1990, *British Standard Methods of Test for Soils for Civil Engineering Purposes*. British Standards Institution, London.

Case, J. and Chilver, A. H. (1959). *Strength of Materials: an Introduction to the Analysis of Stress and Strain*. Edward Arnold, London.

Hight, D. W. (1980). Private communication to author.

Hight, D. W. (1982). 'Simple piezometer probe for the routine measurement of pore pressure in triaxial tests on saturated soils'. Technical note, *Géotechnique*, 32:4:396.

Lumb, P. (1968). 'Choice of strain-rate for drained tests on saturated soils'. Correspondence, *Géotechnique*, 18:4:511.

Olson, R. E. (1964). 'Discussion on the influence of stress history on stress paths in undrained triaxial tests on clay. Laboratory shear testing of soils'. *ASTM STP* No. 361, pp 292–293. American Society for Testing and Materials, Philadelphia, USA.

Sandroni, S. S. (1977). 'The strength of London Clay in total and effective shear stress terms'. PhD Thesis, Imperial College of Science and Technology, University of London.

Sangrey, D. A., Pollard, W. S. and Eagan, J. A. (1978). 'Errors associated with rate of undrained cyclic testing of clay soils. Dynamic Geotechnical Testing'. *ASTM Special Technical Publication* No. 654. American Society for Testing and Materials, Philadelphia, USA.

Wesley, L. D. (1975). 'Influence of stress path and anisotropy on behaviour of a soft alluvial clay'. PhD Thesis, Imperial College of Science and Technology, University of London.

Chapter 19

Further triaxial shear strength tests

19.1 SCOPE

This chapter describes several triaxial test procedures for the measurement of the shear strength of soils, additional to the basic tests described in Chapter 18. Some of these procedures are not often used but may be appropriate in particular circumstances. The procedures are generally indicated in outline only, and make reference to Chapters 17 and 18 for many of the details. The apparatus required is the same as that already described for the tests given in Chapter 18.

There are numerous other ways of determining soil shear strength properties for particular applications, most of which need more advanced procedures requiring additional equipment. Such procedures are beyond the scope of this volume.

19.2 MULTISTAGE TRIAXIAL TEST

19.2.1. Introduction

A multistage quick-undrained triaxial test, in terms of total stress, on a single specimen of soil such as 'boulder clay', was described in Volume 2, Section 13.6.5. Multistage effective stress tests have been reported for instance by Lumb (1964), Kenney and Watson (1961), Watson and Kirwan (1962) and Ruddock (1964). These tests were shown to give drained and consolidated undrained effective shear strength parameters (c_d, ϕ_d or c', ϕ') that are practically indistinguishable from those obtained by the normal procedure using a set of three specimens, for many types of soil. However other factors such as compressibility, dilatancy, pore pressure changes and voids ratio changes do not compare so well with conventional tests. Multistage triaxial tests on partially saturated, fairly permeable soils were described by Ho and Fredlund (1982 (a)). The use of stress paths in multistage testing was described by Janbu (1985).

Some form of stress path plotting is necessary for multistage tests, because it is essential to keep the stress conditions and soil behaviour under observation as the test proceeds, especially immediately before the peak deviator stress is reached. Stress paths and their use are described in Chapter 21.

The main advantage of multistage testing is in the saving of time and material when only one large specimen is available for testing, and when small specimens are impracticable. This applies particularly to soils containing relatively large-scale features such as gravel in boulder clays, or discontinuities such as fissures, which cannot be adequately represented in a small specimen. Multistage tests should in general be restricted to soils such as these that are also of low sensitivity and have a stable structure, and require a relatively small strain and volume change to induce failure. Tests can be performed using either consolidated-drained or consolidated-undrained procedures, usually on specimens that are first saturated, or have a high degree of saturation initially. Three stages of loading are normally applied, the final stage being extended beyond the point of peak deviator stress.

An essential factor in multistage testing is the selecting of an appropriate criterion for 'failure' at each intermediate stage. This should take into account the type of soil and its behaviour, and practical in-situ considerations especially those relating to limitations of strain. Excessive deformation should be avoided during the first two stages of a three-stage test, and these stages should be discontinued just before reaching peak deviator stress. Some general guidance is given in Section 19.2.4, and each type of soil should be treated on its merits rather than following a rigid procedure.

Multistage triaxial testing in general should be regarded not as standard practice but as an expedient to fall back on for certain types of soil when there is no practicable alternative due to the limited number of available specimens.

19.2.2 Test Procedures

GENERAL

The following outline procedures relate to both CU and CD triaxial tests on undisturbed specimens, except where either type of test requires special treatment. Remoulded or compacted specimens may also be used, if enough time is allowed between preparation and testing to enable pore pressures within the sample to equalise.

With a large specimen of clay soil of low permeability, drainage from both ends is desirable so as to halve the drainage path (without having to use side drains), thereby reducing the theoretical consolidation time by a factor of 4. In a drained test the consequent shorter time required for the compression stage usually outweighs the disadvantage of not being able to make pore pressure checks as the test proceeds. To allow double drainage during consolidation preceding an undrained test, the arrangement referred to below is necessary.

PROCEDURAL STEPS

(1) Set up the specimen as for a normal CU or CD test, as shown in Fig. 18.1, Section 18.2.3. Valve a is connected to the pore pressure transducer. Valves b and d are connected to a common back pressure system, or if a back pressure is not necessary, to open burettes with the water levels at the specimen mid-height. With this arrangement pore pressures can be measured during compression (valve a open, valves b and d closed) after allowing drainage from both ends during consolidation (valve a closed, b and d open) without having to change over any connecting lines.

(2) Saturate the specimen by one of the procedures given in Section 18.6.1, to obtain an acceptable B value. Alternatively apply a cell pressure to confirm that the sample initially has an acceptable degree of saturation.

(3) Consolidate to the first (lowest) of the effective confining pressures selected for the test (Section 18.6.2).

(4) Calculate the time to failure (Section 18.6.2, step 9), and estimate the appropriate rate of strain depending on the type of test and drainage conditions.

(5) Start the compression stage in the usual way. As the test proceeds, calculate data for plotting the following against strain (%), applying all the necessary corrections.

CU test: Deviator stress (kPa)
(Section 18.7.3)   Pore pressure (kPa)
                   Principal stress ratio (σ_1'/σ_3')
CD test: Deviator stress (kPa)
(Section 18.7.4)   Volume change (%)

For both types of test, also calculate data for making the stress path plot of t against s', where

$$t = (\sigma_1' - \sigma_3')/2$$
$$s' = (\sigma_1' + \sigma_3')/2$$

(see Section 21.1.4(2)).

If necessary run the test a little slower than normal to enable these graphs to be kept up to date as the test proceeds.

(6) When failure is approached according to the appropriate criterion, stop the compression machine and close the drainage valve or valves (in a drained test). Impending failure may be indicated by the development of a surface of failure within the specimen, or by flattening of the stress/strain curve, or by one of the other criteria discussed in Section 19.2.4.

(7) Immediately reverse the motor to reduce the axial force rapidly until it just reaches zero, and record the corresponding axial strain reading. Sketch the configuration of the specimen. Allow enough time for the pore pressure to reach equilibrium before proceeding further. Removal of the axial force prevents 'creep' of the soil structure between loading stages, and allows a small elastic recovery of strain.

(8) Raise the cell pressure to the level required for the next stage, and wait for the pore pressure to stabilise. Maintain the back pressure at the same level as previously unless it is essential to lower it in order to achieve the next effective confining pressure.

(9) Consolidate the specimen as before. Calculate the new consolidated dimensions.

(10) Determine the time to failure for the next stage of loading from the new consolidation data, and calculate the new rate of strain. Failure is usually approached at a much smaller strain than in the first stage; for example if failure in the first stage was almost reached at between 5 and 10% strain, only 1–2% strain may be needed for a subsequent stage. Furthermore as the effective pressure is increased the value of c_{vi} usually decreases. The machine speed for stages 2 and 3 is therefore likely to be considerably slower than for stage 1.

(11) Wind up the machine platen to remake contact between the cell piston and the top cap, then reset the strain dial gauge and the load ring dial gauge to zero or to a new datum reading.

(12) Resume application of the axial load at the appropriate rate of strain, with simultaneous calculation and plotting as in step (5). Continue until 'failure' is again approached as in step (6).

(13) Calculations are made as follows.

Strain: Use the new specimen length to calculate strain referred to the new zero datum, and add this to the strain at the end of unloading to obtain cumulative strain. This gives strain due to application and removal of deviator stress only.

Deviator stress: Use the new specimen area and volume and strain measured from the new datum, to calculate the (uncorrected) deviator stress. To apply the membrane correction use the cumulative strain.

Volume change: Calculate further volume changes as percentages of the new volume and plot these percentages starting from a new datum point.

Pore water pressure: Plot the recorded values. The pore pressure at the start of each stage is usually equal to, or a little above, the back pressure used for consolidation, which should be unchanged wherever possible.

(14) Repeat steps (7) to (13) in order to apply a total of three stages of loading (if practicable). A fourth stage may sometimes be possible. Continue the final stage beyond peak deviator stress.

(15) Reduce the applied load and pressures to zero, and remove the specimen for sketching, final weighing and measurements, as for a normal triaxial test. Complete the usual calculations.

19.2.3 Graphical Plots

The graphical data to be presented are similar to those for a normal set of CU or CD tests, except that the curves for the second and subsequent compression stages start from the strain corresponding to the unloaded condition of the previous stage. The form of the graphs for the compression stages of a CU test is indicated in Fig 19.1. From the stress–strain curves for the first two stages the 'failure' conditions are extrapolated. These values, together with the actual failure data for the final stage, can be used for constructing the Mohr circles to which a failure envelope is drawn for determining the values of c', ϕ'. However, the stress path plot described in Section 19.2.4(7) below is generally more useful.

19.2.4 Failure Criteria

In a typical triaxial compression test the maximum total applied strain should not exceed about 25%. The available length of the cell piston is one limiting factor, but the severe distortion of the specimen itself at a strain much in excess of 20% can result in dubious values of calculated axial stress. In a three-stage test the axial strain in each stage should not therefore exceed about 8%, but it may be permissible to extend the first stage to a larger strain because stiffening of the soil due to further consolidation often results in smaller strains to failure in subsequent stages.

The failure criterion for each specimen should be selected on the basis of soil type and in-situ conditions. Some of the criteria which might be applied are as follows.

(1) Visible failure, e.g. the development of a slip surface in the specimen.

(2) Flattening out, or reaching the peak, of the deviator stress/strain curve (with all corrections applied).

(3) Reaching a pre-determined strain, such as 10, 15, 20% for three successive stages. For

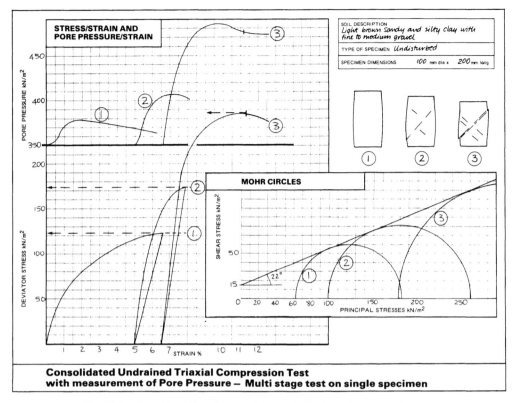

Fig. 19.1   *Graphical data from multistage CU triaxial compression test*

a plastic type of soil, strains of 16, 18 and 20% may be more appropriate (Anderson, 1974).

(4) In a CU test, the condition of maximum principal stress ratio (σ_1'/σ_3') provides a useful indication. In overconsolidated soils it occurs before maximum deviator stress. This ratio should be plotted against strain as the test proceeds.

(5) In a CU test the change in pore water pressure can also be used as a guide. The maximum excess pore pressure (in a non-dilating soil) occurs at about peak principal stress ratio.

(6) In a CD test the volume change behaviour provides a similar guide and the 'critical state' condition, in which the volume remains constant, is relevant.

(7) In addition to the plots referred to in (4), (5) and (6) above, the stress path method of plotting (described in Chapter 21) is particularly useful for undrained effective stress tests. For an overconsolidated soil such as boulder clay it was demonstrated by Watson and Kirwan (1962), using a plot of σ_1' against σ_3', that the slope of the failure envelope can be estimated from a test at a single cell pressure after the maximum effective stress ratio has been reached. This is illustrated in Fig. 19.2, in which the data from the same test as in Fig. 19.1 are plotted on the MIT stress field (Section 21.1.4(2)). In stage 1 the maximum stress ratio is reached at point S, and the succession of readings from S to U lies along the failure envelope. However, it is difficult to derive the slope of the envelope from these closely grouped points. Normally the first stage is terminated at a point such as *T*, before the

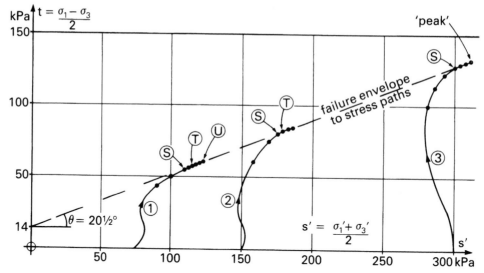

Fig. 19.2. Stress path plot of multistage CU triaxial compression test
$\sin \phi' = \tan \theta = 0.374;\ \phi' = 22°$

specimen has been subjected to a deformation large enough to reach the 'peak' deviator stress. A second stage is needed to enable the failure envelope to be drawn, and this is also terminated at a similar point T. A third stage is applied, and this can be extended far enough to achieve failure and define the peak strength envelope more positively, as shown in Fig. 19.2.

It is usually desirable to make more than one type of plot in order to provide enough evidence for deciding when to terminate one stage and proceed to the next.

19.3 UNDRAINED COMPRESSION TESTS

19.3.1 General

TYPES OF TEST

This section briefly describes four relatively simple types of triaxial compression test which have certain limited applications. These are:

(1) The unconsolidated, undrained test, with measurement of pore water pressure, referred to as the UU test.

(2) The consolidated quick-undrained (C–QU) test.

(3) The consolidated constant volume (CCV) test, in which the pore pressure is held constant.

(4) A consolidated-undrained test for 100 mm diameter specimens of compacted clay.

The first was originally described by Bishop and Henkel. It is similar to the quick-undrained test (Volume 2, Section 13.6.3) except that pore pressures and volume changes are measured, and therefore strain must be applied at a slow enough rate to allow these readings to be representative.

The second is sometimes used as an improvement on the standard quick-undrained test for certain conditions of sample. The specimen is first consolidated to an effective stress equivalent to the mean effective stress in the ground at the level from which it was taken. An equal or greater confining pressure is applied for the compression test, which is run as a 'quick' test without allowing further drainage to take place.

The third procedure was outlined by Bishop and Henkel, and the fourth is based on a method described by D. W. Hight (1996, pers. comm.).

APPLICATIONS

The UU test is applicable to undisturbed specimens in which no change in moisture content from the in-situ value can be permitted. It is also used for the determination of effective shear strength parameters for compacted fill materials, using large specimens (e.g. 100 mm diameter). Smaller specimens of partially saturated intact undisturbed cohesive soil can also be tested in this way. The cohesion intercept, c', is particularly sensitive to small variations in moisture content (Bishop and Henkel, 1962). Tests can be carried out over a range of moisture contents to enable Mohr envelopes for the required value to be interpolated.

The C-QU test procedure is useful for determining the undrained shear strength of soils that have suffered disturbance or moisture change during sampling. An example is a stiff fissured clay in which fissures may have opened up. It was used by Hanrahan (1954) in tests on peat, from which a relationship between shear strength and water content was obtained. There is a change in moisture content during the consolidation stage. This type of test is intended only for the measurement of total stress, and is run too quickly for the measurement of pore pressures.

In the CCV test, errors in pore pressure measurement due to incomplete saturation can be avoided by holding the pore pressure constant by cell pressure adjustments.

The CU test outlined in Section 19.3.5 is intended primarily for compacted clay specimens. This procedure avoids the usual back pressure saturation process in which the soil has access to water at very low effective stresses.

19.3.2 Unconsolidated Undrained (UU) Test on Clay

This procedure, which is based on that described by D. W. Hight (1996, pers. comm.), is for measuring the effective shear strength properties of a 100 mm diameter undisturbed specimen of clay during undrained shear.

APPARATUS

The apparatus is similar to that required for the routine effective stress triaxial tests described in Chapter 18.

Two connections to the base pedestal (as in Fig. 16.2) are required, so that the porous disc and transducer port can be flushed with de-aired water during the test. The top drainage connection is not used. The arrangement is as shown in Fig. 19.3.

Hight's procedure calls for a pore pressure probe at the mid-height of the specimen, in addition to the usual pore pressure transducer at the base. Since this device is beyond the scope of this volume, the procedure given below makes use only of base pore pressure measurements.

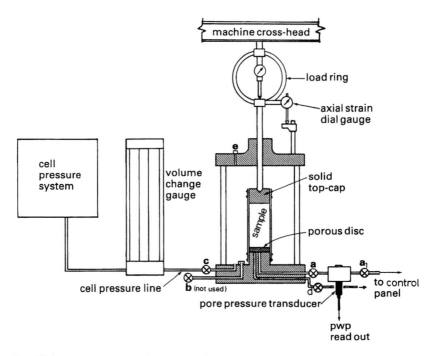

*Fig. 19.3. Arrangement of apparatus for unconsolidated undrained (UU) triaxial
compression test with pore pressure measurement*

SPECIMEN PREPARATION

The test specimen of undisturbed clay, 100 mm diameter and 200 mm high, is prepared as
described in Section 18.4.2, and Section 9.2.5 or 9.3.3 of Volume 2. If the clay contains
sand lenses the engineer responsible for the tests should be notified immediately, so that a
decision can be taken on whether to use a replacement specimen or to proceed with the
laminated specimen.

The specimen is set up in the triaxial cell as described in Section 18.4.7. Filter paper side
drains may be fitted if approved by the engineer, to accelerate pore pressure equalisation,
but the filter paper should not encroach within 10 mm of the lower porous disc where pore
pressure is measured. The way it is fitted is similar to that shown in Fig. 18.10.

SATURATION

The cell pressure is applied to the specimen in three or more steps in order to saturate the
specimen. The pressures to apply depend on the total overburden pressure at the depth at
which the sample was taken, denoted by σ_{vo} (kPa), as summarised in Table 19.1

The value of σ_{vo} should be specified by the engineer, but in the absence of this
information it may be calculated from the equation

$$\sigma_{vo} = \rho g z \quad \text{kPa}$$

where ρ is the soil density (Mg/m^3), z the relevant depth below the surface (m) and
$g = 10 \text{ m/s}^2$ (approximately).

Table 19.1. CELL PRESSURES FOR SATURATION STAGE

Depth z (m)	Cell pressure for saturation stage (kPa)		
	Step 1	Step 2	Step 3
less than 5	$2.0\,\sigma_{vo}$	$2.5\,\sigma_{vo}$	$3.0\,\sigma_{vo}$
5–10	$1.5\,\sigma_{vo}$	$2.0\,\sigma_{vo}$	$2.5\,\sigma_{vo}$
10–40	$1.0\,\sigma_{vo}$	$1.5\,\sigma_{vo}$	$2.0\,\sigma_{vo}$
greater than 40	800	1200	1600

The procedure is as follows:

(1) Apply the first cell pressure (step 1) as soon as possible after setting up the specimen in the triaxial cell. Shortly afterwards, quickly flush the base pedestal and transducer housing with de-aired water.

(2) Record the response of the pore pressure transducer at the instant of application of the cell pressure and at time intervals similar to those used for consolidation tests (e.g. $\frac{1}{4}$, $\frac{1}{2}$, 1, 2, 4, 8 minutes).

(3) Maintain the pressure for at least 4 hours, or until the pore pressure reading has stabilised (i.e. is not changing at a rate greater than 5 kPa/hour). If the pore pressure has not stabilised within 8 hours of application of the cell pressure, the engineer should be notified.

(4) Apply the second and third steps of cell pressure, and record the pore pressure response, allowing stabilisation of pore pressure as above, but with not less than 2 hours between applying one step and the next.

(5) If, when the pore pressure reading has stabilised under the third step of cell pressures, the measured pore pressure is less than 100 kPa, then increase the cell pressure until the pore pressure does exceed 100 kPa. Allow the pore pressure reading to stabilise.

(6) Increase the cell pressure by a further 50 kPa, and calculate the resulting value of the pore pressure coefficient B, using the equation in Section 18.6.1.

(7) Leave the specimen under the applied cell pressure until the pore pressure reading has stabilised, as defined above. Notify the engineer if stabilisation has not been reached within 2 hours of the last step of cell pressure being applied, or if the required cell pressure is likely to exceed the capacity of the equipment.

UNDRAINED SHEAR

(8) The rate of strain given in Hight's procedure is 1% per hour. But without pore pressure measurement at the mid-height, to ensure pore pressure equalisation the rate of strain should be not greater than that derived from consolidation data on a similar specimen, as described in Section 18.6.3. The rate of strain should not exceed 1% per hour.

(9) Take readings during the shear stage as described in Section 18.6.3, but at greater frequencies, as shown in Table 19.2.

Table 19.2. MAXIMUM INTERVALS BETWEEN READINGS DURING
UNDRAINED SHEAR

Time period	Maximum interval between readings (minutes)	No. of readings
First 20 minutes	$\frac{1}{2}$	40
Next 40 minutes	1	40
Next 2 hours	3	40
Next 7 hours	10	42
Thereafter	30	

(10) Observe the specimen carefully, and if appropriate record the point at which a shear zone or zones develop, as described in Section 17.4.1.

(11) Continue the test until an axial strain of at least 10% is reached.

(12) Complete the test as described in steps (13) to (23) of Section 18.6.3. If shear surfaces develop, measure and record their inclination, and the movement along them, as described in Section 17.4.1.

CALCULATING, PLOTTING, REPORTING

Calculations and plotting are as described in Sections 18.7.1 and 18.7.3.

Corrections for changes in area due to movement along slip planes (if they appear) are applied in accordance with Section 17.4.1. Corrections for membrane restraint, distinguishing between conditions before and after the formation of any shear planes, are made as described in Section 17.4.2.

Results including graphical plots are reported as described in Section 18.7.5. for the CU test, except that the consolidation stage does not apply. For the saturation stage, graphical plots of cell pressure and pore pressure readings against time, and pore pressure against cell pressure, should be included.

Some typical graphical plots from the shearing stage are shown in Fig. 19.4.

19.3.3 Consolidated Quick-Undrained (C-QU) Test

GENERAL

The apparatus for this test is shown in Fig. 19.5. The load frame is the same as used for a quick-undrained triaxial test (Volume 2, Section 13.6.3). A burette open to atmosphere is connected via the drainage line to the top of the specimen.

If the soil is not fully saturated, air as well as water will emerge via the drainage line. Air can be separated from the expelled water by connecting the drainage line first to a bubble trap, as shown in Fig. 20.4. In either case the water surface in the open burette should be maintained at a constant level corresponding to the mid-height of the specimen, especially when readings are being taken.

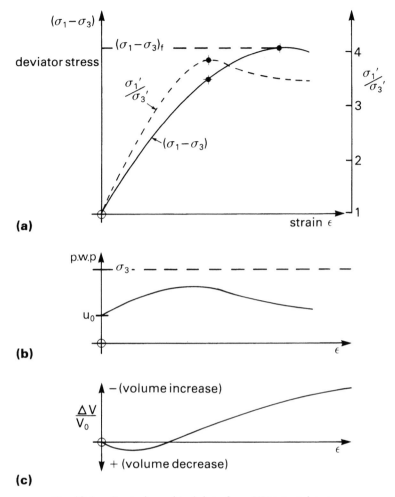

Fig. 19.4. *Typical graphical data from UU triaxial test*

Measurement of pore water pressure at the base during consolidation is desirable but not essential. If the pore pressure facility is used the initial pore pressure is measured as soon as the specimen is set up and sealed onto the cell base.

If pore pressure measurements are not made, double drainage can be used by draining from the base of the specimen as well as from the top, either to separate burettes, or to a single burette. Consolidation is then theoretically four times faster than with drainage from one end only, without having to use side drains.

CONSOLIDATION PRESSURE

The effective confining pressure used for consolidation is often equal to the vertical effective stress σ_v' at the level from which the sample was taken. This is calculated using the equation given in Section 19.3.2 above.

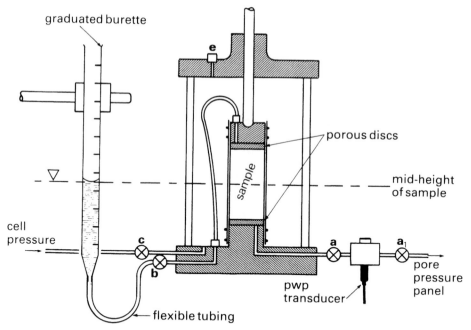

Fig. 19.5 Apparatus for consolidated quick-undrained (C-QU) triaxial compression test, with consolidation to atmosphere

The mean effective in-situ stress, p_0', is more appropriate for isotropic consolidation, where

$$p_0' = \tfrac{1}{3}(\sigma_v' + 2\sigma_h')$$

(21.6)

(See Section 21.1.4.)

The horizontal effective stress σ_h' is equal to $K_0\sigma_v'$, where a value of K_0 (Section 15.3.4) may be assumed if it has not been determined.

CONSOLIDATION PROCEDURE

The consolidation cell pressure is applied, and enough time is allowed to enable the pore pressure to reach equilibrium. Consolidation is started by opening valve b (Fig. 19.5), and readings of the water level in the burette (and pore pressure if appropriate) are taken at time intervals usual for a consolidation test. The level of the water surface in the burette should be maintained constant, at about the mid-height of the specimen, during consolidation.

Completion of consolidation is assessed from the plot of burette readings against square-root time, in the same way as shown in Fig. 18.21, Section 18.6.2.

Valves a and b are then closed because no further drainage is permitted and pore pressure readings are not needed since this is a total stress test. The cell pressure is raised to the desired level, or may remain unchanged for one specimen of a set.

COMPRESSION TEST

The compression machine speed control is set to apply a rate of strain of not more than 1% per minute to the specimen, so that the necessary readings can be recorded and plotted as the test proceeds. Readings of the load ring or force transducer are observed at regular intervals of strain as in the quick-undrained test (Section 13.6.3). The stress–strain curve may be plotted on a similar skew-grid sheet to that shown in Fig. 13.41 (Vol. 2) and the maximum deviator stress is calculated as described in step 21 of Section 13.6.3.

A set of specimens taken from the same depth are all consolidated to the same effective stress, but a different cell pressure is applied to each specimen. This enables the undrained shear strength over a range of confining pressures to be obtained.

In addition to the usual test data (Volume 2, Section 13.6.3, step 23), the sample depth and the cell pressure used for consolidation are also reported. Estimated values of the coefficient of consolidation, c_v, derived from square-root time/volume change graphs from the consolidation readings as explained in Section 15.5.5, may be reported if appropriate.

19.3.4   Consolidated Constant Volume (CCV) Test

In a routine CU test on a specimen that is not completely saturated some volume change may occur during the shear stage due to the compressibility of the air in the voids. This can result in errors in the measured pore pressures and shear strength. These errors can be avoided if the pore pressure is maintained constant at the initial value by adjusting the cell pressure as axial compression is applied, without allowing drainage. This is the principle of the consolidated constant volume (CCV) test.

Initial preparation and consolidation are the same as for the CU test (Section 18.6.3). The rate of strain is also derived in the same way. After the end of the consolidation stage arrangements are made for controlling the cell pressure during the compression stage. This requires constant attention to ensure that the pore pressure remains constant at the value at the start of compression, with reference to a pore pressure transducer read-out. Adjustment of cell pressure can be made by using the air pressure regulator of an air–water system.

However a more sensitive control is obtainable by using a hand control cylinder, as follows. Referring to Fig. 18.4 in Section 18.2.3, the cell pressure system is disconnected from valve c and a connection is made from valve o on the pore pressure panel to valve c on the cell base. These valves are opened after pressurising the control cylinder to equate with the cell pressure. Valve k must be kept closed, and valve a open. During the compression test the pore pressure is maintained constant by operating the control cylinder to adjust the cell pressure.

Readings are taken at regular intervals of strain in the same way as in the CU test, except that cell pressure readings are recorded instead of pore pressures. Calculation of deviator stress must make allowance for changes in the upward force on the cell piston due to changes in cell pressure. Effective stresses for plotting Mohr circles are calculated as for the CU test, but here σ_3 is variable and u remains constant.

When testing specimens that tend to dilate when sheared it is particularly important to ensure that no air is entrapped in the pore pressure system, otherwise errors may occur due to capillary effects at the base of the specimen. For fully saturated soils the shear strength parameters derived from this test are practically identical to those obtained from the CU test.

19.3.5 Consolidated Undrained Test on Compacted Clay

This procedure, which is based on that described by D. W. Hight (1996, pers. comm.), is for measuring the effective shear strength properties of a 100 mm diameter specimen of compacted clay, during undrained shear after isotropic consolidation.

APPARATUS

The apparatus is similar to that required for the routine effective stress triaxial tests described in Chapter 18, with the following modifications:

(1) High air-entry value porous discs are used. They are set into the base pedestal, as shown in Fig. 16.4(b), and into the top cap.

(2) Two connections to the base pedestal (as in Fig. 16.2) are required, so that the porous disc and transducer port can be flushed with de-aired water during the test.

(3) A top cap which permits unrestrained failure on an inclined shear surface is necessary. The principle, using ball bearings and a hardened steel plate, is shown in Fig. 19.6.

(4) Hight's procedure calls for a pore pressure probe at the mid-height of the specimen, in addition to the usual pore pressure transducer at the base. Since this device is beyond the scope of this volume, the procedure given below makes use only of base pore pressure measurement.

SPECIMEN PREPARATION

The test specimen of compacted clay, 100 mm diameter and 200 mm high, is prepared as described in Section 18.4.4, and Section 9.5.6 of Volume 2. If the clay contains gravel-size particles it should be enclosed in two membranes. After compaction the specimen should be sealed and left to mature for at least 24 hours, or for several days if the soil is of very low permeability.

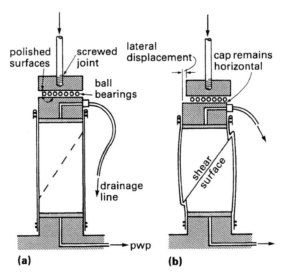

Fig. 19.6. Top cap permitting lateral displacement of top of specimen: (a) initial assembly; (b) unrestrained lateral movement with slip along inclined shear surface

The specimen is set up in the triaxial cell as described in Section 18.4.7. Filter paper side drains may be fitted if required, to accelerate consolidation.

Incremental back pressure saturation is not applied.

APPLICATION OF CELL PRESSURE

The cell pressure required for the test, σ_c, is applied in four equal steps (i.e. $0.25\sigma_c$, $0.5\sigma_c$, $0.75\sigma_c$, σ_c). The value of σ_c should be specified by the engineer, but in the absence of this information it may be calculated from the equation

$$\sigma_c = \rho g z \quad \text{kPa}$$

where ρ is the soil density (Mg/m^3), z is the relevant depth below the surface (m) and $g = 10\,\text{m/s}^2$ (approximately).

The procedure is as follows:

(1) Apply the first cell pressure step $(0.25\sigma_c)$ as soon as possible after setting up the specimen. Shortly afterwards, quickly flush the base pedestal and transducer housing with de-aired water.

(2) Maintain the pressure for at least 4 hours, or until the pore pressure reading has stabilised (i.e. is not changing at a rate greater than 5 kPa per hour).

(3) Apply the second and subsequent steps of cell pressure, allowing stabilisation of pore pressure as above, but with not less than 2 hours between applying one step and the next.

(4) When the pore pressure reading has stabilised under the final pressure increment (cell pressure $= \sigma_c$), the specimen is ready for the isotropic consolidation stage.

CONSOLIDATION

(5) Adjust the back pressure to 100 kPa with the drainage valve closed, and increase the cell pressure if necessary to achieve the specified effective stress. Allow the pore pressure reading to stabilise again.

(6) Open the drainage valve and take readings during consolidation as described in Section 18.6.2.

(7) At the end of consolidation, check the value of the pore pressure coefficient B by increasing the cell pressure by 25 kPa with the drainage valve closed (see Section 18.6.1(1), steps (2) and (3)).

(8) Restore the consolidation stress and wait for pore pressure stabilisation before starting the shear stage.

UNDRAINED SHEAR

(9) The rate of strain given in Hight's procedure is 1% per hour. But without pore pressure measurement at the mid-height the rate of strain should be not greater than that derived from consolidation data, as described in Section 18.6.3, with a maximum of 1% per hour.

(10) Take readings during the shear stage as described in Section 18.6.3, but at greater frequencies, as given in Table 19.2 (Section 19.3.2).

(11) Observe the specimen carefully, and record the point at which a shear zone or zones develop, as described in Section 17.4.1.

(12) Continue the test until an axial strain of at least 10% is reached.

(13) Complete the test as described in steps (13) to (23) of Section 18.6.3. Measure and record the inclination of shear surfaces, and the movement along them, as described in Section 17.4.1.

CALCULATIONS, PLOTTING, REPORTING

Calculations and plotting are as described in Sections 18.7.1 and 18.7.3.

Corrections for changes in area due to movement along slip planes (if they appear) are applied in accordance with Section 17.4.1. Corrections for membrane restraint, distinguishing between conditions before and after the formation of any shear planes, are made as described in Section 17.4.2.

Results are reported as described in Section 18.7.5 for the CU test.

REFERENCES

Anderson, W. F. (1974). 'The use of multi-stage triaxial tests to find the undrained strength parameters of stony boulder clay'. *Proc. Inst. Civ. Eng.* Technical Note No. TN 89.
Ho, D. Y. F. and Fredlung, D. G. (1982). 'A multistage triaxial test for unsaturated soils'. *Geotechnical Testing J.*, Vol. 5, No. 1/2, pp 18 to 25.
Janbu, N. (1985). 'Soil models in offshore engineering'. 25th Rankine Lecture, *Géotechnique*, 35 : 3 : 241.
Kenney, T. C. and Watson, G. H. (1961). 'Multiple-stage triaxial test for determining c' and φ' of saturated soils'. *Proc. 5th Int. Conference on Soil Mechanics and Foundation Eng.*, Paris, Vol. 1, pp 191–195.
Lumb, P. (1964). 'Multi-stage triaxial tests on undisturbed soils'. *Civ. Eng. & Public Works Review*, May, 1964.
Ruddock, E. C. (1966). Correspondence, *Géotechnique*, 16 : 1 : 78.
Watson, G. H. and Kirwan, R. W. (1962). 'A method of obtaining the shear strength parameters for boulder clay'. *Trans. Inst. of Civil Engineers of Ireland*.

Chapter 20

Triaxial consolidation and permeability tests

20.1 INTRODUCTION

20.1.1 General

The consolidation characteristics, and the permeability, of a soil specimen (undisturbed or reconstituted) can be measured in a triaxial cell, and in some ways this method is preferable to the more conventional procedures described in Volume 2. The type of specimen used and its arrangement in the cell are similar for both types of test, except that a drainage connection to each end of the specimen is needed for permeability measurements.

Triaxial consolidation is similar to the consolidation stage of an effective stress triaxial compression test (Section 18.6.2). Permeability measurements can be made on a specimen that has been set up in a triaxial cell for an effective stress compression test, if required.

The test procedures given in this chapter are those covered in Clauses 5 and 6 of BS 1377:Part 6:1990, together with some additional related procedures. A permeability test similar to the BS method is given in ASTM D 5084.

20.1.2 Triaxial Consolidation (Pore Pressure Dissipation) Tests

PRINCIPLES

In the simplest form of triaxial consolidation, which is covered in this chapter, the conditions are isotropic, i.e. the specimen is subjected to increments of equal all-round pressure ($\sigma_1 = \sigma_2 = \sigma_3$). This causes the specimen to consolidate laterally as well as vertically, and it is therefore a three-dimensional process. Because pore pressure is measured as an essential part of the procedure, it is sometimes referred to as a pore pressure dissipation test.

The all-round confining pressure is applied in increments, each of which is held constant until virtually all the excess pore pressure due to that increment has dissipated, i.e. the specimen consolidates.

In a typical test, three or more increments of pressure are applied, to provide three or more stages of consolidation. Each stage is carried out in two phases:

(1) the undrained phase, in which the cell confining pressure is increased so that it exceeds the back pressure by an amount equal to the desired effective stress for consolidation, causing the pore pressure to build up and eventually reach a steady value;

263

(2) the drained phase, in which this excess pore pressure is allowed to dissipate against the back pressure until the pressures virtually equalise (consolidation).

During the consolidation process, water drains out from one end of the specimen (usually the top), and the volume of expelled water is measured; this is equal to the change in volume of a saturated specimen. Pore pressure is monitored at the undrained end of the specimen (usually the base). Readings of volume change and pore pressure are recorded at suitable intervals of time to enable consolidation curves to be drawn.

It is usual to increase the effective stress from one stage to the next by a factor of 2 (as in the oedometer consolidation test)—for example 50, 100, 200, 400 kPa.

The relationship between voids ratio and effective isotropic stress can be derived from a series of incremental loadings. Values of the volumetric coefficients of consolidation and compression for three-dimensional isotropic consolidation can be determined for each pressure increment.

ADVANTAGES AND DISADVANTAGES

Some of the advantages of this type of test over a standard oedometer test are as follows.

(1) A larger specimen can be tested, the usual size being 100 mm diameter and 100 mm high.

(2) A large specimen size enables soils containing discontinuities (e.g. fissures) to be tested satisfactorily, so that field conditions can be more closely represented.

(3) Direct measurement of pore pressure is possible, not only during consolidation but also during the application of external load.

(4) The test can be related to the range of pore pressures likely to apply in practice.

(5) The coefficient of consolidation can be obtained directly from pore pressure measurements. This is advantageous because a geometrical curve-fitting procedure is not then needed. It is necessary only to compare a point on the laboratory curve with a point on the theoretical curve.

(6) Volume changes in partially saturated soils due to undrained loading can be measured.

(7) Direct measurement of permeability can be carried out on the same specimen (see Section 20.3).

(8) The horizontal confining stress, as well as the vertical stress, can be controlled to a known value.

(9) Provision can be made for either vertical (axial) or horizontal (radial) drainage.

(10) Errors due to the deflection of the oedometer load frame, and from side friction at the cell walls, are eliminated.

(11) Consolidation can take place under any of the following conditions:
    (a) Isotropic (equal all-round pressure). (Sections 20.2.1 and 20.2.2.)
    (b) General anisotropic loading.
    (c) No lateral strain (K_0 condition), as in the standard oedometer test.

The application of triaxial consolidation testing to three-dimensional settlement analysis was discussed by Davis and Poulos (1968).

There are some disadvantages compared with oedometer tests, as follows, but they are generally outweighed by the advantages.

(1) A higher level of operator skill, and more attention to procedures are necessary.

(2) Larger specimens mean longer testing times.

(3) Several major items of equipment are in use for long periods.

(4) Head losses in drainage layers and pipelines may give rise to erroneous values of the coefficient of consolidation when the rate of drainage from the specimen is high.

(5) The usual errors inherent in the triaxial apparatus must be allowed for, by applying the relevant corrections described in Section 17.4.

STRESS CONDITIONS

In three-dimensional consolidation (which excludes the K_0 condition), the boundary conditions and stress changes within the soil differ from those which apply during one-dimensional consolidation, because the specimen is not rigidly confined. The radial boundary moves as the volume changes, and in the early stages this movement is greatest near the drained face (Lo, 1960). The equations that are used for calculating the coefficient of consolidation from one-dimensional consolidation do not properly represent these conditions. Values of the volumetric coefficients of compression and consolidation derived from isotropic consolidation tests, which are denoted by m_{vi} and c_{vi} respectively, are therefore not the same as the values of m_v and c_v derived from one-dimensional consolidation tests. Theoretical relationships are given in Section 20.2.1(10) and (11).

PORE PRESSURE ANOMALY

A feature of three-dimensional consolidation with radical drainage which might be surprising when it occurs is the apparent increase in measured excess pore water pressure early in a consolidation stage, as illustrated in Fig. 20.1. This is similar to the 'Mandel–Cryer

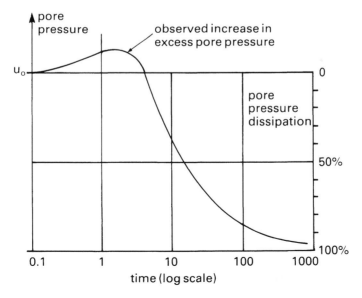

Fig. 20.1   Pore pressure increase sometimes observed in early stage of triaxial consolidation with radial drainage

effect' (Schiffman, Chen and Jordan, 1969), which has been observed both in laboratory tests and in-situ, and can be predicted from a theoretical analysis. It is explained physically in terms of the compatibility of strains which must hold good everywhere in the specimen when both drainage and deformation occur at a boundary. A similar effect has been observed by the author in Rowe cell consolidation tests under 'free-strain' loading.

PRESENTATION

Procedures for an isotropic consolidation test with vertical drainage is described in detail in Section 20.2.1. These procedures are referred to subsequently in the descriptions of a test with radial drainage (Section 20.2.2).

20.1.3   Triaxial Permeability Tests

PRINCIPLES

The triaxial test apparatus provides facilities for maintaining a flow of water through a specimen under a known difference of pressure, and for measuring the rate of flow, while the specimen is subjected to a known effective stress. From these measurements the soil permeability can be calculated. Repeat tests after consolidation under a number of effective stress values can be carried out on the same specimen without having to remove and reset it.

ADVANTAGES

Some of the advantages of measuring permeability in a triaxial cell are as follows:

(1) The specimen can be first saturated under the application of a back pressure, thus reducing or eliminating obstructions to flow due to bubbles of gas. The presence of air leads to unreliable volume change measurements, and to the accumulation of bubbles at the sample outlet. Bjerrum and Huder (1957) reported a sixfold increase in the measured permeability of a sandy clay when the pore pressure was raised from zero to about 800 kPa.

(2) Saturation by back pressure can be used, and is achieved much more quickly than saturation by prolonged flooding or circulation only. This applies particularly to compacted soils.

(3) The test can be carried out under effective stresses and at pore pressures which relate to field conditions.

(4) Small rates of flow can be measured easily.

(5) Constant head or falling head procedures can be used.

(6) A wide range of hydraulic gradients can be applied and accurately measured.

(7) Soils of intermediate permeability, such as silts, which are difficult to test by either of the standard constant head or falling head procedures described in Volume 2, Chapter 10, can be accommodated, as well as clays.

(8) Undisturbed test specimens can be set up easily, and there are no cell wall effects which might give non-uniform flow conditions.

TEST METHODS

Four methods for carrying out permeability tests in a standard triaxial cell are described in Section 20.3. Either the constant head or the falling head principles may be used, depending on circumstances. The methods comprise:

(1) Using two independent constant pressure systems, additional to the system providing the confining pressure, to maintain a flow of water (Section 20.3.1).

(2) Using one additional pressure system and draining to an open burette (Section 20.3.2).

(3) Using two burettes (Section 20.3.3).

(4) Making use of the volume characteristics of a pressure transducer to measure very small rates of flow through materials having very low permeabilities (Section 20.3.4).

Method (1) is covered in Clause 6 of BS 1377:Part 6:1990. A similar procedure is described in ASTM D 5084, where the apparatus is referred to as a flexible wall permeameter.

PORE PRESSURE DISTRIBUTION

A specimen undergoing a triaxial permeability test is represented in Fig. 20.2(a). The cell confining pressure σ_3 is greater than the inlet and outlet water pressures p_1 and p_2. The distribution of pore pressure along the length of the sample is initially assumed to be linear as shown in diagram (b). The effective stress σ' at any height is the amount by which the cell pressure σ_3 exceeds the pore pressure at that point, and increases from the inlet at the base to the outlet at the top.

Voids ratio is related to effective stress, and so must be greater at the inlet than at the outlet (diagram (c)). Permeability depends on voids ratio and must vary in a similar manner. In a steady-state flow condition the rate of flow of water is the same across all horizontal sections, i.e. $q = $ constant. From Darcy's law (Volume 2, Section 10.3.2)

$$i = \frac{Q}{Akt} = \frac{q}{Ak} \qquad (10.4)$$

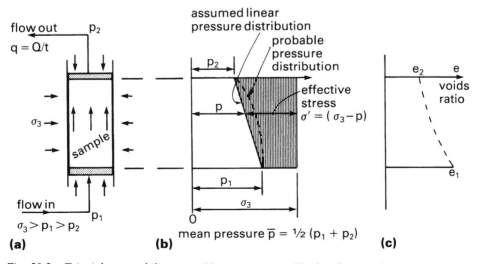

Fig. 20.2 *Triaxial permeability test: (a) arrangement, (b) distribution of pore pressure along sample axis, (c) probable variation of voids ratio*

For a local increase in voids ratio, both the area A and the permeability k increase, and therefore the gradient i decreases. Hence the hydraulic gradient in the specimen must be less than the mean value near the inlet end, and greater than the mean near the outlet end. Therefore the distribution of pore water pressure throughout the specimen height is not uniform, but of the form shown by the dashed curve in Fig. 20(b). If the curve is assumed to be parabolic the average pore pressure is equal to $\frac{1}{3}(p_2 + 2p_1)$, and the average effective stress is equal to

$$\sigma_3 - \tfrac{1}{3}(p_2 + 2p_1)$$

However for many practical purposes the average pore pressure is taken to be equal to $\frac{1}{2}(p_1 + p_2)$, and the mean effective stress to be equal to

$$\sigma_3 - \tfrac{1}{2}(p_1 + p_2).$$

20.2   ISOTROPIC CONSOLIDATION TESTS

20.2.1   Isotropic Consolidation with Vertical Drainage (BS 1377:Part 6:1990:5)

PRINCIPLES

This test is carried out in a triaxial cell without the need for a load frame. A special permeability cell is available for this purpose (see Section 16.2.4). Specimens up to 100 mm high can be used. But a cell of the type normally used for triaxial compression tests (as described in Section 16.2.2) may be used, provided that the piston can be rigidly restrained against the upward force from the cell pressure. Normally a cell for accepting 100 mm diameter specimens would be used, and the specimen proportions in terms of height:diameter ratio could range from 1:1 to 2:1. The advantage of 1:1 specimens is that the time required for the test is appreciably reduced.

BS 1377 allows for specimens of any size from 38 mm diameter upwards. However, it is desirable to use specimens that are as large as practicable because large specimens are more representative of natural soil conditions than small specimens.

SELECTION OF TEST CONDITIONS

Before a test is started, the following test conditions need to be specified:

(1) Size of test specimen

(2) Drainage conditions for the test

(3) Whether voids ratios are to be calculated and plotted

(4) Method of saturation, including cell pressure increments and differential pressure if appropriate; or whether saturation should be omitted

(5) Sequence of effective pressure increments and decrements

(6) Criteria for terminating each primary consolidation and swelling stage

(7) Whether secondary compression characteristics are to be determined.

PRE-TEST CHECKS

The relevant checking procedures ('complete' checks and 'routine' checks) on the apparatus and ancillary items, which are specified in Clause 5.2.4 of BS 1377:Part 6:1990, are the same as those described in Sections 18.3.2–18.3.6.

ENVIRONMENT

The environmental requirements for these tests are the same as those given in Section 18.5.3 for triaxial compression tests.

APPARATUS

Apart from the possible use of the permeability cell referred to above, the apparatus needed is the same as that used for the routine triaxial tests described in Chapter 18. The essential features are shown in Fig. 20.3. Similar connections are made to cell pressure, back pressure and pore water pressure systems. Drainage takes place from the top of the specimen to the back pressure system which includes a volume-change indicator. Pore pressure is measured at the base. The piston of a standard cell must be rigidly restrained against the upward force from the cell pressure.

If changes in specimen volume are to be measured, a volume-change gauge should be connected in the cell pressure line and the cell should first be calibrated for volume change against pressure and time.

If a back pressure is not to be applied, drainage from the top of the specimen can be taken to an open burette as shown in Fig. 19.5 (Section 19.3.3). If the soil is not fully saturated, air as well as water will emerge via the drainage line. Air can be separated from the expelled water by connecting the drainage line first to a bubble trap, as shown in Fig. 20.4. In either case the water surface in the open burette should be maintained at a

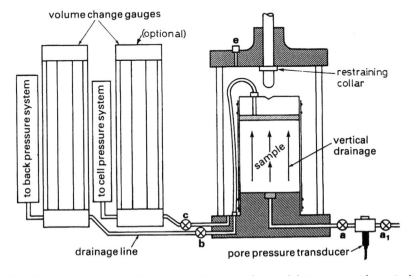

Fig. 20.3   Arrangement of apparatus for triaxial consolidation test with vertical drainage

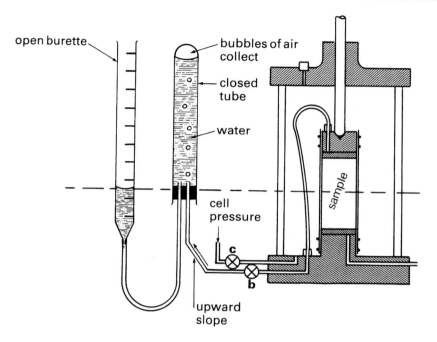

Fig. 20.4   Bubble trap for separating air from water expelled from a partially saturated specimen

constant level corresponding to the mid-height of the specimen, especially when readings are being taken.

PREPARATION AND SATURATION

(1) Prepare and check the cell and ancillary equipment as described in Section 18.3, making sure that the systems are leak-free, air-free and completely filled with de-aired water.

(2) Prepare, measure and weigh the specimen. The initial height H_0 and volume V_0 are used in subsequent calculations.

(3) Set up the specimen on the cell pedestal as described in Section 18.4.7, and enclose it in one or two rubber membranes, after carefully checking them for flaws. Side drains are not used.

(4) Assemble the cell and fill with de-aired water, keeping the air bleed e open.

(5) Measure the initial pore water pressure at the base of the specimen, then close air bleed e.

(6) If appropriate, saturate the specimen by applying alternate increments of cell pressure and back pressure, or by using one of the other methods given in Section 18.6.1. The saturation procedures given in BS 1377 are saturation by increments of cell pressure and back pressure (Clause 5.4.3), and saturation at constant moisture content (Clause 5.4.4). In Section 18.6.1 these are described respectively under methods (1) and (4). Calculate relevant values of the pore pressure parameter B after each increment, using the equation

$$B = \frac{\delta u}{\delta \sigma_3}$$

Apply sufficient increments to obtain a value of B which indicates that saturation has been achieved, as stated in Section 18.6.1(1), step (10). If appropriate, calculate the total volume of water taken up by the specimen (see Section 18.6.1(1), step (13)).

(7) After completion of the saturation stage keep valve b closed and record the final readings of pore pressure and volume-change indicator.

CONSOLIDATION PROCEDURE

(8) With valves b and c closed (Fig. 20.3), increase the pressure in the cell pressure line to give the required effective pressure (σ_c') for the consolidation stage, as given by the equation

$$\sigma_c' = \sigma_c - u_b$$

where σ_c is the cell pressure and u_b the back pressure. The back pressure should be maintained at a constant value throughout all stages if practicable. If a reduction of back pressure is unavoidable in order to obtain a desired high value of effective stress, it should not be reduced below the level of the pore pressure at the end of the saturation stage, or 300 kPa, whichever is the greater.

(9) Open valve c to admit the increased pressure to the cell. Open valve a and observe the pore pressure until a steady value is reached, which is recorded (u_i). If necessary plot readings of pore pressure against time to establish when equilibrium is reached. Also record the reading of the volume-change indicator.

This is the 'build-up' phase, and the resulting excess pore pressure to be dissipated during consolidation is equal to $(u_i - u_b)$. The build-up phase (steps (8) and (9) above) may be carried out in two or more steps if necessary. A value of B can be obtained after the application of each cell pressure increment.

(10) The 'consolidation phase' is started by opening valve b and starting the clock, allowing drainage into the back pressure system. Take readings of pore water pressure and drainage volume change (and cell volume change if used) at the usual geometrically spaced time intervals (Section 18.6.2, steps (4) and (5)). The stage is completed when at least 95% of the excess pore pressure has been dissipated, or a steady state has been reached. The pore pressure dissipation U (%) at any time t from the start of consolidation is given by the equation

$$U = \frac{u_i - u}{u_i - u_b} \times 100 \qquad (15.28)$$

where u is the pore pressure at that time (Section 15.5.5). A value of U equal to or exceeding 95% indicates that the consolidation stage can be terminated.

(11) When it is evident that consolidation is complete, record the readings of pore pressure (u_f) and of the volume-change indicator, then close valve b. Calculate the total change in volume of the specimen, ΔV_c, during the consolidation stage.

(12) Repeat steps (8)–(11) for each consolidation stage at successfully higher effective pressures, for as many stages as are needed. The back pressure should preferably remain

constant, and in any case should not be reduced below 300 kPa to avoid air bubbling out of solution.

(13) The rebound (swelling) characteristics, if required, can be obtained by following the above procedure but reducing the cell pressure in a series of appropriate decrements. During the drained phase the pore pressure increases, usually more quickly than it dissipated.

(14) When equilibrium has been established on the final unloading stage, close valve b before reducing the cell pressure and back pressure to zero.

(15) If the test is to be terminated immediately after completing the final consolidation stage (step 12) without allowing swelling, omit step (13) and follow step (14).

(16) Drain and dismantle the cell, and remove the specimen as quickly as possible for final measurements, as described in Section 18.6.3, steps (15)–(22) (except (18)), or by using the rapid procedure given in Section 18.6.5.

PLOTTING AND CALCULATION

Symbols are similar to those used in Section 18.7.1, Tables 18.3 and 18.4.

(1) Calculate the initial values of moisture content, density, dry density, voids ratio and degree of saturation for the specimen, as described in Section 18.7.1.

(2) If saturation by increments of back pressure was applied, plot the value of the pore pressure coefficient B for each increment ($B = \delta u / \delta \sigma_3$) against pore pressure or cell pressure.

(3) During the saturation stage, if it is assumed that the water entering the sample only replaces air in the voids, there is no change in volume or height or diameter of the specimen, therefore

$$V_s = V_0 \quad \text{and} \quad H_s = H_0 \quad \text{and} \quad D_s = D_0$$

If measurement of the change in volume during saturation (ΔV_s) is wanted, it is derived from cell volume change readings as explained in Section 18.7.1 using Equation (18.9).

(4) Calculate the value of B for each undrained ('build-up') phase.

(5) Plot the pore pressure at the end of each build-up phase against the cell pressure, as in Fig. 20.5(a). The build-up of pore pressure may also be plotted against time if required.

(6) For each consolidation stage the measured volume change is usually plotted against square-root time, and pore pressure dissipation (%) against log time (see Fig. 20.5(b) and (c)).

(7) From each pore pressure dissipation plot read off the value of t_{50} (minutes) corresponding to 50% pore pressure dissipation, as shown in Fig. 20.5(c). These are used for calculating values of c_{vi} (steps (11)–(13) below).

(8) The sample height H at the end of any consolidation stage is calculated from the equation

$$H = H_0 \left[1 - \frac{1}{3} \frac{\Delta V}{V_0} \right] \tag{20.1}$$

where ΔV is the cumulative volume change, measured on the drainage line, from the start of the first stage of consolidation to the end of the stage in question. The mean height $\bar{H}$ during a stage (required for step 11) is equal to $(H_1 + H_2)/2$, where H_1 and H_2 are the

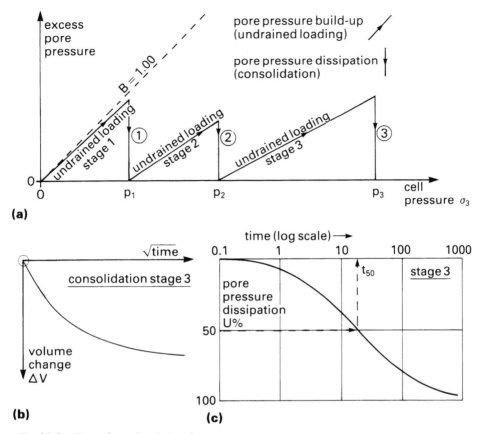

Fig. 20.5   *Typical graphical data from triaxial consolidation test: (a) build-up and dissipation of pore pressure, (b) volume-change against square-root time for a consolidation stage, (c) pore pressure dissipation against log time for a consolidation stage, showing derivation of* t_{50}

heights at the beginning and end of the stage (i.e. H_1 is the calculated height at the end of the previous stage).

(9) The voids ratio e at the end of a consolidation stage, if required, is calculated from the equation

$$e = e_S - (1 + e_S)\,\frac{\Delta V}{V_s}$$ (20.2)

where ΔV is defined as above.

(10) The coefficient of volume compressibility, m_{vi}, for isotropic consolidation, for each consolidation stage is calculated from the equation

$$m_{vi} = \frac{\delta e}{\delta p'} \times \frac{1000}{1 + e_1}\ \text{m}^2/\text{MN}$$ (20.3)

where δe is the change in voids ratio during the stage $(e_1 - e_2)$, e_1 is the voids ratio at beginning of stage, e_2 is the voids ratio at end of stage, and $\delta p'$ is the effective stress increment (kPa) for the stage $(p_1' - p_2')$.

If values of voids ratios are not calculated, values of m_{vi} can be obtained from the equation

$$m_{vi} = \frac{\Delta V_2 - \Delta V_1}{V_0 - \Delta V_1} \times \frac{1000}{p_2' - p_1'} \tag{20.4}$$

where ΔV_1 is the cumulative change in volume of the specimen up to the end of the previous consolidation stage (cm^3), ΔV_2 is the corresponding change up to the end of the consolidation stage being considered (cm^3) and V_0, p_1', and p_2' are as defined above.

An approximate relationship of m_{vi} to the equivalent value of m_v derived from a one-dimensional oedometer consolidation test is

$$m_{vi} = 1.5 m_v \tag{20.5}$$

(11) The theoretical equation for calculating the value of the coefficient of consolidation, c_{vi}, for isotropic consolidation is

$$c_{vi} = \frac{0.379 \bar{H}^2}{t_{50}} \tag{20.6}$$

Using customary units, if the mean height $\bar{H}$ is measured in millimetres and t_{50} is in minutes, then this equation becomes

$$c_{vi} = \frac{0.199 \bar{H}^2}{t_{50}} \, \text{m}^2/\text{year} \tag{20.7}$$

For a specimen initially 100 mm high, for practical purposes this becomes

$$c_{vi} = \frac{2000}{t_{50}} \tag{20.8}$$

(12) If a temperature correction to the calculated value of c_{vi} is appropriate, a correction factor derived from the curve given in Fig. 14.18 of Volume 2 (Section 14.3.16) should be applied.

(13) The calculated values of c_{vi} can be adjusted to give equivalent values derived from an oedometer consolidation test (c_v) by multiplying by a factor equal to

$$f_{cv} = \frac{1}{1 - B(1 - A)(1 - K_0)} \tag{20.9}$$

(Rowe, 1959) where A and B are the Skempton pore pressure coefficients and K_0 is the coefficient of earth pressure at rest. The values of A and K_0 may not be known for the test specimen, but the following values are typical. For a normally consolidated clay, A is likely to lie between 0.4 and 1, and K_0 around 0.5. If B is close to unity the above multiplying factor lies in the range 1.2–1.33. Typical 'equivalent oedometer' values of c_v are therefore equal to about $1.25 \times c_{vi}$.

(14) Voids ratio e is plotted against effective pressure on a log scale (log p'), to give an $e/\log p'$ curve similar to that obtained from an oedemeter consolidation test (see Fig. 20.6).

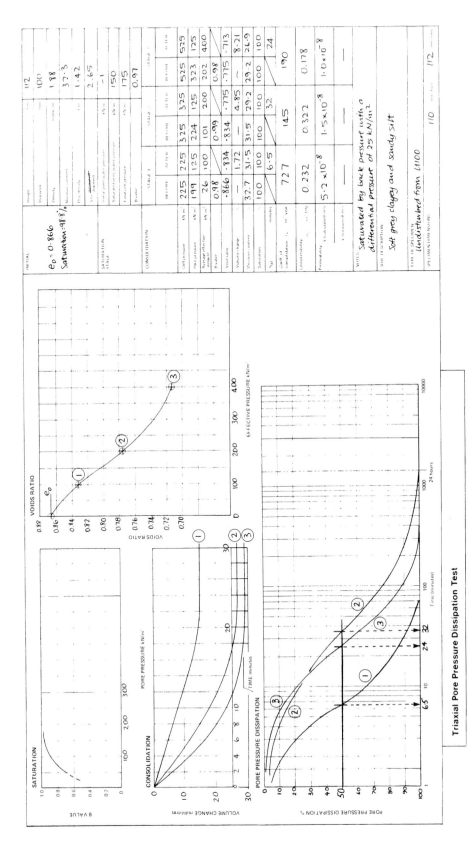

Fig. 20.6 Test data and graphical plots from a three-stage triaxial consolidation test

The first point on this graph corresponds to e_0 and the initial effective pressure after saturation, i.e. before the application of the first consolidation pressure increment.

(15) A calculated value of the vertical permeability at any stage can be obtained from the equation given in Volume 2, Section 10.5.4:

$$k = c_v \times m_v \times 0.31 \times 10^{-9} \, \text{m/s} \qquad\qquad (20.10)$$

TEST RESULTS

A typical set of results from a three-stage test is shown in Fig. 20.6. These plots form part of the test report, which should comprise the following. Items marked * are additional to those called for in Clause 5.7 of BS 1377:Part 6:1990.

Statement (if appropriate) that the test was carried out in accordance with Clause 5 of BS 1377:Part 6:1990, using isotropic consolidation in a triaxial cell

Sample identification, reference number and location

Type of sample

Method of preparation of the test specimen

Comments on the condition, quality and possible disturbance of the soil sample, and any difficulties experienced during preparation on the test specimen

Visual soil description, including soil fabric and any unusual features

Location and orientation of test specimen within original sample

Initial and final specimen details:
    dimensions
    densities and dry densities
    moisture contents
    particle density, including whether measured or assumed
    voids ratios
    degree of saturation

Method of saturation, including pressure increments and differential pressure applied, if appropriate

Volume of water taken into the specimen during saturation

Cell pressure, pore pressure and B value at the end of saturation

Data for each pressure stage:
    cell pressure and back pressure
    effective stress at beginning and end of each stage
    pore pressure change and B value for each undrained loading

Data for each consolidation phase:
    voids ratio, if required, and percentage pore pressure dissipation
    volume change
    values of the coefficients m_{vi}, c_{vi}, to two significant figures
Graphical plots:
    *B value against pore pressure or cell pressure
    Voids ratio/log p' curve

Volume change/square root time curve for each effective stress

Percentage pore pressure dissipation/log time curve for each effective stress

*Values of t_{50} derived for each stage

*Calculated values of moisture content and degree of saturation at the end of each stage

*Calculated permeability values, and measured values if direct measurements were made.

20.2.2 Isotropic Consolidation Test with Horizontal Drainage

PRINCIPLES

A triaxial consolidation test can be carried out with drainage horizontally (radially) to a porous boundary. The specimen is set up as shown in Fig. 20.7, with a layer of thin porous material between it and the rubber membrane, and an impervious membrane separating the top surface from the top drainage disc. The circumferential porous material must make contact with the porous disc. The bottom drainage disc is omitted. Pore pressure is measured at the centre of the base by means of a porous ceramic insert about 10 mm diameter. All external connections are as shown in Fig. 20.3.

Normal filter paper drains are inadequate as a circumferential drain for this type of test (Rowe, 1959). A sheet of Vyon porous plastic material, 1.5 mm thick, is suitable, slotted to

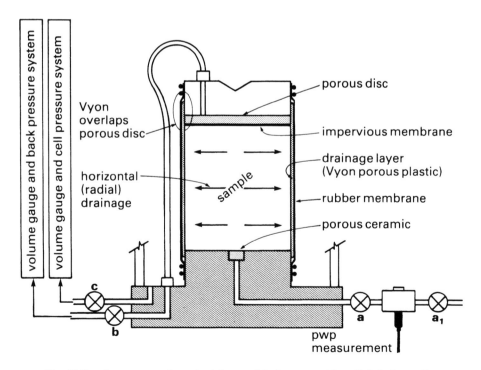

Fig. 20.7 *Arrangement for triaxial consolidation test with radial drainage (base port to valve d not shown)*

allow deformation as the specimen consolidates and to minimise the effects of restraint. The porous material should be saturated and de-aired by boiling, and care must be taken when fitting to avoid entrapping air.

PROCEDURE

The test procedure, which is not included in BS 1377, is similar to that outlined in Section 20.2.1, and similar graphical plots are made. Calculations are also similar except for the determination of the coefficient of consolidation (denoted by c_h), for which two methods of calculation are outlined below.

(1) *Square-root time method*

This method makes use of Equation (15.29), as in Section 18.7.2. But here, with drainage to a radial boundary only, the value of λ is 64 (from Table 15.4). The value of c_h is therefore calculated from the equation

$$c_h = \frac{\pi D^2}{64 t_{100}} \tag{20.11}$$

where D is the specimen diameter. t_{100} is obtained from a square-root time/volume change curve for the pressure increment, in the same way as described in Section 18.7.2.

The value of c_h is independent of the sample height. In engineering units, from Equation (15.30) (Section 15.5.5), c_h is equal to

$$\frac{1.652 D^2}{64 t_{100}}$$

i.e.

$$c_h = \frac{0.026 D^2}{t_{100}} \ \mathrm{m^2/year} \tag{20.12}$$

For a specimen initially 100 mm diameter, this becomes

$$c_h = \frac{260}{t_{100}} \ \mathrm{m^2/year} \tag{20.13}$$

(2) *Log time method*

The factor given in Section 22.2.3, Table 22.3, for consolidation in the Rowe cell with radial-outward drainge is used.

The value of t_{50} is read off from the pore pressure dissipation/log time curve, and is used in the equation

$$c_h = 0.131 \times 0.173 \times \frac{D^2}{t_{50}} \tag{Table 22.3}$$

$$= \frac{0.023 D^2}{t_{50}} \ \mathrm{m^2/year} \tag{20.14}$$

In Equations (20.12) and (20.14), D represents the geometric mean of the diameters at the beginning and end of the consolidation stage, i.e. $D^2 = D_1 D_2$. The diameter D_2 at the end of each increment is calculated from Equation (20.1) with diameters substituted for heights.

For a specimen initially 100 mm diameter, Equation (20.14) becomes

$$c_h = \frac{230}{t_{50}} \text{ m}^2/\text{year} \tag{20.15}$$

which is practically the same as Equation (20.13).

REPORTING RESULTS

Test results are reported in a manner similar to that for an isotropic consolidation test with vertical drainage (Section 20.2.1), except that reference to a BS procedure does not apply. The initial statement should make it clear that the drainage is horizontal (radial), and details of the porous drainage material should be included.

20.3 ANISOTROPIC CONSOLIDATION

20.3.1 Consolidation test with σ_v greater than σ_h

PRINCIPLES

Consolidation tests under anisotropic stress conditions, in which the vertical stress (σ_v) is not equal to the horizontal stress (σ_h) provided by the cell pressure, are not included in BS 1377 : 1990. However, it is possible to carry out anisotropic consolidation tests for the condition in which the axial stress exceeds the horizontal stress ($\sigma_v > \sigma_h$) in an ordinary triaxial cell with no special additional equipment.

In the test outlined below an axial stress is applied to the specimen, in addition to the cell pressure, so as to maintain the ratio of the horizontal and vertical total principal stresses (σ_h/σ_v) at a constant value which is less than unity.

APPARATUS AND PREPARATION

A normal triaxial cell fitted with connections to a drainage line and a pore pressure measuring system is used. Volume-change indicators are incorporated in the back pressure and cell pressure lines. Axial load may be applied either by a dead-weight hanger, or by a triaxial load frame with force measuring device. Several increments of both vertical and horizontal stresses may be applied, maintaining the same ratio of σ_h/σ_v.

The cell and specimen are prepared and set up in the same way as for a CD triaxial test (Sections 18.3 and 18.4). If a dead-weight hanger is to be used, a dial gauge for measuring axial deformation is clamped to the upper end of the piston as shown in Fig. 16.16, Section 16.4.2.

The specimen may first be saturated by increments of back pressure, or the desired initial confining pressure can be applied directly to the specimen, in one stage or several increments, and the B value measured.

With the valve on the drainage line closed, the axial force is increased to give the required principal stress ratio σ_h/σ_v, denoted by K which is less than 1. The force is calculated as indicated below.

DEAD-WEIGHT LOADING

The following symbols are used:

Consolidated specimen length	L_c	(mm)
Consolidated specimen area	A_c	(mm^2)
Area of piston	a	(mm^2)
Force exerted by load ring	P	(N)
Axial strain	ε	(%) (−)
Corresponding area of cross-section of specimen	A	(mm^2)
Cell pressure	σ_h	(kPa)
Volume change (drained test only) (water out of sample +)	ΔV	(cm^3)
Downward force on top cap	F	(N)
Force due to effective mass of piston and top cap	Q	(N)

The forces P, F and Q are calculated from the load dial reading and calibration constant.

Mass of dead-load hanger	m_h	(g)
Mass of weights applied to hanger	m	(g)

The mass of the top cap and piston can usually be neglected, but if they are significant an additional compressive (+) stress equal to

$$\frac{(m_p - m_w) \times 9.81}{1000} \text{ kPa}$$

should be added, where m_p is the mass of top cap and piston (g) and m_w is the volume (cm^3) or mass (g) of water displaced by the top cap and submerged part of the piston.

The forces acting on the specimen are shown diagrammatically in Fig. 20.8 (a). The net downward force F applied to the specimen top cap is given by the equation

$$F = \left[\frac{m_h + m + (m_p - m_w)}{1000} \right] \times 9.81 - \frac{\sigma_h a}{1000} \text{ N}$$

It is assumed below that piston friction is counteracted by the effective mass of the piston and top cap $(m_p - m_w)$.

The axial stress σ_v is equal to

$$\left(\frac{F}{A \times 1000} \right) + \sigma_h \text{ kPa}$$

i.e.
$$\sigma_v = \frac{9.81}{A}(m_h + m) - \sigma_h \frac{a}{A} + \sigma_h \text{ kPa} \tag{20.16}$$

If $\sigma_h/\sigma_v = \text{constant} = K$, i.e. $\sigma_v = \sigma_h/K$, then

$$\sigma_h \left(\frac{1}{K} - 1 + \frac{a}{A} \right) = \frac{9.81}{A}(m_h + m)$$

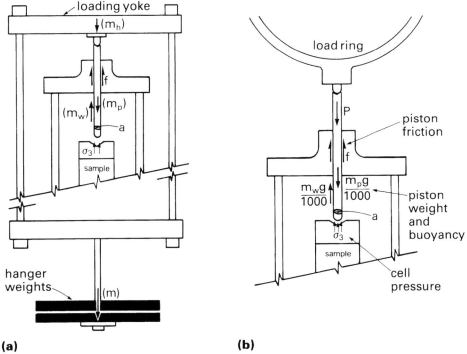

Fig. 20.8   Anisotropic consolidation tests in triaxial cell, illustrating forces acting on the specimen: (a) under dead weight loading, (b) using a load ring (σ_h is equal to the cell pressure σ_3)

Hence

$$m = \frac{A}{9.81} \; \sigma_h \left(\frac{1}{K} - 1 + \frac{a}{A} \right) - m_h \quad \text{kPa} \qquad (20.17)$$

If piston friction is significant, a constant mass to overcome this should be added.

When the pore pressure reading and undrained deformations have reached equilibrium, consolidation is started by opening the drainage valve and starting the clock. Readings of volume change and pore pressure are taken at intervals as in an isotropic consolidation test, with the addition of readings of axial deformation from which axial strain ($\varepsilon\%$) is calculated. The area of cross-section of the specimen (A) at the end of a consolidation stage is calculated from the equation

$$A = A_0 \left[\frac{1 - \dfrac{\Delta V}{V_0}}{1 - \dfrac{\varepsilon\%}{100}} \right] \qquad (17.6)$$

where V_0 is the initial specimen volume. This corrected area is used in Equation (20.17) for calculating the hanger load needed for the next stage of consolidation, after selecting the next cell pressure to be applied. From several stages the voids ratio/log pressure can be plotted, as in the isotropic test, where the pressure is the effective major principal stress.

After completing all consolidation stages, further loading up to failure may be carried out by increasing the weights on the hanger, i.e. under controlled stress conditions. Alternatively the cell may be transferred to a loading frame for loading to failure at constant rate of strain. In the meantime the axial loading as well as the cell pressure must be maintained. A direct measurement of vertical permeability may also be made if required.

LOAD RING LOADING

When using a load frame fitted with an external load measuring device, the force, or load dial reading, needed to give the required principal stress ratio ($K = \sigma_h/\sigma_v$) is determined as described below. The symbols are the same as those used above, with the addition of the following:

Force exerted by load ring: P newtons

Frictional force in cell bushing opposing downward movement of piston: f newtons

These forces are indicated in Fig. 20.8 (b).

The net downward force F on the specimen top cap is given by the equation

$$F = P + \frac{(m_p - m_w) \times 9.81}{1000} - \frac{\sigma_h a}{1000} - f \ \text{N} \qquad (20.18)$$

The force due to the effective mass of piston and top cap is represented by Q, where

$$Q = \frac{m_p - m_w}{1000} \times 9.81 \ \text{N}$$

Axial stress
$$\sigma_v = \left(\frac{F}{A} \times 1000 \right) + \sigma_h$$

i.e.
$$\sigma_v = \frac{1000}{A} (P + Q) - \sigma_h \frac{a}{A} - \frac{1000 f}{A} + \sigma_h \ \text{kPa} \qquad (20.19)$$

Putting $\sigma_v = \sigma_h / K$

$$\sigma_h \left(\frac{1}{K} - 1 + \frac{a}{A} \right) = \frac{1000}{A} (P + Q - f) \ \text{kPa}$$

Hence
$$P = \frac{A}{1000} \sigma_h \left(\frac{1}{K} - 1 + \frac{a}{A} \right) - Q + f \ \text{N} \qquad (20.20)$$

If the effective mass of the piston and top cap counteracts the piston friction (i.e. $Q = f$), then

$$P = \frac{A}{1000} \sigma_h \left(\frac{1}{K} - 1 + \frac{a}{A} \right) \ \text{N} \qquad (20.21)$$

If a load ring reading in dial divisions is used, and the mean calibration is C_r N/div, the dial reading required is equal to P/C_r divisions. The reading due to the upward force on the piston from the cell pressure cannot be eliminated by adjusting the dial gauge if the cell pressure is to be increased in subsequent stages of the test.

The axial force is set to the required value and the pore pressure and undrained deformations are allowed to reach equilibrium. Consolidation is started by opening the drainage valve and starting the clock. Readings are taken as described above. The axial force must be kept constant by winding up the machine platen to compensate for axial deformation of the specimen.

Further stages of consolidation may be applied by selecting suitable cell pressures and calculating the corresponding axial forces to maintain the same stress ratio by using Equation (20.20) or (20.21).

20.3.2   Other Anisotropic Conditions

Other types of anisotropic tests, such as those in which σ_h is greater than σ_v, and the K_0 condition of no lateral yield ($K_0 = \sigma_h{}'/\sigma_v{}'$), are beyond the scope of this book.

20.4   MEASUREMENT OF PERMEABILITY

20.4.1   Triaxial Permeability Test with Two Back Pressure Systems

PRINCIPLE

In this test the specimen in the triaxial apparatus is subjected to known conditions of effective stress under the application of a back pressure, and a constant hydraulic gradient is applied across it, i.e. it is a constant head test. The coefficient of permeability, k, is determined by measuring the volume of water passing through the specimen (usually downwards) in a known time.

The procedure is suitable for soils of low and intermediate permeability, i.e. clays and silts. It is covered in Clause 6 of BS 1377 : Part 6 : 1990, and in ASTM D 5084.

TEST CONDITIONS

Before a test is started, the following test conditions need to be specified:

(1) Size of test specimen

(2) Direction of flow of water

(3) Effective stress at which permeability measurement is to be carried out

(4) Method of saturation, including cell pressure increments and differential pressure if appropriate; or whether saturation should be omitted

(5) Whether voids ratios are to be calculated.

PRE-TEST CHECKS

The checking procedures for the apparatus are the same as described in Sections 18.3.2–18.3.6. In this case there are two back pressure systems to be flushed and checked.

The drainage lines in the back pressure systems should be calibrated for head loss in the pipelines due to flow of water, as described in Section 17.3.7. The calibration is applied as in Section 17.4.6

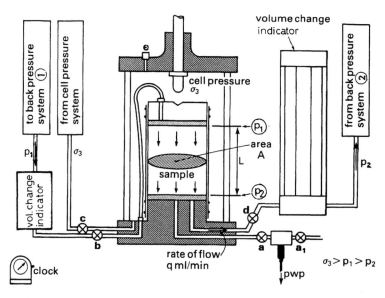

Fig. 20.9   Arrangement of apparatus for triaxial permeability test using two back pressure systems

ENVIRONMENT

The environmental requirements for these tests are the same as those given in Section 18.5.3 for triaxial compression tests.

APPARATUS

The apparatus is the same as described in Section 20.2.1, with the addition of a second constant pressure system to supply a second back pressure. One constant pressure system is connected to the base of the specimen, one to the top, in addition to the system used for applying the cell confining pressure, as shown in Fig. 20.9. The cell pressure must always be greater than the other pressures.

An installation designed for multiple tests in permeability cells is shown in Fig. 16.7 (Section 16.2.4.).

If the cell is fitted with two outlets from the base pedestal, one outlet is connected to the pore pressure system via valve a as usual, and the other is connected to one of the back pressure systems via valve d, as shown in Fig. 20.9. The port between valve d and the base pedestal should be flushed, de-aired and filled with de-aired water in the same way as the pore pressure connection. The procedure in ASTM D 5084 recommends that two drainage lines should be connected to the top cap as well as to the base pedestal, to facilitate removal of air and saturation of the specimen.

If there is only one outlet from the base pedestal, and a pore pressure transducer is used, the additional pressure system is connected to valve a_1 as shown in Fig. 20.10.

Wherever possible a volume-change indicator should be incorporated in both the inflow and the outflow pressure systems. When the rates of flow observed through the two gauges are practically equal, steady state flow condition is confirmed. When only one

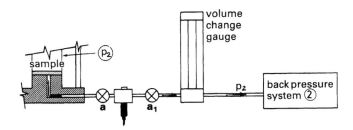

Fig. 20.10   *Connections to triaxial cell with one outlet from base pedestal for permeability test using pore pressure transducer*

volume-change indicator is available, it should be fitted in the pressure system connected to the sample inlet, so that water entering the gauge will be freshly de-aired water from the source of pressure. If fitted on the specimen outlet line, any bubbles of gas remaining in the specimen would find their way into the volume-change indicator, giving uncertain readings and necessitating tedious de-airing of the burettes afterwards.

The scale intervals on the burette of the volume-change indicator should provide the required discrimination. Where pressure differences are very small, it might be necessary to take account of the pressure variations which occur due to movement of the paraffin/water interface, because paraffin has a lower density than water. (See Section 16.5.6).

A recent alternative to conventional volume-change indicators is the volume-change device described by Araruna, Harwood and Clarke (1995), used in conjunction with a flow pump (Fig. 16.32, Section 16.6.6). Further details of the arrangement and procedure for permeability tests are given by Araruna, Clarke and Harwood (1995). They illustrate the use of two of their volume-change units, one on the outflow back pressure line to measure the rate of flow through the specimen, and one on the cell pressure line for determination of changes in specimen volume.

PROCEDURE

(1) Flush and de-air the pore pressure and back pressure systems, and their connections to the specimen, as described in Section 18.3, and then close all valves.

(2) Prepare and measure the specimen, and set it up as described in Section 18.4 between two saturated coarse porous discs.

(3) Saturate the specimen by one of the methods given in Section 18.6.1 to ensure that virtually all the air in the voids is driven into solution.

(4) Consolidate the specimen using top drainage only to the required effective stress (Section 20.2.1) until the pore pressure measured at the base equalises with the back pressure, or until outward drainage of water has ceased. Close valves b and d (Fig. 20.9).

(5) Adjust the pressure (p_2) in the base pressure system to a value equal to the back pressure (p_1) at the top of the specimen, and open valve d (Fig. 20.9) or valve a_1 (Fig. 20.10).

(6) Increase the pressure p_1 in the top pressure system to a value less than the cell pressure σ_3, such that the pressure difference ($p_1 - p_2$) gives the desired hydraulic gradient across the specimen for the permeability test. The pressure difference depends on the hydraulic

gradient that is needed to give a reasonable rate of flow through the specimen. A gradient of 20, sometimes more, might be needed to initiate any flow at all in a clay soil. When a high gradient is to be applied the pressure should be increased carefully after opening valve b (step (8)). The rate of flow should be observed during this operation. Piping or erosion within the specimen must be avoided.

Wherever possible the hydraulic gradient used in the test should be compatible with that likely to occur in the field. However, for soils of low permeability it is usually necessary to apply substantially greater hydraulic gradients in the laboratory so that testing times will not be unduly prolonged. Maximum hydraulic gradients recommended in ASTM D 5084 are as follows:

Soil permeability (m/s)	Maximum hydraulic gradient
10^{-5} to 10^{-6}	2
10^{-6} to 10^{-7}	5
10^{-7} to 10^{-8}	10
10^{-8} to 10^{-9}	20
less than 10^{-9}	30

Excessively large hydraulic gradients should be avoided with soft compressible soils, otherwise consolidation due to seepage pressures can reduce the permeability. The specimen will then decrease in volume and more water will flow out than flow in.

(7) Record the readings of the volume-change indicators on the back pressure lines when they are steady.

(8) Open valve b (Fig. 20.9) to admit the pressure to the specimen, and start the timer. When a steady condition is achieved (step (10)) the average effective stress in the specimen is approximately equal to $\sigma'_1 - (p_1 + p_2)/2$.

(9) Record the reading of both volume-change indicators at half-minute intervals, or at regular intervals appropriate to the rate of flow.

(10) As the test proceeds, calculate the cumulative flow of water, Q (ml), for every reading of each volume-change indicator. Plot each value of Q against elapsed time. Continue the test until the two plots are linear and parallel, which indicates that a steady state has been reached.

(11) If the volume-change indicator nears the end of its travel so that the direction of flow in the burettes must be reversed, this should be done by quickly operating the reversing valve or valves at the same instant as the burette reading is observed and recorded. This reading provides the new datum for subsequent readings, to which the cumulative flow up to the time of reversal must be added.

(12) If a larger differential pressure is desired, p_2 can be reduced (but not enough to allow air bubbles to form) and p_1 increased by the same amount, to maintain the same average effective stress. Some time may be needed for equilibrium to be re-established.

If the flow is interrupted by closing either or both of the inlet and outlet valves b, d or a, some redistribution of pore pressure within the specimen may take place and the steady state condition may not be resumed until some time after restarting the flow.

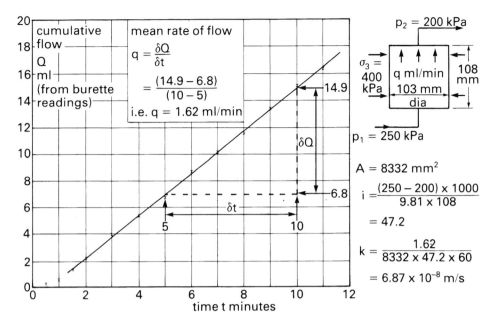

Fig. 20.11   Graphical data from triaxial permeability test, with calculations of rate of flow, hydraulic gradient and permeability

(13) Record the temperature in the vicinity of the triaxial cell to nearest 0.5 °C.

(14) When sufficient readings have been taken and plotted to confirm that a steady state has been achieved, stop the test by closing valves b and d.

(15) If an additional test at a lower effective stress is required, repeat steps (5) to (14) with the pressures p_1 and p_2 increased to the appropriate values.

(16) If an additional test at a higher effective stress is required, consolidate the specimen as in step (4) by applying the appropriate pressures, and repeat steps (5) to (14).

CALCULATIONS

(1) Calculate the area of cross-section of the specimen A (mm^2).

(2) From the linear part of the graph plotted in step (10) above, measure the slope to calculate the rate of flow q ml/minute; i.e. $q = \delta Q/\delta t$ ml/minute.

A typical graph and flow rate calculation are shown in Fig. 20.11.

(3) If the rate of flow is relatively small, the effect of head losses in the pipelines and connections can be neglected and the pressure difference across the specimen is equal to $(p_1 - p_2)$. But if the head loss calibration (described in Chapter 17, Section 17.3.7) indicates that pipeline losses are a significant portion of the pressure difference, these losses must be deducted from the measured pressure difference (Section 17.4.6). The total pipeline loss p_c observed during the test is read off from the calibration curve (of the type shown in Fig. 17.21 (b)), and Δp is equal to $(p_1 - p_2) - p_c$.

(4) The permeability k_v (m/s) of the specimen in the vertical direction is calculated by using the equation in Section 10.3.2 of Volume 2:

$$k_v = \frac{q}{60Ai} \quad \text{m/s} \tag{10.5}$$

where A is the area of cross section of specimen (mm^2), i is the hydraulic gradient across the specimen and q is the rate of flow (ml/minute).

A pressure difference of 1 kPa is equivalent to a head of water $1/9.81$ m, i.e. 102 mm. The mean hydraulic gradient is the difference in head per unit length, i.e.

$$i = \frac{102}{L} \times \Delta p$$

where L is the length of the specimen (mm). Substituting in Equation (10.5),

$$k_v = \frac{qL}{60A \times 102\,\Delta p} \tag{20.22}$$

i.e.

$$k_v = \frac{1.63qL}{A\Delta p} \times 10^{-4} \quad \text{m/s}$$

where $\Delta p = (p_1 - p_2) - p_c$.

If the length L of the specimen is about 100 mm, for any diameter, $i = \Delta p$ approximately, i.e. the mean hydraulic gradient is numerically about equal to the pressure difference in kPa. Equation (20.22) then becomes

$$k_v = \frac{q}{60A\,\Delta p} \tag{20.23}$$

For a specimen 100 mm long and 100 mm diameter,

$$k_v = \frac{q}{60 \times 7854 \times \Delta p} = \frac{2.08q}{\Delta p} \times 10^{-6}\,\text{m/s} \tag{20.24}$$

The derived permeability relates to the mean effective stress in the specimen at which the test was carried out.

If necessary the calculated coefficient of permeability, k_v, is corrected to the corresponding value at 20 °C by multiplying the result obtained as above by the temperature correction factor R_t for water viscosity, which is derived from Fig. 14.18 of Volume 2 (second edition).

TEST RESULTS

The test report should include the following:

Statement (if appropriate) that the test was carried out in accordance with Clause 6 of BS 1377 : Part 6 : 1990, under constant head conditions in a triaxial cell

Sample identification, type, condition, soil description, and method of preparation of the test specimen (all as listed in Section 20.2.1)

Whether the test specimen was undisturbed, or remoulded

Initial details of test specimen:

dimensions

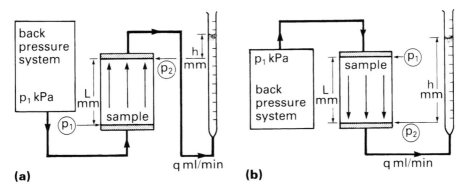

Fig. 20.12   *Arrangement for triaxial permeability test using one back pressure system: (a) upward flow, (b) downward flow*

moisture content

density and dry density

Method of saturation, and relevant details

Value of the pore pressure coefficient, B, achieved

Appropriate data from the consolidation stage, as detailed in Section 20.2.1

Final density and overall moisture content

Coefficient of vertical permeability, k_v (m/s), at 20 °C, to two significant figures

Mean effective stress during the test

Pressure difference across the specimen, and hydraulic gradient, during the test.

20.4.2   Triaxial Permeability Test with One Back Pressure System

PRINCIPLE

This procedure, which is not covered in BS 1377, requires only one back pressure system. If run as a constant head test the principle is similar to that described in Section 20.3.1. A modified procedure applicable when applied pressures are small makes use of the falling head principle.

APPARATUS

Permeability can be measured with only one constant pressure system connected to the specimen (in addition to the system providing cell pressure) if the outlet from the specimen is connected to an open burette as shown in Fig. 20.12. If the permeability is high enough to permit displacement of air from the voids in the sample by upward movement of water, the back pressure system should be connected to the base, allowing drainage from the top (Fig. 20.12(a)). If saturation is first achieved by application of back pressure increments, the direction of flow is immaterial and the burette can be connected to valve a_1 in Fig. 20.10 after disconnecting the pore pressure panel. Flow is then downwards as indicated in Fig. 20.12 (b). The outlet pressure can be raised slightly above atmospheric by elevating the burette so that the water level is higher than the exit point from the specimen. For every metre increase in height the pressure is increased by 9.81 kPa.

When hydraulic gradients exceeding unity are to be applied, upward flow can lead to instability and piping especially in non-cohesive soils. Downward flow gives stable conditions and is generally to be preferred.

Bubbles of air or gas are likely to emerge from the specimen at these low pressures. If the bubbles are allowed to emerge through the burette the water level readings would be affected. Air can be removed by fitting an air trap, filled with water initially, as indicated in Fig. 20.4. Tubing leading to it should slope upwards so as not to form another air trap. The burette readings then measure the total volume (air + water) emerging from the specimen, and this should be equal to the volume of water entering.

PROCEDURE

(1) Small rate of flow

If the rate of flow is small, the test is carried out in a manner similar to that described in Section 20.3.1, except that flow measurements are obtained by reading the burette. A volume-change indicator could be included in the inlet pressure line for comparison. The condition of constant head is applicable if the burette is progressively lowered so that the level of the water it contains is maintained constant at the initial level. The connecting tube should be long enough to allow for this movement.

Notation is the same as that used in Section 20.3.1. The hydraulic gradient i across the specimen (without allowing for pipeline losses) is given by the equation

$$i = \frac{102p_1 - h}{L} \tag{20.25}$$

If the pressure head due to the height of water in the burette is small (say less than 5% of p_i), Equation (20.25) approximates to

$$i = \frac{102p_1}{L}$$

The rate of flow q ml/minute is obtained from a graph of burette readings over a period of time, as described in Section 20.4.1. The coefficient of permeability k_v m/s is calculated from Equation (20.22) if the above approximate is valid, i.e.

$$k_v = \frac{qL}{60A \times 102p_1} = \frac{qL}{6120Ap_1} \tag{20.26}$$

(2) Large rate of flow

If the rate of flow is relatively large, such that during the test period the total flow exceeds the capacity of the burette, an overflow arrangement similar to that shown in Volume 2, Fig. 10.23, should be provided. The outlet water level then remains constant, and the flow is measured by collecting the water in a measuring cylinder. Cumulative flow should be recorded so that the graphical method for determining rate of flow under steady conditions (Fig. 20.11) can be applied. A correction for pipeline losses may be necessary.

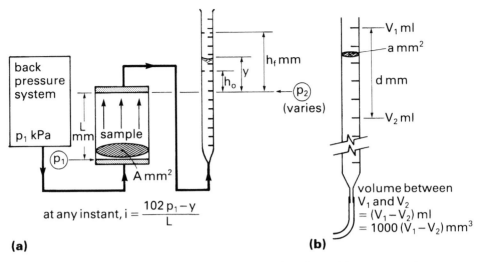

Fig. 20.13   *Triaxial permeability test under falling head condition: (a) pressures and hydraulic gradient, (b) measurements for determining cross-sectional area of burette*

(3) *Falling head test*

If the inlet pressure is not much greater than the outlet pressure under the head of water in the burette, and a constant level overflow is not used, the variation in outlet pressure as the water level in the burette rises might be significant (Fig. 20.13 (a)). The conditions are then those of a falling head test (Volume 2, Section 10.7), and the calculation of permeability is based on Equation (10.15) of Section 10.3.6, modified as follows.

$$k = 3.84 \left[\frac{aL}{At} \log_{10} \left(\frac{102p_1 - h_0}{102p_1 - h_f} \right) \right] \times 10^{-5} \ \ \text{m/s} \qquad (20.27)$$

In this equation,

a = area of cross-section of burette (mm^2) (Fig. 20.13 (b))

$$= \frac{V_1 - V_2}{d} \times 1000 \ \text{mm}^2$$

h_0 = height of water level in burette above outlet end of specimen initially (mm)

h_f = corresponding height after time t (mm)

t = duration in minutes

L, A, and p_1 are as in procedure (1) above.

Equation (20.27) need be used only if the outlet pressure due to the burette water level is more than about 10% of the inlet pressure, i.e. if

$$\frac{9.81h}{1000} > 0.1p_1 \ \ \text{kPa}$$

i.e.

$$h \ \text{mm} > 10p_1 \ \ \text{kPa}$$

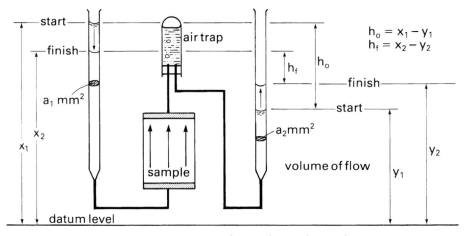

Volume of flow in time t = Q ml = $\dfrac{a_1(x_1-x_2)}{1000} = \dfrac{a_2(y_2-y_1)}{1000}$

Change in head in time t = $(h_o - h_f)$ mm

Equivalent area of 'standpipe' = $\dfrac{1000\,Q}{(h_o - h_f)}$ mm^2

Fig. 20.14   Outline of apparatus for triaxial permeability test using two burettes,
with representative calculations

For example, if the inlet pressure p_1 is 100 kPa, the height h could be up to 1 m before this refinement need be applied, to an accuracy within 10%.

20.4.3   Triaxial Permeability Test with Two Burettes

PRINCIPLE AND APPARATUS

A simple form of triaxial permeability test can be carried out on soils of intermediate permeability by using two burettes, as shown in Fig. 20.14. An air trap should be included on the outlet line if the specimen is not fully saturated. Careful application of a partial vacuum to the outlet (upper) end of the specimen might be necessary initially to remove bubbles which could otherwise collect there and impede the flow.

FALLING HEAD METHOD

The difference in level between the water in the two burettes is measured initially (h_0 mm), and again after flow has taken place under steady conditions after a known time t minutes ($h_f$mm). Since this is a falling head test the permeability is calculated from Equation (10.15) of Volume 2, Section 10.3.6, modified as follows

$$k = 3.84\left[\frac{1000Q}{h_0 - h_f}\frac{L}{At}\log_{10}\left(\frac{h_0}{h_f}\right)\right] \times 10^{-5} \;\; \text{m/s} \qquad (20.28)$$

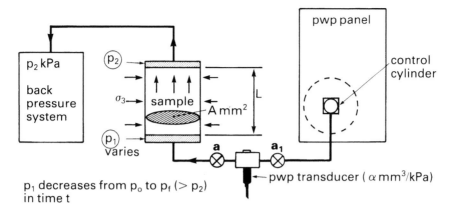

Fig. 20.15   *Outline of apparatus for measurement of very small permeabilities using a pressure transducer. (Pore pressure line must be completely free of air and leaks)*

where Q is the volume of water (ml) measured in either burette, flowing through the specimen in time t minutes; h_0, h_f are as defined (mm); and L, A and t are as before.

The area of the standpipe tube, a, in Equation (10.15) is equivalent to $1000\,Q/(h_0 - h_f)\,\text{mm}^2$.

CONSTANT HEAD METHOD

If the burettes are connected to the triaxial cell by long lengths of flexible tubing, both water levels can be maintained constant by raising the inlet burette and lowering the outlet burette. The test is then carried out under a constant head of h_0 mm, and the hydraulic gradient is h_0/L_0.

If the average rate of flow is q ml/minute, the permeability is calculated from Equation (10.5) (Volume 2, Section 10.3.2) putting $i = h_0/L$,

$$k = \frac{qL}{60Ah_0}\ \ \text{m/s} \tag{20.29}$$

20.4.4 Measurement of Very Small Rates of Flow

The following method is based on a procedure described by Remy (1973), and can be used for measuring very low permeabilities when the rate of flow of water is too small to measure satisfactorily by the usual means. It makes use of the volume change characteristics of a pressure transducer, or a mercury null indicator.

APPARATUS

The specimen is set up in a triaxial cell with a pore pressure transducer connected to the base pedestal, as shown in Fig. 20.15. The pore pressure system must be properly de-aired and all valves, couplings and tubing must be checked to ensure that they are completely free of leaks. No perceptible loss of pressure should be indicated by the transducer over several hours after pressurising and closing valves a and a_1.

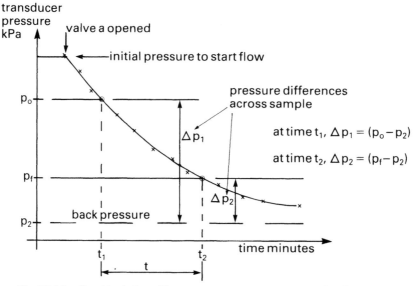

Fig. 20.16   *Graphical plot of base pore pressure against time for determination of very small permeability*

PROCEDURE

The specimen is first saturated in the usual way by the application of back pressure to the top end, until a *B* value of near unity is obtained, followed by consolidation to the required effective stress until equilibrium is established. The pressure at the base of the specimen is then increased above the back pressure value by using the control cylinder to give a suitable pressure gradient across the specimen. An increase of 100 kPa or more might be necessary. Valve a_1 is closed, and from that moment readings of the pore pressure transducer are taken at regular intervals of time. A graph is drawn of pressure readings against time, of the form indicated in Fig. 20.16.

CALCULATIONS

During a time interval *t* (minutes), the pressure at the base falls from p_0 to p_f (kPa). The volume of water flowing into the specimen can be calculated from the deformation properties of the pressure transducer, expressed as mm^3 per kPa, and denoted by $\alpha \, mm^3/$ kPa. This is equivalent to a standpipe in a falling-head permeameter having an area of cross-section of $a \, mm^2$, where $a = \alpha/102 \, mm^2$. The permeability of the specimen can be calculated from Equation (10.15) of Section 10.3.6 for a falling head test, as follows.

$$k = \left[3.84 \, \frac{\alpha}{102} \, \frac{L}{At} \, \log_{10} \left(\frac{p_0 - p_2}{p_f - p_2} \right) \right] \times 10^{-5} \text{ m/s} \qquad (20.30)$$

As an example, if for a 'soft' transducer the value of α is of the order of $10^{-2} \, mm^3$/kPa, the above equation becomes approximately

$$k = \left[3.8 \, \frac{L}{At} \, \log_{10} \left(\frac{p_0 - p_2}{p_f - p_2} \right) \right] \times 10^{-9} \text{ m/s} \qquad (20.31)$$

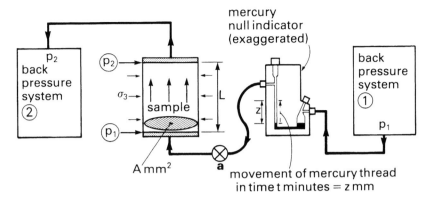

Fig. 20.17 *Measurement of very small flow using a mecury null-indicator*

USE OF MERCURY NULL INDICATOR

An alternative method when a pressure transducer is not available is to measure the movement of the mercury thread in a mercury null indicator to determine the volume of flow under the application of a constant applied pressure (Fig. 20.17). Assuming a mercury thread of 1.5 mm diameter, a movement of z mm gives a volume displacement of $1.767z/1000$ cm^3 ($=$ ml). If the constant head condition can be assumed, the permeability is calculated from the equation

$$k = \frac{1.767z}{1000} \times \frac{L}{60A \times 102(p_1 - p_2)t} \quad \text{m/s}$$

i.e.

$$k = \frac{2.89zL}{A(p_1 - p_2)t} \times 10^{-7} \quad \text{m/s} \tag{20.32}$$

The accuracy of these very small volume changes depends on their being absolutely no leaks anywhere in the pipelines, valves and connections between the measuring device and the specimen.

CAPILLARY TUBE METHOD

An accurate method for measuring small rates of flow during a long period of time under a constant head was developed at Manchester University (Wilkinson, 1968). The inlet and outlet pressure systems connected to the specimen each include a one metre length of precision-bore thick walled glass tubing, 1.5 mm bore, mounted horizontally just above bench level (Fig. 20.18). After de-airing a single small bubble of air is introduced through a capillary tube which can be opened to atmosphere, with the aid of a screw control cylinder. A scale mounted alongside the tube enables the velocity of the bubble, and hence the rate of flow, to be measured as water flows through the specimen. Steady state conditions can be assumed when the velocities in the two tubes are practically equal. If the scales are marked in mm^3, or ml, the rate of flow (mm/minute) can be observed directly. One metre length represents 1.767 ml.

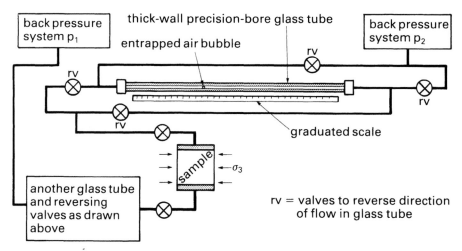

Fig. 20.18 *Apparatus for measuring small rates of flow in constant head permeability tests (capillary tube inlet for air bubble not shown)*

The valves and connecting pipework shown in Fig. 20.18 enable the direction of movement of the bubble to be reversed during a test if necessary. The insides of the tubes must be thoroughly cleaned to ensure that they are completely free from grease and other contaminants which might impede the flow.

REFERENCES

Araruna, J. T., Harwood, A. H. and Clarke, B. G. (1995). 'A practical, economical and precise volume change measurement device'. Technical Note, *Géotechnique*, 45:3:541–544.

Araruna, J. T., Clarke, B. G. and Harwood, A. H. (1995). 'Quick, accurate, consistent measurements of permeability of clays'. *International Conference on Advances in Site Investigation Practice*, ICE, London, March, Session VI, pp 1–12.

ASTM Designation D 5084, *Standard Test Method for Measurement of Hydraulic Conductivity of Porous Materials Using a Flexible Wall Permeameter*. American Society for Testing and Materials, Philadelphia, USA.

Bjerrum, L. and Huder, J. (1957). 'Measurement of the permeability of compacted clays'. *Proc. 4th Int. Conference of Soil Mechanics and Foundation Eng.*, London. Vol. 1, pp 6–8.

Cornforth, D. H. (1974). 'One-Dimensional consolidation curves of a medium sand'. Technical Note, *Géotechnique*, 24:4:678.

Davis, E. H. and Poulos, H. G. (1968). 'The use of elastic theory for settlement prediction under three-dimensional conditions'. *Géotechnique*, 18:1:67.

Lambe, T. W. and Whitman, R. V. (1979). *Soil Mechanics*, (S.I. version). Wiley, New York.

Lo, K. Y. (1960). Correspondence, *Géotechnique*, 10:1:36.

Rémy, J. P. (1973). 'The measurement of small permeabilities in the laboratory'. Technical Note, *Géotechnique*, 23:3:454.

Rowe, P. W. (1959). 'Measurement of the coefficient of consolidation of lacustrine clay'. *Géotechnique*, 9:3:107.

Schiffman, R. L., Chen, A. T. F. and Jordan, J. C. (1969). 'An analysis of consolidation theories'. *J. Soil Mechanics & Foundation Div. ASCE* Vol. 95, No. S M 1, pp 285–312.

Wilkinson, W. B. (1968). 'Permeability and consolidation measurements in clay soils'. PhD thesis, Manchester University.

Chapter 21

Stress paths in triaxial testing

21.1 INTRODUCTION

21.1.1 Scope

This chapter provides an introduction to the concept and use of stress paths as applied to the laboratory testing of soils. The 'stress path method' was described by Lambe (1967) as a systematic approach to stability and deformation problems in soil mechanics.

The relevance of stress path plots to routine laboratory triaxial tests is described, and leads to the presentation of test data by means of stress paths. The application of stress paths to geotechnical problems is beyond the scope of this volume.

21.1.2 Principles

STRESS CHANGES IN SOILS

During a laboratory test on a soil specimen, or as load is applied to a mass of soil in the ground by a foundation, each element of soil experiences changes in its state of stress. A stress path gives a continuous representation of the relationship between the components of stress at a given point as they change. Use of a stress path provides the geotechnical engineer with an easily recognisable pattern which assists him in identifying the mechanism of soil behaviour. It also provides a means of selecting and specifying the sequence of stresses to be applied to a specimen in a test for a particular purpose.

FACTORS AFFECTING SOIL BEHAVIOUR

Soils in general are not elastic materials and their behaviour in-situ depends on many factors (see Section 15.4.1). These include the magnitude of the imposed stress changes; the way in which they change; and the previous history of loading, whether due to natural causes (geomorphological) or to changes imposed by man (e.g. previous loading, excavation, alteration of ground water level). It is therefore desirable to trace the states of stress of an element of soil throughout its loading history, and a stress path provides a convenient and easily understood means of conveying that information.

ADVANTAGES OF STRESS PATHS

Use of the stress path method in the laboratory enables the field stress changes, past, present and future, to be modelled much more realistically than by using conventional test procedures alone. The conventional approach may be good enough in many instances, but for certain problems a closer approach to field conditions is necessary.

There are numerous ways in which stress paths can be plotted, some of which are outlined in Section 21.1.4. The names assigned to these types of plot may not be universally accepted but are used here for convenience.

21.1.3 Definitions

STRESS PATH A curve drawn through a series of points on a plot of stresses.

STRESS POINT A point on a plot of stresses representing the selected components of stress at any instant.

STRESS FIELD The two-dimensional surface used for the plot of stresses. (Three-dimensional 'stress space' is not used here.)

VECTOR CURVE or STRESS TRAJECTORY Alternative names for certain kinds of stress path.

EFFECTIVE STRESS PATH (ESP) A stress path plotted in terms of effective stresses.

TOTAL STRESS PATH (TSP) A stress path plotted in terms of total stresses.

21.1.4 Types of Plot

REPRESENTATION OF STATE OF STRESS

The state of stress in a soil element can be represented in many ways. It is completely defined by the three mutually perpendicular principal stresses σ_1, σ_2, and σ_3; the orientation of these stresses; and the pore pressure u (equal to the pore water pressure u_w in saturated soils). When representing the axially symmetrical conditions of the triaxial compression test the two horizontal principal stresses are equal, i.e. $\sigma_2 = \sigma_3$.

uuA familiar method of representing the state of stress at a given instant is by means of a Mohr circle. But this is not a satisfactory way of indicating changes of stress. Other methods use a single point (the stress point) to represent each circle, and the selection of the particular point, and of the axes of reference, influence the type of stress path plot. Several types are outlined below, after a review of the Mohr diagram.

MOHR CIRCLES

A Mohr circle represents the state of stress (shear stress and normal stress) on any plane within a soil specimen that is subjected to axisymmetrical stress, as explained in Volume 2, Section 13.3.4. Either total stresses or effective stresses can be represented, but for the following illustration only total stresses are considered.

The state of stress in a test specimen subjected only to an equal all-round cell pressure, σ_3, is represented on the Mohr diagram (Fig. 21.1) by the point A on the major principal stress axis, where $OA = \sigma_3$. During a compression test the vertical stress σ_1 is steadily increased and successive states of stress can be represented by Mohr circles of increasing size, b, c, d, e, f in Fig. 21.1. The limiting condition is represented by circle f, which just touches the failure envelope at F. Point F represents the state of stress on the theoretical surface of failure, on which the ratio of shear stress reaches its maximum attainable value.

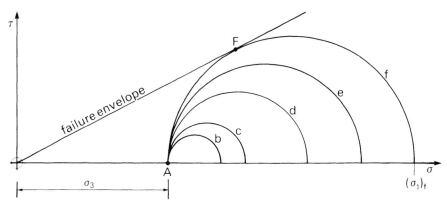

Fig. 21.1   Mohr circles representation of increasing total axial compressive stress in triaxial compression test

STRESS PATHS

A diagram like Fig. 21.1 becomes very confusing even if used for only one test specimen. The presentation is greatly clarified if circles are replaced by points, which is done in the various stress path methods of plotting outlined below.

(1) *Vector curve*

Probably the earliest reference to any form of stress path plotting was made by Taylor (1944), as quoted by Holtz (1947). Taylor devised a 'vector curve' connecting stress points representing the stress on a plane at $(45° + \phi/2)$ to the horizontal (Fig. 21.2 (a)). These points were obtained by drawing a tangent, inclined at an angle ϕ, to the stress circles,

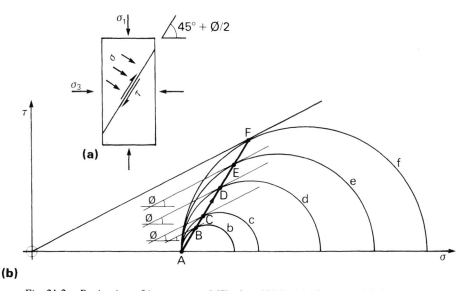

Fig. 21.2   Derivation of 'vector curve' (Taylor, 1994): (a) theoretical failure plane considered, (b) representation of stresses on that plane

touching them at points B, C, ..., F, as shown in Fig. 21.2(b). Alternatively the values of σ and τ can be calculated from each value of σ_1 and σ_3. Joining these points from the starting point A gave the 'vector curve' A B C ... F up to failure represented by the point F. This method of plotting was also described by Casagrande and Wilson (1953).

Disadvantages of this method of plotting are that it presupposes a value of ϕ for the soil, and the derivation of each point is rather cumbersome. This method is not in common use.

(2) MIT stress field

A method of plotting derived from Mohr circles is that developed by Prof. T. W. Lambe of the Massachusetts Institute of Technology (Lambe, 1964), and known as the MIT stress path plot. Other references to this method are by Lambe (1967), Simons and Menzies (1977), Lambe and Marr (1979) and Lambe and Whitman (1979).

The stress point used is that representing the maximum shear stress at any stage, i.e. the topmost point of each Mohr circle (point J in Fig. 21.3 (a)). The stress path is therefore the locus of points of maximum shear stress experienced by a soil element as the state of stress varies. This is illustrated in terms of total stress in Fig. 21.3 (b), in which the line A B C ... P is the stress path.

From Fig. 21.3 (a) it can be seen that the abscissa (horizontal axis) of the topmost point J of the Mohr circle is equal to $\frac{1}{2}(\sigma_1 + \sigma_3)$ and that the ordinate (vertical axis) is equal to the radius of the circle, i.e. $\frac{1}{2}(\sigma_1 - \sigma_3)$. Parameters s and t are therefore defined as

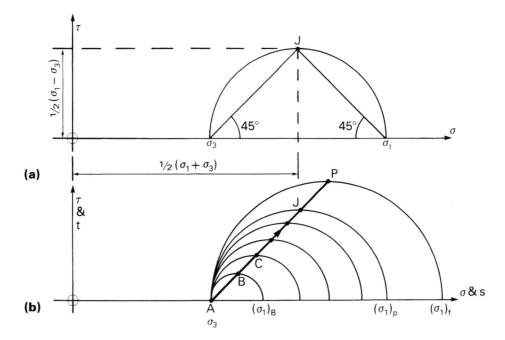

Fig. 21.3 Derivation of MIT stress field (Lambe, 1964): (a) definition of stress point, (b) stress path as locus of stress points

$$s = \frac{\sigma_1 + \sigma_3}{2} \tag{21.1}$$

$$t = \frac{\sigma_1 - \sigma_3}{2} \tag{21.2}$$

and these parameters are used in the MIT stress path plot.

In terms of effective stress, similar relationships apply and s' is substituted for s, i.e.

$$s' = \frac{\sigma_1' + \sigma_3'}{2} \tag{21.3}$$

The deviator stress is the same as for total stresses because the pore water pressure cancels out, and t is used for both total and effective stress path plots.

The symbols s and t used here are those defined by Atkinson and Bransby (1978). Unfortunately some authors use the symbols p and q in this sense, which can cause confusion with the Cambridge notation defined below.

(3) Cambridge stress field

Roscoe, Schofield and Wroth (1958) at the University of Cambridge, England, developed the use of the mean of the three principal effective stresses (σ_1, σ_2, σ_3) instead of the mean of the major and minor principal stresses (Fig. 21.4 (a)). This method of plotting is known as the Cambridge stress path plot, in which the parameter p' is defined in terms of effective stress by the equation

$$p' = \frac{\sigma_1' + \sigma_2' + \sigma_3'}{3} \tag{21.4}$$

The parameter q is defined as being equal to the deviator stress:

$$q = \sigma_1' - \sigma_3' = \sigma_1 - \sigma_3 \tag{21.5}$$

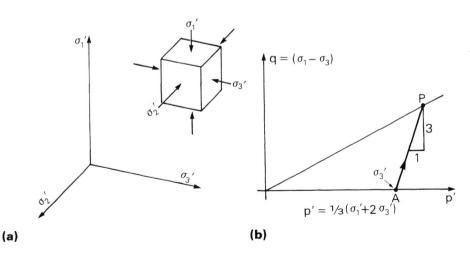

(a)   **(b)**

Fig. 21.4   *Cambridge stress field (Roscoe et al, 1958): (a) principal stresses,*
(b) stress path for increasing major principal stress

In the triaxial test two of the principal effective stresses are equal to the horizontal effective stress, and Equation (21.4) is expressed as

$$p' = \frac{\sigma_1' + 2\sigma_3'}{3} \tag{21.6}$$

The mean total stress is similarly defined as

$$p = \frac{\sigma_1 + 2\sigma_3}{3} \tag{21.7}$$

The stress path shown in Fig. 21.3 (b) is drawn in Fig. 21.4 (b) as the line A P, which rises at a slope of 3 on 1. The concept that the volumetric behaviour of soil is dependent upon a mean effective stress (written as $\frac{1}{3}(\sigma_1' + \sigma_2' + \sigma_3')$) is central to this type of stress path plot. The introduction of the intermediate principal effective stress leads to a more fundamental representation of the state of stress with regard to yield and the elastic behaviour of overconsolidated soils. By plotting volume changes, in terms of voids ratio e (or as $v = 1 + e$) in the plane perpendicular to the (p', q') surface, a three-dimensional space for representing stresses and deformations is obtained which has been extensively used for 'critical state' analysis at Cambridge and elsewhere. This concept is beyond the scope of this book.

(4) Axial symmetry

For the condition of axial symmetry, Henkel (1960) used the stress plane in which the stresses in a triaxial test lie, as shown in Fig. 21.5. The equal stresses σ_2 and σ_3 on the

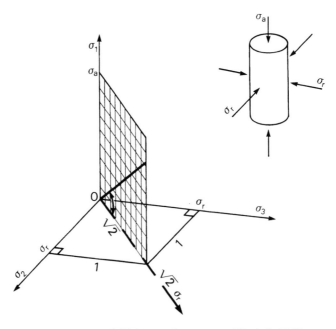

Fig. 21.5   Stress field for axial symmetry (Henkel, 1960)

horizontal axes were denoted by σ_r (the radial stress), and the vertical (axial) stress by σ_a. The combined representation of σ_2 and σ_3 in the stress plane is equal to $\sqrt{2}\sigma_r$, and the stress field consisted of a plot of σ_a (vertical axis) against $\sqrt{2}\sigma_r$ (horizontal axis). The resulting plot is a distortion of the principal effective stress plot referred to below, and is no longer used.

(5) *Principal effective stresses*

An alternative method used at Imperial College (Skempton and Sowa, 1963) is to plot σ_1' against σ_3', as shown in Fig. 21.6, or σ_1 against σ_3 for a total stress plot. This provides a convenient way of plotting data from tests in the Bishop–Wesley stress path cell (Bishop and Wesley, 1975), in which axial and radial stresses are applied directly, but the use of this apparatus lies beyond the scope of this book.

(6) *Deviator stress field*

This type of stress field provides a convenient plot for the $\bar{B}$ test (Section 21.4). Deviator stress $(\sigma_1 - \sigma_3)$ is plotted against confining pressures σ_3', as indicated in Fig. 21.7. (Bishop and Henkel, 1962, Part IV.6). A point P on the stress path is obtained by rotating the diameter of the Mohr circle through $90°$ about its left-hand end, as indicated.

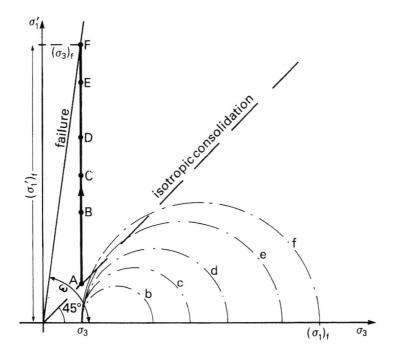

Fig. 21.6 Stress field using principal effective stresses (Skempton and Sowa, 1963)

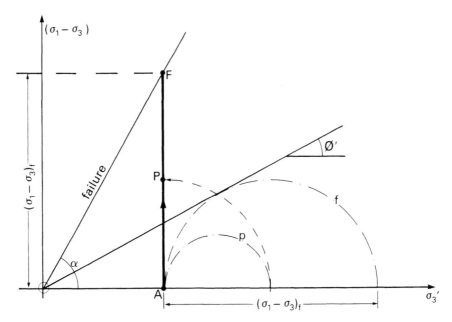

Fig. 21.7 *Stress field using deviator stress (Bishop and Henkel, 1962)*

21.1.5 Applications

METHODS ADOPTED

The stress path plots most often used are the MIT plot and the Cambridge plot, and both provide a two-dimensional representation of the three-dimensional stress space. The MIT plot is the more convenient to use for many laboratory tests, and is marginally preferable for the consideration of failure conditions. The Cambridge plot is more fundamentally representative and is advantageous when considering problems of deformation and yield. Most of the applications of stress paths referred to in this chapter make use of the MIT plot.

COMMENT

In view of the ease with which stress paths of the kinds referred to above can be plotted, and failure envelopes derived, it seems remarkable that the cumbersome method of plotting Mohr circles has persisted for so long. For an undrained test in which pore pressure is measured, a stress path gives a much clearer indication of behaviour than the combination of conventional plots of stress and pore pressure change against strain, to which Mohr circles add little. Even for drained and quick-undrained tests, the representation of failure by single points makes for an easier interpretation of the failure envelope, and enables a regression analysis to be applied if appropriate. The author recommends the adoption of stress path plots to supplement or replace Mohr circle diagrams in routine tests as well as for special applications.

A stress path plot is the alternative method of presentation of graphical test data given in BS 1377 : Part 8 : 1990.

21.2 CHARACTERISTICS OF STRESS PATH PLOTS

21.2.1 General Relationships

A stress path represents graphically a change from one set of stress conditions to another. A stress path plotted in terms of total stresses (t against s in the MIT plot, or q against p in the Cambridge plot) is referred to as a total stress path (TSP); and one plotted in terms of effective stresses (t' against s', or q' against p') is called an effective stress path (ESP). The direction of change should always be indicated by an arrow, as shown in Fig. 21.3 (b) and in subsequent diagrams. When stresses are reversed the same stress path might not necessarily be retraced.

The relationship between a point in the MIT stress field and the principal stresses σ_1 and σ_3 is illustrated in Fig. 21.3 (a). The two lines through the point P at 45° to the horizontal intersect the horizontal axis where $s = \sigma_1$ and $s = \sigma_3$. Conversely if σ_1 and σ_3 are known, the stress point P can be located. The same applies to the principal effective stresses σ'_1 and σ'_3.

The stress path characteristics discussed below in Sections 21.2.2 to 21.2.6 relate mainly to the MIT stress field. The most useful relationships, together with corresponding relationships for the Cambridge, 'principal stress' and 'deviator stress' fields, are summarised in Table 21.1, Section 21.2.7.

21.2.2 Stress Paths for Triaxial Compression

Typical stress paths derived from 'routine' triaxial compression tests are outlined below.

TOTAL STRESS TEST

The stress path in the MIT stress field for a total stress triaxial compression test is shown by the line AP in Fig. 21.3 (b). It rises at 45° from the point A at which s is equal to the cell confining pressure σ_3, and reaches its highest point at P corresponding to the maximum deviator stress. If deformation is continued beyond this point and the deviator stress remains constant, the stress path remains at P. If the deviator stress decreases the loading path is re-traced, e.g. from P to Q in Fig. 21.11 (a).

DRAINED TEST ON NORMALLY CONSOLIDATED CLAY

In a drained triaxial compression test that is run slowly enough to prevent changes in pore pressure the parameters s, s', and t on the MIT plot all increase at the same rate as the vertical stress increases. The total stress path (TSP) shown as AP in Fig. 21.8 (a) rises at 45°, and the effective stress path (ESP) is parallel to it, displaced to the left by a horizontal distance equal to the pore pressure u_b (i.e. the back pressure) as shown by $A_1 P_1$. If drainage is to atmosphere and there is no back pressure the ESP and TSP coincide. Maximum deviator stress is represented by the point P on the ESP, and by P_1 on the TSP.

Tests on additional specimens of the same soil, carried out at different effective confining pressures, would give additional points similar to P through which a line representing the failure envelope can be drawn (see Section 21.2.3).

In the Cambridge plot the ESP and TSP are both inclined at a slope of 3 on 1, and are also separated by a horizontal distance equal to the pore pressure (Fig. 21.8 (b)).

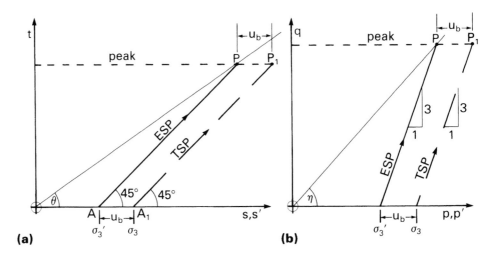

Fig. 21.8 *Stress paths of total and effective stresses for a drained triaxial compression test: (a) MIT field, (b) Cambridge field*

DRAINED TEST ON OVERCONSOLIDATED CLAY

After reaching a peak value the shear resistance of an overconsolidated clay decreases if deformation continues to be applied (see Fig. 15.7, Section 15.4.2), and the effective stress path retraces the loading path (similar to PQ in Fig. 21.11 (a)). Further deformation causes failure in localised zones, and the shear resistance continues to decrease towards a constant value at the 'residual' state (Section 15.4.4). However the residual condition is unlikely to be achieved within the limitations of a triaxial test.

UNDRAINED TEST ON NORMALLY CONSOLIDATED CLAY

In an undrained test on a normally consolidated clay the pore pressure increases as the deviator stress is applied, as shown in Fig. 15.12 (a) and (b), Section 15.5.2. The increasing pore pressure increases the difference between effective stress and total stress, and so the

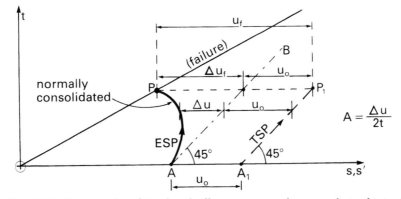

Fig. 21.9 *Stress paths of total and effective stresses for an undrained triaxial compression test on normally-consolidated clay*

effective stress path follows a curve which deviates to the left of the $45°$ line A B, as indicated by the curve A P in Fig. 21.9. The corresponding total stress path is the line $A_1 P_1$ at $45°$ to the axis.

The horizontal distance between the ESP and the TSP at any point is equal to the pore pressure u at that instant. It is made up of two components — the initial pore pressure u_0, and the excess pore pressure Δu due to the application of the deviator stress. Since the parameter t is equal to $\frac{1}{2}(\sigma_1 - \sigma_3)$ the value of the pore pressure coefficient A at any point is equal to $\Delta u/2t$. At failure, the excess pore pressure is Δu_f and $A_f = \Delta u_f/2t_f$.

A similar kind of curve is obtained on the Cambridge stress field, from which at any point the value of A is equal to $\Delta u/q$. A particular property of the Cambridge plot is that undrained effective stress paths for an isotropic *elastic* soil are vertical, since p' remains constant.

UNDRAINED TEST ON OVERCONSOLIDATED CLAY

Deviator stress and pore pressure changes in a heavily overconsolidated clay during a triaxial compression test are indicated in Fig. 15.12 (d) and (e), Section 15.5.2. When the pore pressure begins to decrease due to the effect of dilatancy the resulting stress path curves to the right, as shown by the line A B F in Fig. 21.10. At the point B it crosses over the $45°$ line representing zero pore pressure change, after which Δu becomes negative; and at failure A_f is negative. If the initial pore pressure u_0 is not high enough the effective stress path may cross the TSP line and the actual pore pressure would become negative — a condition which has to be avoided.

The curve AG in Fig. 21.10 represents a lightly overconsolidated clay in which the pore pressure does not fall below the initial value u_0.

COMPACTED CLAY

A partially saturated compacted clay can be described as 'quasi-overconsolidated'. The undrained effective stress path typically displays a reverse curvature as indicated by the dashed curve AH in Fig. 21.10.

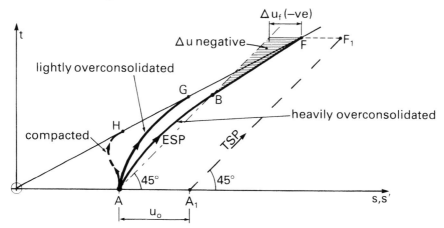

Fig. 21.10 Stress paths for undrained tests on heavily overconsolidated, lightly overconsolidated and compacted clay specimens

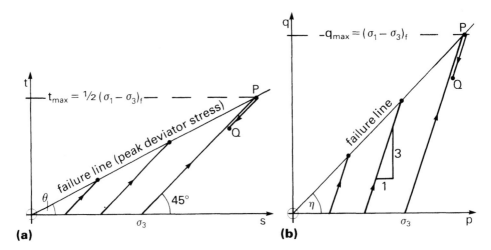

Fig. 21.11   Failure envelope of total stress on (a) MIT stress field, (b) Cambridge stress field

21.2.3   Derivation of Shear Strength Parameters

PRINCIPLE

The shear strength parameters can be derived from a set of triaxial tests by using stress path plots, without having to draw Mohr circles. The line drawn through the set of points representing the appropriate failure criterion gives a failure envelope, as shown in Fig. 21.11 in terms of total stresses on the MIT stress field (diagram (a)) and on the Cambridge stress field (diagram (b)). A failure envelope in terms of effective stresses for a set of undrained tests is shown in Fig. 21.12, where it is referred to as the K_f line. Although the inclination and intercept of these envelopes are not equal to ϕ and c (or ϕ' and c', etc.) those parameters can be derived very easily, as explained below for the three stress fields most often used.

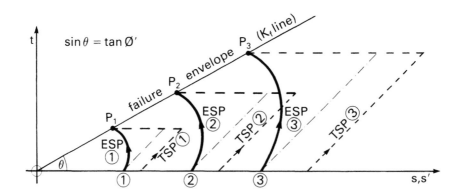

Fig. 23.12   Failure envelope of effective stress paths from set of undrained tests (MIT stress field)

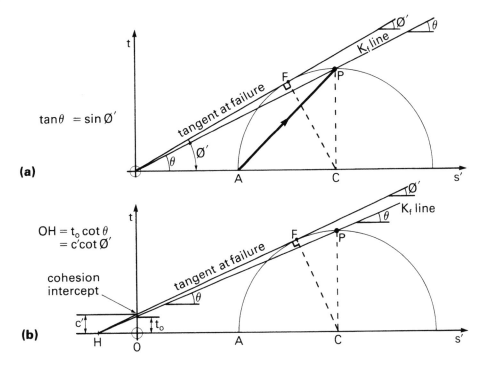

Fig. 21.13 *Derivation of shear strength parameters from* K_f *line: (a) with no cohesion intercept, (b) with a cohesion intercept*

MIT STRESS FIELD

A Mohr circle of effective stress representing failure of a specimen in a triaxial compression test is shown in Fig. 21.13 (a), its centre being at C. The tangent to the circle from the origin is the failure envelope (assuming no cohesion intercept), and its inclination is the angle ϕ'. It touches the circle of failure at F. The topmost point of the circle is denoted by P, and the line AP is the stress path to failure. The line OP, inclined to the horizontal at some angle θ smaller than ϕ', is the equivalent failure envelope on the stress path plot, i.e. the K_f line. The relationship between the angles ϕ' and θ is as follows. In Fig. 21.13 (a),

$$\tan \theta = \frac{PC}{OC}$$

and

$$\sin \phi' = \frac{FC}{OC}$$

But since PC and FC are both radii of the circle, they must be equal and therefore

$$\tan \theta = \sin \phi' \tag{21.8}$$

If the failure envelope gives an intercept on the t axis, denoted by t_0, the value of the apparent cohesion, c', can be derived as follows. In Fig. 21.13 (b) it can be seen that the

Mohr envelope and the stress path envelope intersect on the horizontal axis at H, and therefore

$$OH = t_0 \cot \theta$$

also

$$OH = c' \cot \phi'$$

Hence

$$c' = \frac{t_0 \cot \theta}{\cot \phi'}$$

Substituting from Equation (21.8),

$$c' = t_0 \frac{1/\sin \phi'}{\cot \phi'}$$

i.e.

$$c' = \frac{t_0}{\cos \phi'} \tag{21.9}$$

Values of c' and ϕ' are calculated from the intercept and slope of the K_f line by using Equations (21.8) and (21.9).

CAMBRIDGE STRESS FIELD

In the Cambridge stress path plot (Fig. 21.8 (b)) the relationship between the inclination of the failure envelope, η, and the angle of shear resistance, ϕ', is given by the equation

$$\sin \phi' = \frac{3 \tan \eta}{6 + \tan \eta} \tag{21.10}$$

The converse relationship is

$$\tan \eta = \frac{6 \sin \phi'}{3 - \sin \phi'} \tag{21.11}$$

If the stress path failure envelope intersects the q axis at q_0 the apparent cohesion c' is obtained from the equation

$$c' = \frac{3 - \sin \phi'}{6 \cos \phi'} q_0 \tag{21.12}$$

DEVIATOR STRESS FIELD

In the deviator stress field Fig. 21.7 the inclination of the failure envelope, α, is related to ϕ' by the equation

$$\tan \alpha = \frac{2 \sin \phi'}{1 - \sin \phi'} \tag{21.13}$$

If the stress path failure envelope intersects the vertical (y) axis at y_0, the apparent cohesion c' is given by the equation

$$c' = \frac{1 - \sin \phi'}{2 \cos \phi'} y_0. \tag{21.14}$$

The intersection of the envelope with the horizontal (x') axis is at a distance x_0 from the origin, where

$$x_0 = \frac{c'}{\tan \phi'} \tag{21.15}$$

21.2.4 Strain Contours

A stress path plot in itself provides no information on the stress–strain characteristics of the soil, for which reference has to be made to the curve of deviator stress against strain. However, strains can be indicated on a stress path plot by writing the axial strain against each point plotted. When several curves of a set are plotted together, contours of equal strain can be sketched in by interpolating between the marked points, as indicated in Fig. 21.14. The contours may approximate to straight lines radiating from a point near the origin, but over a wide stress range they are more likely to be curved slightly downwards.

21.2.5 Summary of Relationships

Some of the more useful relationships in the MIT stress field, most of which have been explained above, are summarised in Table 21.1. Also included are corresponding relationships for three other types of stress field that are often used.

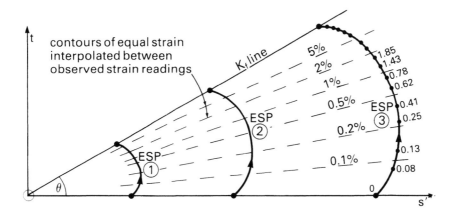

Fig. 21.14 *Stress paths for a set of undrained tests, with strain contours*

Table 21.1 USEFUL RELATIONSHIPS FOR STRESS PATHS

Parameter or function	MIT stress field	Cambridge stress field (axisymmetric)	Principal stress field	Deviator stress field
Abscissa – total stress	$s = \frac{1}{2}(\sigma_1 + \sigma_3)$	$p = \frac{1}{3}(\sigma_1 + 2\sigma_3)$	σ_3	(not used)
effective stress	$s' = \frac{1}{2}(\sigma_1' + \sigma_3')$	$p' = \frac{1}{3}(\sigma_1' + 2\sigma_3')$	σ_3'	$x = \sigma_3'$
Ordinate – total stress	$t = \frac{1}{2}(\sigma_1 - \sigma_3)$	$q = \sigma_1 - \sigma_3$	σ_1	(not used)
effective stress	$t' = \frac{1}{2}(\sigma_1' - \sigma_3')$	$q' = \sigma_1' - \sigma_3'$	σ_1'	$y = \sigma_1' - \sigma_3'$
Major principal effective stress	$\sigma_1' = s' + t$	$\sigma_1' = p' + \frac{2}{3}q$	σ_1'	$\sigma_1' = x + y$
Minor principal effective stress	$\sigma_3' = s' - t$	$\sigma_3' = p' + \frac{2}{3}q$	σ_3'	$\sigma_3' = x$
Slope of line from origin to stress point	$\tan\theta = \dfrac{t}{s'}$	$\tan\eta = \dfrac{q}{p'}$	$\tan\omega = \dfrac{\sigma_1'}{\sigma_3'}$	$\tan\alpha = \dfrac{y}{x}$
Mohr–Coulomb failure envelope in compression (K_f line)	$\tan\theta = \sin\phi'$	$\eta = \dfrac{6\sin\phi'}{3 - \sin\phi'}$	$\tan\omega = \dfrac{1 + \sin\phi'}{1 - \sin\phi'}$	$\tan\alpha = \dfrac{2\sin\phi'}{1 - \sin\phi'}$
Cohesion intercept	$t = c'\cos\phi'$	$q = \dfrac{6\cos\phi'}{3 - \sin\phi'}c'$	$\sigma_1' = \dfrac{2\cos\phi'}{1 - \sin\phi'}c'$	$y = \dfrac{2\cos\phi'}{1 - \sin\phi'}c'$
Earth pressure coefficient (for K_0 line put $K = K_0$)	$K = \dfrac{1 - \tan\theta}{1 + \tan\theta}$	$K = \dfrac{3 - \tan\eta}{3 + 2\tan\eta}$	$K = \dfrac{\sigma_3'}{\sigma_1'}$	$K = \dfrac{1}{1 + \tan\alpha}$
Drained compression—slope of stress path	1 on 1	3 on 1	vertical	vertical
Isotropic consolidation—slope of stress path (K_1 line)	along s' axis	along p' axis	1 on 1	along σ_3' axis

21.3 DERIVATION OF STRESS PATHS FROM TRIAXIAL TESTS

CALCULATION AND PLOTTING

Stress paths can be plotted easily from the data observed during triaxial tests in which pore pressure are measured. A set of stress paths from a set of test specimens enables the parameters c', ϕ', to be derived without having to draw a separate Mohr circles diagram.

When the object of the test is to determine only the shear strength parameters c', ϕ', the test can be carried out exactly as described in Chapter 18. Values of s' and t for plotting the effective stress path are calculated from corrected deviator stress values and pore pressure readings as follows.

At any point on the stress/strain curve, the corrected deviator stress $(\sigma_1 - \sigma_3)$ is denoted by q; and the pore water pressure is u, the cell pressure σ_3 (constant), and the effective cell pressure is σ_3' $(= \sigma_3 - u)$. All stresses are in kPa.

In the MIT stress field

$$t = t' = \frac{\sigma_1 - \sigma_3}{2} = \frac{1}{2} q \tag{21.2}$$

i.e.

$$s = \tfrac{1}{2}(\sigma_1 + \sigma_3) = \tfrac{1}{2}[(q + \sigma_3) + \sigma_3)]$$
$$s = \tfrac{1}{2}q + \sigma_3 \tag{21.16}$$

and

$$s' = \tfrac{1}{2}q + \sigma_3' \tag{21.17}$$

In the Cambridge stress field,

$$q' = q = (\sigma_1 - \sigma_3) \tag{21.5}$$
$$p = \tfrac{1}{3}(\sigma_1 + 2\sigma_3) \tag{21.7}$$

Combining the above

$$p = \tfrac{1}{3}q + \sigma_3 \tag{21.18}$$

and

$$p' = \tfrac{1}{3}q + \sigma_3' \tag{21.19}$$

In any type of triaxial test (compression or extension) the plotting of a stress path as the test proceeds gives a clear picture of the soil behaviour. For some special tests the stress path is established first and test conditions are adjusted so that the prescribed path is followed as closely as possible. An example is the test for the pore pressure coefficient $\bar{B}$, described in Section 21.4.

Part of a test data sheet showing the calculation of the first few values of s' and t' from a CU triaxial test is given in Fig. 21.15 (a). The stress path curves from a set of samples are plotted in Fig. 21.15 (b).

GENERAL COMMENTS

When the stress path is to be used to derive pore pressure or strain relationships at stress levels below failure, the rate of strain applied must be slow enough to ensure that equalisation of pore pressure within the specimen can take place by every reading. This requires a much slower rate of strain than that normally applied when 'failure' is the only significant criterion. The 'time to failure', t_f, derived from Table 18.1 (Section 18.6.2)

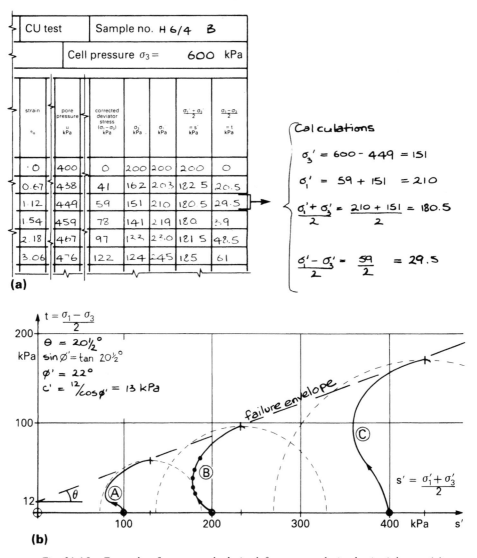

Fig. 21.15   *Example of stress path derived from an undrained triaxial test: (a)*
calculation of s', t' (b) stress path plots

indicates the time needed from the start of compression to the first significant pore
pressure reading (Blight, 1963). For instance if readings are needed at intervals of 0.5%
strain, the strain used for calculating the rate of strain (see Section 18.6.3) should be 0.5%,
not the estimated strain at failure.

A stress path indicates at a glance the relationships between pore pressure change and
applied stresses. The value of the parameter A at any one point on the ESP can be
calculated from the equation

$$A = \frac{\Delta u}{2t} \quad \text{or} \quad A = \frac{\Delta u}{q} \qquad (21.20)$$

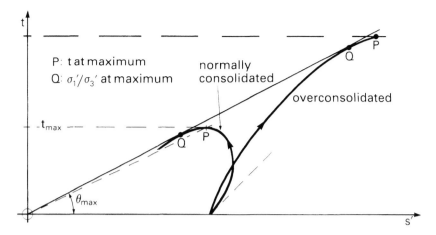

Fig. 21.16   Representation of two failure criteria on stress path plot

The failure envelope for the criterion of maximum principal effective stress ratio (Section 15.4.2) is represented by the line of greatest slope from the origin to the stress path (points Q, Fig. 21.16). This line might not be the same as that for maximum deviator stress (points P) for which the maximum values of t apply.

PRESENTATION OF TEST DATA

For many applications the value of the data obtained from routine triaxial tests, as well as from specially derived 'stress path' tests, could be enhanced if presented in a manner which clearly shows the relationships referred to above. The traditional presentation of triaxial test data described in Chapter 18 can easily be modified to include stress path plots, on the lines indicated in Fig. 21.15 (b). This is in accordance with Clause 7.5(f) of BS 1377: Part 8: 1990. If Mohr circles at failure are also required it is better not to superimpose them on the stress path plot, but to plot them as a separate diagram.

21.4   TEST FOR PORE PRESSURE COEFFICIENT $\bar{B}$

21.4.1   Introduction

SCOPE

This test, often referred to as the $\bar{B}$ test, (pronounced 'bee-bar'), is a type of stress path test that has been used to supplement routine tests for many years. It is used to determine the values of the pore pressure coefficient $\bar{B}$ for a partially saturated soil under conditions in which the vertical and horizontal principal stresses change simultaneously, such as in an earth embankment while under construction. The test is usually carried out on recompacted specimens of clay to be used in the embankment, but undisturbed specimens from foundation strata can also be tested in the same way.

The test procedure is given in some detail as an example of a stress path test using normal triaxial equipment and procedures. The principles and applications were originally described by Bishop (1954).

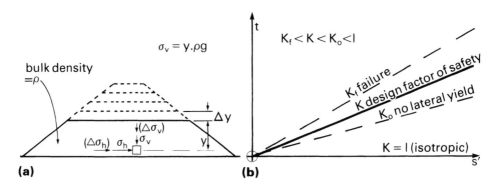

Fig. 21.17 *Stress changes during construction of earth embankment:* (a) *soil element considered,* (b) *limiting stress path for design factor of safety*

PRINCIPLE

An element of soil near the base of a partly constructed earth embankment is indicated in Fig. 21.17 (a). During construction the total vertical stress on that element progressively increases as the embankment is raised. The stress increment $\Delta\sigma_v$ due to an increase in height Δy is equal to $\Delta_y . \rho g$, where ρ is the bulk density of the fill. It is assumed that the principal effective stress ratio σ'_v/σ'_h remains constant at a value denoted by $1/K$. If no lateral yield occurs, the value of K would be equal to K_0, but since the embankment is not confined laterally, K is rather less than K_0. It is required that the value of K should never be less than that corresponding to the minimum factor of safety specified by the designer (see Section 21.4.2). In terms of the MIT stress path (s', t) plot, the path to be followed is the heavy line path marked K in Fig. 21.17 (b). Since the vertical stress σ_v is the major principal stress it is expressed as σ_1, and σ_h is expressed as σ_3.

An increase in vertical stress of $\Delta\sigma_1$ causes the pore water pressure to increase by Δu. For the condition of constant stress ratio referred to above, the pore pressure change is related to the change in major principal stress by the equation

$$\Delta u = \bar{B} . \Delta\sigma_1 \tag{15.15}$$

The relationship between $\bar{B}$ and the pore pressure coefficients A and B is explained in Section 15.3.3.

The value of $\bar{B}$ can be derived from a triaxial rest in which the above conditions are met, by determining the ratio $\Delta u/\Delta\sigma_1$ for the desired stress increment.

At any instant the height of the embankment above the point being considered is denoted by y, and the pore pressure at that point by u. The principal total and effective stresses are calculated as follows:

$$\sigma_1 = y . \rho g \tag{21.21}$$

$$\sigma'_1 = y\rho g - u \tag{21.22}$$

$$\sigma'_3 = K\sigma'_1 = K(y\rho g - u) \tag{21.23}$$

$$\sigma_3 = \sigma'_3 + u = Ky\rho g + (1 - K)u \tag{21.24}$$

where ρ is the mean bulk density of the fill.

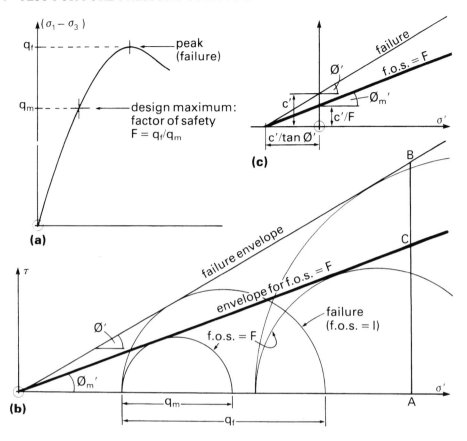

Fig. 21.18   *Shear strength properties of embankment material: (a) stress–strain curve, (b) Mohr circles and envelopes for failure and for design factor of safety (no cohesion intercept), (c) envelopes with cohesion intercept*

Measurement of the pore pressures during construction of the embankment enables its stability to be monitored. Comparison of measured pore pressure at a certain stage with the maximum permissible value derived from the test provides a measure of the factor of safety. If the pore pressures become excessive, construction is stopped until they have dissipated to a safe value.

21.4.2   Factor of Safety

A stress–strain curve for a drained triaxial compression test on embankment material is represented in Fig. 21.18 (a). The maximum deviator stress at failure is denoted by q_f, but the working stresses would usually be such that a lower value, q_m, will never be exceeded. The ratio of failure stress to maximum working stress is known as the 'factor of safety' (f.o.s.). It is denoted by F, and is defined by the equation

$$F = \frac{q_f}{q_m} \tag{21.25}$$

A set of Mohr circles representing the limiting working condition (f.o.s. $= F$) can be drawn in the same way as for the failure condition, as shown in Fig. 21.18 (b). The envelope rises at an angle ϕ'_m to the horizontal. At any point A on the horizontal axis the ratio AB/AC is equal to F, i.e. $F = \tan \phi/\tan \phi'_m$, or

$$\tan \phi'_m = \frac{\tan \phi}{F} \tag{21.26}$$

If the failure envelope gives a cohesion intercept c', the f.o.s. envelope has an intercept equal to c'/F, as shown in Fig. 21.18 (c). The two envelopes intersect on the horizontal axis at a distance $c'/\tan \phi'$ to the left of the origin.

The object of the $\bar{B}$ test is to apply stresses so as always to maintain the condition of the limiting factor of safety F, i.e. to follow the envelope OC, and to observe the resulting pore pressure response. This would be very difficult to do from the Mohr circle plot, but a plot using the deviator stress field (Section 21.1.4 (6)) makes it easier to keep the sample close to the desired factor of safety.

21.4.3   Defining the Stress Path

The effective shear strength parameters c', ϕ', for the soil to be tested are obtained from a set of standard consolidated-drained triaxial compression tests as described in Chapter 18. The size of specimens used should be the same as those to be used in the $\bar{B}$ test. Recompacted specimens should all be prepared and matured in exactly the same way, in accordance with the specified moisture content and density criteria. The effective confining pressures used for the strength tests should give a reasonable spread over the appropriate stress range.

The slope of the stress path at failure ($\tan \alpha$) in the deviator stress field is calculated from Equation (21.13) in Section 21.2.3. Using the design value of the factor of safety F, the value of ϕ'_m corresponding to F is calculated from Equation (21.26). The slope of the stress path envelope representing ϕ'_m, inclined at an angle β, is calculated from the same equation as for α, i.e.

$$\tan \beta = \frac{2 \sin \phi'_m}{1 - \sin \phi'_m}$$

The two envelopes are represented in Fig. 21.19 (a). The intercept $c'/tan\phi'$ is also calculated if appropriate (Fig. 21.19 (b)). Typical calculations are given in Fig. 21.20.

The envelopes representing the factor of safety condition, and failure, are drawn to a convenient scale on a large sheet of graph paper to accommodate the maximum stresses likely to be applied, as indicated in Fig. 21.21. The two axes representing deviator stress $(\sigma_1 - \sigma_3)$ and minor effective principal stress (σ'_3) must be drawn to a common scale.

The rate of strain to be applied during the test is calculated as in Section 18.6.3, but the 'strain at failure' should be replaced by the strain at which the first significant readings are to be taken. Suitable intervals between readings depend on the type of soil and its behaviour, and might be as little as 0.2% strain. It is essential to allow pore pressures within the specimen to equalise as the test proceeds, and the calculated machine speed will

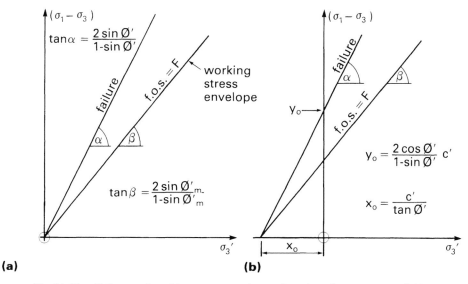

Fig. 21.19 *Failure and working stress envelopes plotted on deviator stress field:*
(a) no cohesion intercept, (b) with cohesion intercept

be much slower than in a standard undrained test for determining the failure conditions only. Once the test has been started it should continue without interruption until it is finished, and shift working may be necessary to maintain continuous supervision, unless a data-logging system is available.

Shear strength parameters for sample: $c' = 0$, $\phi' = 34°$

Factor of safety $F = 1.5$

$c'/F = 0$ kPa

$\tan \phi'_m = \frac{1}{F} \tan \phi' = \frac{1}{1.5} \tan 34° = 0.450$

$\therefore \phi'_m = 24.2°$

Slope of stress path envelope for maximum working stress:

$\tan \beta = \frac{2 \sin \phi'_m}{1 - \sin \phi'_m} = \frac{0.820}{0.590} = 1.390$

$\therefore \underline{\beta = 54.3°}$ Intercept $c'/\tan \phi' = 0$

Slope of stress path envelope at failure:

$\tan \alpha = \frac{2 \sin \phi'}{1 - \sin \phi'} = \frac{1.118}{0.441} = 2.536$

$\therefore \underline{\alpha = 68.5°}$

Fig. 21.20 *Example of calculations required before starting $\bar{B}$ test*

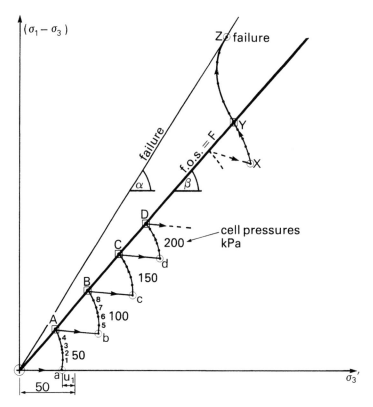

Fig. 21.21 Working stress path plotted on deviator stress field, showing method of plotting test data

21.4.4 Apparatus

A hydraulic stress-path cell can be used with advantage for the $\bar{B}$ test. However a standard triaxial cell can be used instead, connected to the usual ancillary equipment as shown in Fig. 19.3 (Section 19.3.2). The essential features are:

Pore water pressure is measured at the base of the specimen.

A porous disc of high air-entry value is fitted to the base pedestal so that pore water pressure measurements are not affected by the presence of air.

No drainage is allowed from the top of the specimen and a solid top cap is used.

A back pressure system is not required.

A volume-change indicator is incorporated in the cell pressure line. Although this is an undrained test the specimen volume will change because the soil is not saturated and the air voids are compressible.

The triaxial cell must be carefully calibrated for changes in volume with increasing pressure and with time.

The cell piston area must be known.

The strain dial gauge should read to 0.002 mm.

A layer of oil should be introduced at the top of the water in the cell to minimise piston leakage; or if leakage is unavoidable, to enable the rate of leakage to be estimated.

21.4.5 Test Procedure

The test specimen is prepared, matured, measured and set up in accordance with the standard procedures given in Chapter 18. A porous disc with a 'high air-entry value' (Section 16.7.3) should be used at the base of the specimen in place of the normal porous disc. It should be sealed on to the cell base pedestal as described in Section 16.2.2 and shown in Fig. 16.4.

Before starting the test, preparations should be made for rapidly carrying out the calculations given in Section 21.4.6, because it is essential that the stress path diagram should be kept up to date as the test proceeds. Use of a programmable calculator, already programmed to carry out these tasks, is the easiest approach.

The test is carried out as follows. The specimen is *not* saturated before starting the test.

(1) Record the initial pore water pressure.

(2) Fill the triaxial cell with de-aired water, making sure that no pockets of air are entrapped.

(3) Wind up the machine base platen until the piston just makes contact with the specimen top cap. Close the air bleed e. Set the machine drive unit to give the appropriate rate of strain.

(4) Observe the zero settings of the load ring dial gauge, and set the strain dial gauge to zero or to a convenient datum reading. Record the initial volume gauge reading.

(5) Increase the pressure in the cell to the first stage pressure (e.g. 50 kPa).

(6) Restore contact between the piston and top cap if it has been lost due to compression of the specimen and deflection of the load ring. When the conditions are steady, record readings of the load ring dial, strain dial, volume-change indicator and pore water pressure. (See Section 17.3.1 regarding cell volume-change corrections.)

(7) Plot the stress point $\sigma_3' = $ (cell pressure $-$ pwp) on the horizontal axis of the stress path diagram (point a, Fig. 21.21). In this instance

$$\sigma_3' = (50 - u_1)\,\text{kPa}$$

(8) Start the motor of the compression machine to apply the deviator stress to the specimen. At suitable intervals of strain (e.g. 0.2%) take readings of:

   strain dial gauge

   load ring dial gauge

   volume-change indicator

   pore water pressure

   clock time.

(9) Calculate the deviator stress $(\sigma_1 - \sigma_3)$ and effective cell pressure (σ_3') as described in Section 21.4.6. Plot these values on the stress path diagram as the test proceeds, as in Fig. 21.21 (points 1, 2, 3, ...).

(10) When the plotted stress path intersects the f.o.s. line (Fig. 21.21), switch off the motor. The factor of safety has then been reduced to the design limiting value. As the line is approached a close watch should be maintained to prevent overshooting the line.

(11) Record the readings listed in step 8, and plot the exact position of the stress point (point A).

(12) Increase the cell pressure by a further increment. When conditions are steady record the readings listed above and plot the new stress point (b). The deviator stress might be reduced slightly because of a small compression of the specimen and relaxation of the load ring.

(13) Resume application of deviator stress, and plot readings as in step 9 until the stress path reaches the f.o.s. line again, at point (B).

(14) Continue with the further steps similar to steps 12 and 13 to obtain the stress path similar to b–B–c–C–d–D...in Fig. 21.21. As the cell pressure increases the specimen becomes progressively more saturated and there is less separation between the points for successive stages.

(15) When the upper limit of the desired stress range has been reached the deviator stress can be increased with no further change of cell pressure until failure occurs, as represented by X–Y–Z in Fig. 21.21.

(16) Remove the axial load and cell pressure, take out the specimen and weigh and measure it (including moisture content determination) in the usual way.

The stress path derived from a typical test is shown in Fig. 21.22. The first few sets of readings, covering the first two cell pressure increments, are given in Fig. 21.23.

21.4.6   Calculations

AXIAL STRAIN

For each point on the stress path plot calculate the cumulative axial strain, based on the initial specimen length. If the cumulative axial compression at any instant, relative to the initial strain dial reading (step 4 in Section 21.4.5), is denoted by x mm, the axial strain $\varepsilon\%$ is equal to $x/L_0 \times 100\%$, where L_0 is the initial specimen length (mm).

VOLUME CHANGE

The specimen volume is changed during the test by two different processes which are applied alternately:

(a) increase of cell pressure

(b) application of deviator stress.

Corrections to measured volume changes have to be made to allow for:

(1) cell expansion due to pressure change

(2) cell expansion with time due to 'creep'

(3) piston displacement

(4) cell leakage.

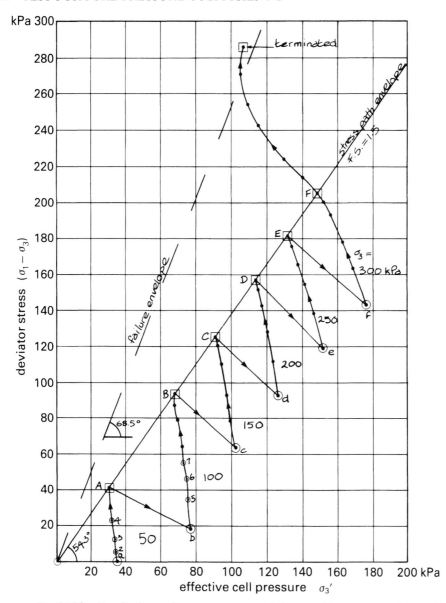

Fig. 21.22   Results from a B *test on compacted clay: working stress envelope and observed stress path*

The sense of these corrections ($+$ or $-$) is explained in Section 17.4.4. Items 2 and 4 can be neglected if proved by calibration checks to be insignificant. Volume changes and corrections can be understood easily if cumulative measurements are used and cumulative corrections applied. The calculation procedure recommended is summarised as follows.

(a) *Cell pressure increase stage*

Calculate the cumulative volume change from the volume-change indicator readings.

324

B̄ test data

Sample length L_0 = 200 mm
area A_0 = 8171 mm²
volume V_0 = 1634 cm³

Piston area a = 765 mm²
a/A_0 = 0.0936

Load ring no. 711-3-27
Rate of strain 0.0054 mm/min
Date started 15.9.83

Location Derwent Dam
Sample no. 18/5

REFERENCE POINT	STRAIN			CELL VOLUME								CELL PRESSURE	PORE PRESSURE	AXIAL LOAD				DEVIATOR STRESS					PRINCIPAL STRESSES		
				R.H.			corrections			volume change													effective		total
	dial gauge div =10⁻¹ mm	diff mm	ε %	gauge cm³	diff cm³	cum V_w cm³	piston $-\Delta V_p$ cm³	cell ΔV_c cm³	total ΔV_t cm³	ΔV_s cm³	$\Delta V/V_0$ %	σ_3 kPa	u kPa	load dial div	ring calibration N div	load N	un-corrected $\frac{1000N}{A_c}$ kPa	piston corrections $\frac{a}{A}\times\sigma_3$ kPa	net stress σ_{net} kPa	area & vol $\sigma_{net}\times\frac{100-\varepsilon}{100}\times\frac{V}{V_0}$ kPa	membrane correction σ_m kPa	corrected deviator stress $(\sigma_1-\sigma_3)$ kPa	σ_3' kPa	σ_1' kPa	σ_1 kPa
0	100.0	0	0	74.3	0	0	0	0	0	0	0	0	−1	0	8.10	0	0	0	0	0	0	0	−1	−1	0
a	100.0	0	0	60.5	18.8	18.8	0	5.0	5.0	13.8	0.8	50	15	3.0	8.10	24.3	2.97	4.7	−1.8	–	–	0	35		
1	110.0	0.10	0.05	58.2	21.1	21.1	0.1	5.0	4.9	16.2	1.0	50	15	4.9		39.7	4.86	4.7	0.2	–	–	0.2	35		
2	150.0	0.30	0.15	57.9	21.4	21.4	0.2	5.0	4.8	16.6	1.0	50	16	9.8		79.4	9.71	4.7	5.0	5.0	–	5.0	34		
3	160.0	0.60	0.3	57.4	21.9	21.9	0.5	5.0	4.5	17.4	1.1	50	16	17.1		138.5	17.0	4.7	12.3	12.4	–	12.4	34		
4	200.0	1.00	0.5	55.4	25.9	25.9	0.8	5.0	4.2	19.7	1.2	50	18	37.5		303	37.2	4.7	32.5	32.6	–	32.6	32		
A	220.0	1.20	0.6	54.5	24.8	24.8	0.9	5.0	4.1	20.7	1.3	50	19	46		371	45.4	4.7	40.7	41.0	–	41.0	31	72	91
b	226.0	1.26	0.63	45.3	34.0	34.0	0.9	9.1	8.2	25.8	1.6	100	23	28	8.10	227	27.8	9.4	18.4	18.5	–	18.5	77		
5	260.0	1.60	0.8	42.2	37.1	57.1	1.2	9.1	7.9	29.2	1.8	100	25	44		356	43.6	9.4	34.2	34.3	–	34	75		
6	280.0	1.80	0.9	41.3	38.0	58.0	1.3	9.1	7.8	30.2	1.8	100	26	56		453	55.4	9.4	46.0	46.4	0.3	46	74		
7	300.0	2.00	1.0	40.3	39.0	59.0	1.5	9.1	7.6	31.4	1.9	100	27.5	65		526	64.3	9.4	54.9	55.4	0.3	55	72.5		

Fig. 21.23   Examples of readings from a B̄ test

Deduct the cumulative correction for cell volume change due to pressure change. Deduct the cumulative correction for piston displacement, derived from the strain dial reading.

(b) *Deviator stress stage*

Calculate the cumulative volume change from the volume-change indicator reading.

Deduct the cumulative corrections for cell volume change and piston displacement. Deduct the correction for the cell volume change with time (if significant).

At any time during the test it might be necessary to reverse the volume-change indicator and start reading again from a new datum. Differences from this datum are then added to the cumulative value at the moment of reversal.

The cumulative correction for cell expansion, ΔV_c, is read directly from the cell calibration chart of the type shown in Section 17.3.1, Fig. 17.6, against the cell pressure just applied. The correction due to piston displacement, ΔV_p, is equal to $-ax/1000\,\text{ml}$, where a is the area of cross-section of the piston (mm^2) and x is the cumulative displacement (mm). The total cumulative correction ΔV_t is equal to

$$\Delta V_t = \Delta V_c + \Delta V_p$$
$$= \Delta V_c - \frac{ax}{1000}\,\text{ml} \tag{21.27}$$

The cumulative change in volume of the specimen, ΔV_s, is calculated from the equation

$$\Delta V_s = \Delta V_m - \Delta V_t \tag{21.28}$$

where ΔV_m is the cumulative measured volume change obtained as outlined above. If cell leakage is significant the cumulative loss at any time is added to ΔV_t before using the above equation.

DEVIATOR STRESS

The measured deviator stress at any point on the stress path is calculated from the following equation, which includes an allowance for the upward force on the piston from the cell pressure.

$$(\sigma_1 - \sigma_3)_m = \frac{(RC_r \times 1000) - a\sigma_3}{A}\,\text{kPa} \tag{21.29}$$

where R = load ring reading (dial divisions)

C_r = load ring calibration at that reading (N/div)

a = area of piston (mm^2)

σ_3 = cell pressure (kPa)

A = area of cross-section of specimen at that point (mm^2)

$$= 100 A_0\,\frac{1 - \dfrac{\Delta V_s}{V_0}}{100 - \varepsilon\%}\,\text{mm}^2 \qquad\qquad \text{(from Equation (17.6), Section 17.4.1)}$$

The membrane correction (Section 17.4.2) is deducted from $(\sigma_1 - \sigma_3)_m$ to give the corrected deviator stress $(\sigma_1 - \sigma_3)$.

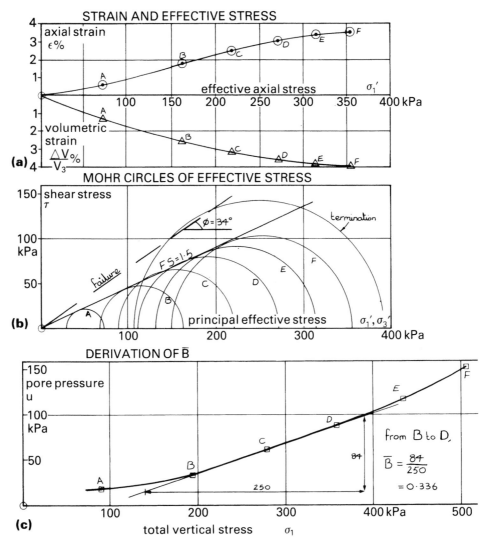

Fig 21.24 *Graphical plots from a* B *test on compacted clay:* (*a*) *axial strain and volume change against major principal effective stress* σ_1, (*b*) *Mohr circles,* (*c*) *plot of pore pressure against total major principal stress* σ_1, *showing derivation of pore pressure coefficient* $\bar{B}$

PRINCIPAL STRESSES

The principal stresses are calculated as follows:

$$\sigma_3 = \text{cell pressure}$$
$$\sigma_3' = \sigma_3 - u$$
$$\sigma_1' = (\sigma_1 - \sigma_3) + \sigma_3'$$
$$\sigma_1 = \sigma_1' + u$$

where u is the pore pressure at that point.

Location Downside Dam								Sample no. 13/5		

Soil description Dark grey sandy & silty clay with some fine to medium gravel

Type of sample Compacted at (omc - 1%) with BS 2.5 kg rammer.   101 mm dia x 200 mm long

Initial density 2.13 Mg/m³	Effective strength parameters
moisture 13.3 %	$c' = 0$ kPa $\varnothing' = 34°$
dry density 1.88 Mg/m³	
	Test parameters
Rate of strain 0.17 % per hour	factor of safety FS = 1.5
Duration of test 29 hours	$c'/F = 0$ kPa $\varnothing'_m = 24°$

READINGS AFTER EACH LOADING STAGE

Stage ref. (see graph)	Strain ε %	$\dfrac{\bar{V}}{V_o}$ %	deviator stress $(\sigma_1-\sigma_3)$ kPa	cell pressure σ_3 kPa	axial stress σ_1 kPa	pore pressure u kPa	effective stress σ_3' kPa	σ_1' kPa	mean coeff. $\bar{B}$
A	0.60	1.3	41	50	91	19	31	72	⎫ 0.12
B	1.65	2.6	94	100	194	32	68	162	⎬
C	2.3	3.1	129	150	279	60	90	219	⎬ 0.34
D	2.9	3.5	158	200	358	87	113	271	⎬
E	3.3	3.8	183	250	433	119	131	314	⎬ 0.44
F	3.6	3.9	206	300	506	152	148	354	⎭
(end)	4.9	4.4	286	300	586	194	106	392	

$\bar{B}$ VALUES

Nominal stress range (σ_1) kPa		Average $\bar{B}$
90	— 200	0.12
200	— 350	0.34
350	— 500	0.44

Fig. 21.25   Summary of results of $\bar{B}$ test

21.4.7   Plotting and Reporting

GRAPHICAL PLOTS

The plotting of the stress path diagram (Fig. 21.22) is described in Section 21.4.5, and it is essential that this is kept up to date as the test proceeds. This plot forms part of the test data.

The following graphical plots are also presented, relating to the points on the f.o.s. line (i.e. points A, B, C...in Fig. 21.22). Typical examples are shown in Fig. 21.24 as stated below.

Axial strain ε% against effective major principal stress σ_1' (Fig. 21.24 (a), upper curve).

Volume change against σ_1' (Fig. 21.24 (a), lower curve).

Mohr circles, including the failure circle (Fig. 21.24 (b)).

Pore water pressure u against total major principal stress σ_1 (Fig. 21.24 (c)). The slope of this curve, i.e. $\Delta u/\Delta\sigma_1$, gives the value of the parameter $\bar{B}$. If the curve is not linear, different values of $\bar{B}$ are derived for different stress ranges and reported accordingly.

PRESENTATION OF RESULTS

The above graphs (Figs 21.22 to 21.24) form part of the test results, which should also include:

Usual sample identification, index properties and measurements of the test specimen

Value of c', ϕ', obtained from a standard test

Specified factor of safety

Calculated values of ϕ'_m and $c'/\tan \phi'$

Rate of strain applied

Duration of test

Value of ϕ' at failure of the test specimen.

For each point on the factor of safety line, tabulate values of:

cell pressure

deviator stress (corrected)

strain

pore water pressure

effective cell pressure

cumulative specimen volume change

average value of $\bar{B}$ for each stated range of stress (σ_1).

A typical summary of results, relating to the test illustrated in Figs. 21.22 to 21.24, is given in Fig. 21.25.

REFERENCES

Atkinson, J. H. and Bransby, P. L. (1978). *The Mechanics of Soils*. McGraw-Hill.
Bishop, A. W. (1954). 'The use of pore pressure coefficients in practice'. *Géotechnique*, 4:4:148.
Bishop, A. W. and Wesley, L. D. (1975). 'A hydraulic triaxial apparatus for controlled stress path testing'. *Géotechnique*, 35:4:657.
Blight, G. E. (1963). 'The effect of non-uniform pore pressures on laboratory measurements of the shear strength of soils'. Symposium on Laboratory Shear Testing of Soils, pp 173–184. *ASTM Special Technical Publication* No. 361, American Society for Testing and Materials, Philadelphia, USA.
Casagrande, A. (1963). 'The determination of the pre-consolidation load and its practical significance'. *Proc. 1st Int. Conference of Soil Mechanics*, Vol. 3, Cambridge, Mass, USA.
Casagrande, A. and Wilson, S. D. (1953). 'Prestress induced in consolidated-quick triaxial tests'. *Proc. 3rd Int. Conference of Soil Mechanics & Foundation Eng.*, Vol. 1, pp 106–110.
Hanzawa, H. (1983). 'Three case histories for short term stability of soft clay deposits'. *Soils and Foundations*, Japan, Vol. 23, No. 2, pp 140–154.
Henkel, D. J. (1960). 'The shear strength of saturated remoulded soils'. *ASCE Speciality Conf. on Shear Strength of Cohesive Soils*, Boulder, Col., USA.
Holtz, W. G. (1947). 'The use of the maximum principal stress ratio as the failure criterion in evaluating triaxial shear tests on earth materials'. *Proc. ASTM*. Vol. 47, pp 1067–1076.
Ladd, C. C. and Foott, R. (1974). 'New design procedures for stability of soft clays'. *J. Geotech. Eng. Div. ASCE*, Vol. 100, No. GT7, pp 763–786.
Ladd, C. C., Foott, R., Ishihara, K., Schlosser, F. and Poulos, H. G. (1977). 'Stress deformation and strength characteristics. State of the art report'. *Proc. 9th Int. Conference on Soil Mechanics & Foundation Eng.*, Tokyo, Vol. 2, pp 421–494.
Lambe, T. W. (1964). 'Methods of estimating settlement'. *J. Soil Mechanics &; Foundation Div. ASCE*. Conference on the Design of Foundations for Control of Settlements, pp 47–71.

Lambe, T. W. (1967). 'Stress path method'. *J. Soil Mechanics & Foundation Div., ASCE*, Vol. 93, No. SM6, pp 309–331.

Lambe, T. W. and Marr, W. A. (1979). 'Stress path method: second edition'. *J. Geotech. Eng. Div. ASCE*, Vol. 105, No. GT6, pp 727–738.

Lambe, T. W. and Whitman, R. V. (1979). *Soil Mechanics* (S. I. Version). Wiley, New York.

Leonards, G. A. (1962). *Foundation Engineering*. McGraw-Hill, New York.

Roscoe, K. H., Schofield, A. N. and Wroth, C. P. (1958). 'On the yielding of soils'. *Géotechnique*, 8:1:22.

Schmertmann, J. H. (1953). 'The undisturbed consolidation behaviour of clay'. *Trans ASCE*, Vol. 120, p 1201.

Simons, N. E. and Menzies, B. K. (1977). *A Short Course in Foundation Engineering*, Butterworths.

Skempton, A. W. and Sowa, V. A. (1963). 'The behaviour of saturated clays during sampling and testing'. *Géotechnique*, 13:4:269.

Taylor, D. W. (1944). Tenth Progress Report to the US Engineering Department, MIT Soil Mechanics Laboratory, USA.

Wroth, C. P. (1975). 'In-situ measurement of initial stresses and deformation characteristics'. *Proc. ASCE Special Conf. on In-situ Measurement of Soil Properties*, Vol. 2, pp 181–230.

Chapter 22

Hydraulic cell consolidation and permeability tests

22.1 THE ROWE CONSOLIDATION CELL

22.1.1 Scope

The type of hydraulic consolidation cell described in this chapter, generally known as the Rowe cell, was developed at Manchester University by Professor P. W. Rowe, and details were published by Rowe and Barden (1966). Their objective was to overcome most of the disadvantages of the conventional oedometer apparatus (described in Volume 2, Chapter 14) when performing consolidation tests on low-permeability soils, including non-uniform deposits.

This design of cell differs from a conventional oedometer in that the test specimen is loaded hydraulically by water pressure acting on a flexible diaphragm, instead of by a mechanical lever system. This arrangement enables specimens of large diameter (up to 254 mm (10 in) diameter in commercial work) to be tested, and allows for large settlement deformations. But the most important features are the ability to control drainage and to measure pore water pressure during the course of consolidation tests. Several drainage conditions (vertical or horizontal) are possible, and back pressure can be applied to the specimen.

In the apparatus known as the Rowe cell, the diaphragm through which the hydraulic pressure applies stress to the test specimen is of the extendable-bellows type. Other types of diaphragm, such as a rolling-seal type, are also available but the principles and methods of test are the same.

The advantages of this type of cell over the conventional oedometer apparatus, together with some illustrations of its application, were described in detail by Rowe (1968) and are summarised in Section 22.1.3. Further applications, especially with regard to the effect of the 'fabric' of non-homogeneous soils, were demonstrated in the Twelfth Rankine Lecture (Rowe, 1972). (See Section 22.1.8.)

Background information on the cell and its development, ancillary equipment, and its advantages, uses and applications are outlined in Sections 22.1.2 to 22.1.8. Aspects of theory relevant to analysis of test data are given in Section 22.2

Preparation of the equipment for a test, sample preparation, and assembly of the cell are described in Sections 22.3 to 22.5. The consolidation test procedures described in Section 22.6 deal first with the most common type of test, which is described in detail. This is the vertical drainage test with measurement of pore water pressure, referred to as the 'basic' test. Variations in procedures for other types of consolidation test follow, and the relevant curve fitting procedures are explained. Permeability tests are described in Section 22.7.

The consolidation and permeability test procedures described here are generally as specified in Clauses 3 and 4 of BS 1377:Part 6:1990. The Rowe cell can also be used for several types of continuous-loading consolidation tests, but these are beyond the scope of this book.

22.1.2 Historical Outline

Developments leading to the present design of the Rowe cell are traced in outline below.

DIAPHRAGM LOADING

An apparatus using the diaphragm loading principle for confined compression tests on sands was described by Rowe (1954). Air pressure was used on a flat flexible membrane in contact with a 10 in diameter specimen. A convoluted diaphragm was adopted for the present design of Rowe cells to permit large settlement deformations of the specimen (Rowe and Barden, 1966).

Cells of 3 in, 6 in, 10 in and 20 in diameter were built to that design at Manchester University, all fitted with pressure transducers for measuring pore water pressure. The 3 in diameter cell became commercially available in 1966, and the 10 in and 6 in sizes followed in 1967. The 20 in cell was intended only for research purposes.

Oedometers using hydraulic loading were designed independently by Bishop, Green and Skinner at Imperial College, London (Simons and Beng, 1969), for testing 3 in and 4 in diameter specimens. Provision was made for pore water measurement and the application of back pressure.

A modified 10 in Rowe cell, using air pressure acting on a flat diaphragm of butyl rubber, was described by Barden (1974) in connection with the consolidation of 'dry' soils, i.e. soils with continuous interconnected air voids.

A subsequent development at Manchester University was the construction of consolidation cells 500 mm square and 1 metre square, using the same diaphragm loading principle, for consolidating large specimens prior to tests in their centrifuge.

PORE PRESSURE MEASUREMENT

Early attempts to measure pore water pressures in consolidation tests were confined to research experiments in the USA. Following the introduction of the pore pressure measuring equipment developed by Bishop and Henkel in 1957, consolidation with pore pressure measurement in the triaxial cell on specimens up to 100 mm diameter became an established procedure in Britain.

Pore pressure measurements were made by Leonards and Girault (1961) in a conventional type of oedometer cell 112 mm diameter, using a manually controlled mercury/water pore pressure device. A major step forward was the application of an electrical pressure transducer to the measurement of pore water pressure by Whitman, Richardson and Healy (1961). They described the advantages of this device and investigated the compliance and response time of the system. Pore pressure transducers and associated monitoring systems were subsequently improved and adapted to soil testing requirements enabling Rowe and Barden to fit them to their cells from the outset.

HORIZONTAL DRAINAGE

Drainage in a horizontal (radial) direction during consolidation was first carried out in a triaxial cell, using filter paper side drains as described by Bishop and Henkel (1957) (see Section 20.2.2). Investigations by Rowe (1959) in a lacustrine clay using this procedure showed that even when filter paper was fitted around the specimen in contiguous strips it did not form an effective drain for that type of soil. In radial drainage tests on 4 in diameter triaxial specimens of clay, Escario and Uriel (1961) reached the same conclusion and used instead a surrounding layer of micaceous sand 5 mm thick, which proved satisfactory. McKinlay (1961) contained the specimen in a porous stainless steel ring for radial drainage tests in a standard oedometer cell.

The Vyon porous plastics sheet material used by Rowe and Barden has been found to be entirely satisfactory as a peripheral drain, having a permeability higher than that of silts. Normally a thickness of 1.5 mm provides enough drainage capacity.

Radial drainage inwards to a central sand drain was achieved in a conventional oedometer by Rowe (1959). Tests using porous ceramic discs to represent horizontal drainage layers in the consolidation of 3 in, 6 in and 10 in specimens with drainage radially inwards to drains of clean fine sand, were carried out by Rowe and Shields (1965). The same authors (Shields and Rowe, 1965) reported that a sand drain diameter not exceeding 5% of the specimen diameter had little effect on the measured compressibility of the specimen, and that a thin-walled mandrel used for forming the drain caused little disturbance of the soil. Singh and Hattab (1979) described several types of mandrel for forming central drains for Rowe cell tests.

WALL FRICTION

The effect of wall friction in a conventional oedometer cell was studied by Leonards and Girault (1961). They found that a coating of Teflon on the cell walls virtually eliminated friction for loads above a certain critical value in that type of cell. Silicone grease has been found to be equally effective, and is now commonly used in the Rowe cell.

CURVE FITTING

Analysis of consolidation by horizontal radial drainage required the extension of the Terzaghi theory to these conditions. The theoretical analysis for radial drainage to drainage wells was developed by Barron (1947), and that for drainage radially outwards to a continuous peripheral drain was given by McKinlay (1961). The outcome of these investigations is explained in Section 22.2.3, and provides the basis for the curve fitting procedures given in Section 22.6.6.

22.1.3 Advantages of the Rowe Cell

The Rowe cell has many advantages over the traditional oedometer consolidation apparatus. The main features responsible for these improvements are the hydraulic loading system; the control facilities and ability to measure pore water pressure; and the capability of testing specimens of large diameter. Advantages under these headings are stated below. Some of these points are amplified in Section 22.1.8.

HYDRAULIC LOADING SYSTEM

(1) The specimen is less susceptible to vibration effects, which a lever loading system can magnify, than in a conventional oedometer press.

(2) Pressures of up to 1000 kPa can be applied easily even to large specimens.

(3) Corrections required for the deformation of the loading system when subjected to pressure are negligible, except perhaps for very stiff soils.

CONTROL FACILITIES

(4) Drainage of the specimen can be controlled, and several different drainage conditions can be imposed on the specimen.

(5) Control of drainage enables loading to be applied to the specimen in the undrained condition, allowing full development of pore pressure. Consequently the initial immediate settlement can be measured separately from the consolidation settlement, which starts only when the drainage line is opened.

(6) Pore water pressure can be measured accurately at any time and with immediate response. Pore pressure readings enable the beginning and end of the primary consolidation phase to be positively established.

(7) The volume of water draining from the specimen can be measured, as well as surface settlement.

(8) The specimen can be saturated either by applying increments of back pressure until a B value of near unity is obtained, or by controlling the applied effective stress, before starting consolidation.

(9) Tests can be carried out under an elevated back pressure, which ensures fully saturated conditions, gives a rapid p.w.p. response, and ensures reliable time relationships (Lowe, Zaccheo and Feldman, 1964).

(10) The specimen can be loaded either by applying a uniform pressure over the surface ('free strain'), or through a rigid plate which maintains the loaded surface plane ('equal strain'). Free strain loading through the flexible diaphragm allows individual pockets of soil to reach their own equilibrium condition.

(11) Fine control of loadings, including initial loads at low pressures, can be accomplished easily.

LARGE SPECIMEN SIZE

(12) Tests on large specimens provide more reliable data for settlement analysis (in reality a three-dimensional problem) than conventional one-dimensional oedometer tests on small specimens.

(13) Large specimens (i.e. 150 mm diameter and 50 mm thick, or larger) have been found to give higher, and more reliable, values of c_v, especially under low stresses, than conventional oedometer test samples (McGown, Barden and Lee, 1974). Better agreement has been reported between predicted and observed rates of settlement, as well as their magnitude (Lo, Bozozuk and Law, 1976). This may be partly due to the relatively smaller effect of structural viscosity in larger specimens, quite apart from the effect of fabric referred to below.

(14) The use of large specimens enables the effect of the soil macro structure (the 'fabric') to be taken into account in the consolidation process, thereby enabling a realistic estimate

of the rate of consolidation to be made (Rowe, 1968 and 1972). A common example is in layered deposits.

(15) Large specimens generally suffer appreciably less disturbance to the microfabric than do small specimens. Excessive disturbance affects the $e/\log p$ plot and tends to obscure the effect of stress history; gives low values to preconsolidation pressure and over-consolidation ratio; and gives high m_v values at low stresses. Tests on high quality large diameter specimens minimise these shortcomings.

(16) Large specimens permit reliable measurements of permeability to be made, both vertically and horizontally, under known stress conditions and taking into account the effect of the soil fabric.

(17) The 250 mm diameter cell is large enough to install a central drainage well, for a realistic assessment of drainage wells in the soil.

22.1.4 Description of Cell and Accessories

DESIGN

Three sizes of Rowe cell are commercially available, based on the 3 in, 6 in and 10 in diameters originally developed by Prof. Rowe. They are referred to in this chapter by the rounded metric equivalents of 75 mm, 150 mm, and 250 mm diameter, and typical cells are shown in Fig. 16.8 (Section 16.2.4). Exact sizes are shown in Table 22.1 together with other relevant dimensions. The diameters of a series of cells produced by ELE are very slightly smaller than the above, so as to give rational values of area, i.e. 45, 180 and 500 square centimetres. These sizes are also listed in Table 22.1.

The following description relates generally to the 250 mm diameter cell, the main features of which are shown in Fig. 22.1. The other sizes are similar in principle and differ mainly in details of construction, as indicated.

The cell itself consists of three parts: the body, the cover, and the base. In an early commercial design the body was machined from aluminium bronze, the base was made of steel and the top of aluminium alloy. The presence of three different metals in contact with chemically aggressive water in a soil sample sometimes gave rise to serious corrosion, due to chemical reaction and electrolytic effects. The design referred to above uses an aluminium

Table 22.1. DIMENSIONS OF ROWE CELLS

Nominal diameters		3 in (75 mm)		6 in (150 mm)		10 in (250 mm)	
Sample diameter:							
exact equivalent	mm	76.2		152.4		254	
new series	mm		75.7		151.4		252.3
Sample area	mm²	4560	4500	18241	18000	50671	50000
Recommended sample height	mm		30		50		90
Sample volume (based on recommended height)	cm³	136.8	135	912.2	900	4560.4	4500

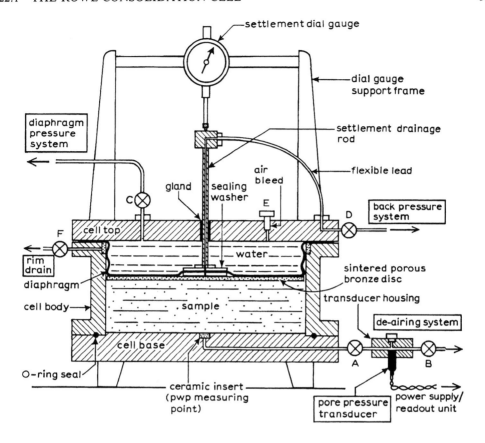

Fig. 22.1 Main features of 250 mm diameter Rowe hydraulic consolidation cell

alloy for all three components, suitably treated to eliminate porosity, together with a smooth plastic lining bonded on to the base and sides.

CELL COMPONENTS

The *cell body* has a flange at each end with bolt holes for securing the base and cover (Fig. 22.2 (a)). On the 75 mm and some 150 mm cells there are no flanges, but the body is clamped between the base and cover by long tie-bolts, as shown in Fig. 22.2 (b). The flanged design enables two bodies to be bolted together when required. Washers are fitted under bolt-heads and nuts. A recess at the upper end, referred to as the rim drain, has an outlet leading to a valve (valve F in Fig. 22.1), but this feature is not usually provided in the smaller cells. The inside face of the body must be smooth and free from pitting or other irregularities, which may be achieved by means of a bonded lining of plastics material.

The *cell cover* is fitted with a convoluted flexible bellows (the diaphragm) of natural or synthetic rubber, the outer edge of which provides a seal between the cover and upper body flange. The diaphragm transmits a uniform load to the soil below it by means of water pressure acting on top (Fig. 22.1). A brass or aluminium alloy hollow spindle passes through a low-friction seal in the centre of the cover. Its lower end passes through the

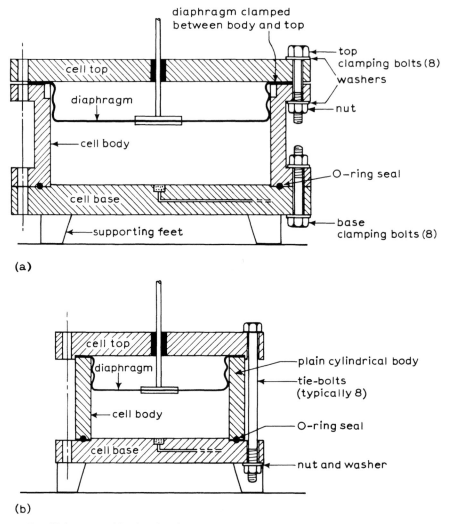

Fig. 22.2   *Assembly details of Rowe cells: (a) 250 mm diameter, (b) 75 mm and some 150 mm diameter*

centre of the diaphragm where it is sealed by two metal washers. Its upper end is connected by flexible tubing to the drainage valve D fitted to the edge of the cover. The blanked-off upper end of the spindle provides a bearing for the anvil of the settlement dial gauge, which is rigidly supported by a bracket assembly bolted to the cover. The cover is also fitted with an inlet valve (C) for connection to a constant pressure water system for applying the vertical load on to the specimen, and an air bleed screw (E).

The *cell base* is fitted with a recess for an O-ring seal to bear against the lower body flange. At the centre of the base is a small circular recess containing a porous ceramic or plastics insert bonded in place. This is the main pore water pressure measuring point, and leads to valve A (Fig. 22.1) on the outer edge, to which is connected the electric pressure transducer housing block and valve B. In the 250 mm cell three additional pore pressure points are provided at distances of 70 mm (0.55 times the cell radius R), 25 mm ($0.1R$), and

114 mm (0.9R) from the centre. Their connecting ports are sealed by blanking plugs on the edge of the base when not in use. The base sits on three metal block supporting feet.

ACCESSORIES

Items used with the cell, and normally supplied with it or obtainable as spares, are as follows:

Rigid circular loading plate with central drainage hole, which can be plugged, and detachable lifting handle, fitted on top of the bronze disc when required to provide 'equal strain' loading.

Porous discs, typically of sintered bronze, placed on top of the specimen immediately beneath the diaphragm to collect water draining vertically from the specimen during consolidation and conduct it to the drainage outlet spindle. A 75 mm diameter disc can usually be regarded as 'rigid', but discs of the larger sizes have some flexibility. (This type of disc is not mentioned in BS 1377.)

Disc of porous plastic material (Vyon) 3 mm thick, to provide flexibility and uniformity of loading pressure on the top surface of the specimen for the 'free strain' (uniform stress) condition.

Porous Vyon sheet, 1.5 mm thick, for forming a peripheral drain.

Dial gauge for measuring vertical settlement (see Instrumentation below).

Spare porous inserts for pore pressure measuring points.

Spare O-ring base seal.

Spare diaphragm.

Flange sealing ring for use when bolting two cell bodies together (see Section 22.4.2).

Typical properties of porous discs and other drainage media are summarised in Table 22.2.

ANCILLARY EQUIPMENT

Other items of equipment necessary for carrying out the tests are detailed in Sections 22.1.5 and 22.1.6. Specifications for these items are the same as those used for triaxial tests, as described in Chapter 16 and referred to in Section 18.2.1.

Table 22.2. PROPERTIES OF DRAINAGE MEDIA FOR ROWE CELLS

Material	Typical details	Typical permeability
Sintered bronze porous disc	3 mm thick	0.77×10^{-6} m/s
Flexible 'Vyon' porous plastic:		
Top disc	3 mm thick	4×10^{-10} m/s
Peripheral drain	1.5 mm thick	
Sand for central drainage well	Washed fine sand 90–300 μm	2×10^{-4} m/s
		Not less than 3×10^4 times that of test specimen; preferably 10^6 times*

*(Gibson and Shefford, 1968).

22.1.5 Instrumentation

MANUAL RECORDING

Instrumentation necessary for the 'basic' Rowe cell consolidation test (see Section 22.1.7(a)), with manual recording of data, is described below.

Vertical settlement of the centre of the specimen is measured with a dial gauge, mounted as described in Section 22.1.4. For the smaller cells a gauge with 10 mm travel reading to 0.002 mm is suitable, but for a 250 mm specimen in which larger settlements are expected a 50 mm travel gauge reading to 0.01 mm is more appropriate.

Drainage water is led via valve D (Fig. 22.1) to a volume-change indicator on the back pressure line (see Section 22.1.6).

Pore pressure at the centre of the base of the specimen is measured by a calibrated electric pressure transducer mounted in the de-airing block (usually of brass) connected to the cell base at valve A (Fig. 22.1). The other valve B is connected to a de-airing and flushing system (see below).

The vertical stress applied to the specimen is measured by means of the pressure gauge on the diaphragm pressure system connected to valve C (see below).

During a 'basic' consolidation test, measurements of the following are observed and recorded at appropriate intervals.

Diaphragm pressure
Back pressure
Pore water pressure
Vertical settlement
Volume of water draining out
Time

ELECTRONIC INSTRUMENTATION

All the above parameters can be easily monitored by electronic instruments. To accomplish this a linear displacement transducer is substituted for the settlement dial gauge, and pressure transducers are fitted to the diaphragm pressure and back pressure lines (outlets C and D, Fig. 22.1). A transducerised volume-change indicator is used on the back pressure line in place of the burette gauge. Pore pressure is measured by the transducer at A as described above.

To operate this system a power supply/signal conditioning and readout unit with five separate channels plus timer is required. Incoming signals can be fed to a data-logging system, or to a computer programmed to process the readings and print out tabulated and graphical data automatically (see Chapter 16, Section 16.6.8).

22.1.6 Ancillary Equipment

ITEMS REQUIRED FOR EACH TEST

The following equipment is necessary for each test being carried out in a Rowe cell. A typical general arrangement of the apparatus for the most common type of test is shown diagrammatically in Fig. 22.3.

(1) Two independently controlled pressure systems, giving maximum pressure up to 1000 kPa, one for loading the diaphragm and one for providing the back pressure. Either

mercury/water, or motorised oil/water, or motorised air/water systems may be used (see Section 16.3), but the air/water system is generally the most satisfactory because of the large volume capacity available. Each pressure system should include a calibrated pressure gauge of 'test' grade, reading to 5 or 10 kPa, on the water line close to the cell.

(2) Calibrated volume-change indicator on the back pressure line, reversible twin tube paraffin type, capacity related to the size of specimen (Section 16.5.6).

(3) Power supply and readout unit for the electric pore pressure transducer. This unit is usually calibrated to give a digital display in pressure units (kPa) (Section 16.6.7).

(4) Timer, reading to 1 second.

(5) Calibrated thermometer reading to 0.5 °C.

(6) Valves and connecting tubing are of the same type as referred to in Section 18.2.1 for triaxial apparatus.

(7) Materials and consumables:

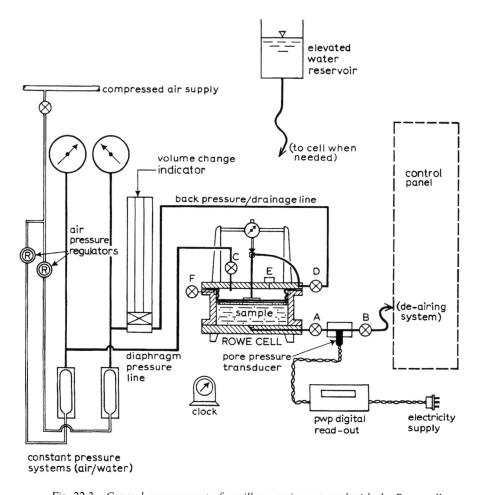

Fig. 22.3   General arrangement of ancillary equipment used with the Rowe cell

Silicone grease   ⎫   (requirements
Vyon porous plastic sheet, 1.5 mm and 3 mm thick   ⎬   depend on
Latex rubber sheet   ⎭   type of test)

(8) For permeability tests, a third pressure system (additional to item 1 above) is needed.

ITEMS REQUIRED FOR PRE-TEST CHECKS

The following items are needed for assembling the cells and checking the systems prior to testing. One set can serve a number of cells.

(9) Pore pressure panel, as described in Section 16.5.5 (Fig. 16.10) for flushing, de-airing and transducer calibration.

(10) System for de-airing water under vacuum (see Volume 2, Chapter 10) and distribution pipework.

(11) Vacuum pump and pipework installation.

(12) Elevated water reservoir, such as a permeability constant-head tank (Chapter 10), the level of which can be adjusted.

(13) Spanners to fit clamping bolts and nuts (two required).

(14) Seals and clamps for blanking off pore pressure points.

(15) Immersion tank to contain the cell when being assembled.

ITEMS REQUIRED FOR SPECIMEN PREPARATION AND MEASUREMENT

The following items are required for the preparation of undisturbed test specimens and their measurement. Details are given in Section 22.4.1.

(16) Extruder for ejecting undisturbed samples from piston tubes directly into the cell. A large compression load frame can be adapted for extruding large diameter samples as outlined in Section 22.4.1 (Fig. 22.14). Saturated soils containing organic matter should preferably be extruded under water, as shown in Fig 22.16.

(17) Cutting shoe of appropriate diameter, for trimming the sample to the required specimen diameter. Shoes of two diameters are needed for each size of cell, one corresponding to the cell diameter and one of slightly smaller diameter for specimens that are to be tested with a peripheral drain (Section 22.5.3).

(18) Standard laboratory small tools and equipment for cutting and trimming undisturbed specimens.

(19) Equipment for measuring and weighing specimens (including balance of high enough capacity for large samples in cells).

(20) Equipment for moisture content measurement.

(21) Apparatus for determining the particle density.

(22) Mandrel and guide jig for forming vertical drainage hole (for tests with radial drainage to the centre only). Details including methods of use are given in Section 22.5.4.

22.1 THE ROWE CONSOLIDATION CELL

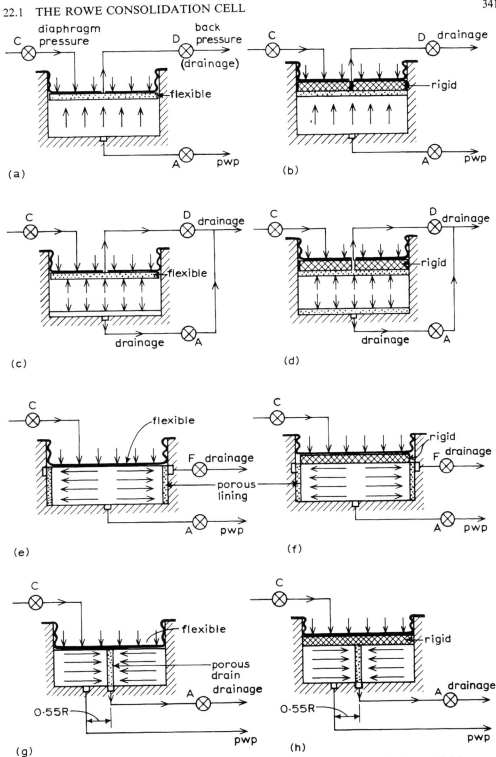

Fig. 22.4 *Drainage and loading conditions for consolidation tests in Rowe cell: (a),*
(c), (e), (g) with 'free strain' loading, (b), (d), (f), (h) with 'equal strain' loading

22.1.7 Types of Test

CATEGORIES OF SPECIMEN

Tests may be carried out in the Rowe cell on the following categories of specimen.

(1) Undisturbed specimen, either extruded from a sampling tube or hand-trimmed from a block.

(2) Remoulded soil, preconsolidated in the cell from a slurry.

(3) Compacted soil, prepared in the cell body by static compression or dynamic compaction.

(4) Specimens that have been remoulded or compacted elsewhere, then trimmed and transferred to the cell as for undisturbed material.

Preparation of these types of specimen is described in Section 22.4.

DESIGNATION OF CONSOLIDATION TESTS

Consolidation tests using incremental loading stages can be carried out on any of the above specimen types under four different conditions of drainage, each of which may be subject to two types of loading ('free strain' or 'equal strain', see below), giving eight different types of test. The eight possible drainage and loading configurations are shown in Fig. 22.4 (a) to (h). In practice not all combinations are used. The test type designations (a) to (h) shown here are referred to as such throughout this chapter.

(a) *Free strain, vertical single drainage*

Vertical drainage to the top surface of the specimen only (single drainage), with pore water pressure measured at the centre of the base under flexible surface loading (uniform pressure distribution, i.e. 'free strain' loading). See Fig. 22.4 (a).

This is the most usual type of test carried out in the Rowe cell, and is the one described here as the 'basic' test.

(b) *Equal strain, vertical single drainage*

As (a), but under a rigid loading plate which maintains a plane surface to the specimen, i.e. the 'equal strain' loading condition. See Fig. 22.4 (b).

(c) *Free strain, vertical double drainage*

Vertical drainage to both top and bottom surfaces of the specimen simultaneously (double drainage), without measurement of pore water pressure, with 'free strain' loading. See Fig. 22.4 (c).

(d) *Equal strain, verticle double drainage*

As (c) but under a rigid plate giving 'equal strain' loading. See Fig. 22.4 (d).

(e) *Free strain, horizontal outward drainage*

Horizontal (radial) drainage to a pervious boundary at the perimeter with the top and bottom faces sealed, but measuring pore water pressure at the centre of the base; 'free strain' loading. See Fig. 22.4 (e).

(f) *Equal strain, horizontal outward drainage*

As (e) but under 'equal strain' loading. See Fig. 22.4 (f).

(g) *Free strain horizontal inward drainage*

Horizontal (radial) drainage to a drainage well at the central axis, under 'free strain' loading. Pore water pressure is measured in the base at a distance of $0.55R$ from the centre (where R is the radius of the cell); other off-centre pore pressure points may be used in addition if required. See Fig. 22.4 (g).

(h) *Equal strain, horizontal inward drainage*

As (g) but under 'equal strain' loading. See Fig. 22.4 (h).

   Methods of analysis and curve fitting procedures vary according to the type of test, and are explained in Section 22.2.3.

DESIGNATION OF PERMEABILITY TESTS

Direct measurement of permeability can be made on any of the above categories of specimen, either as an independent test or on a specimen undergoing a consolidation test, under a known vertical effective stress. Four different types of test are designated as follows, and the flow conditions are indicated in Fig. 22.5.

(j) Permeability test with the flow of water vertically upwards (Fig. 22.5 (j)).
(k) As (j) but with flow vertically downwards (Fig. 22.5 (k)).
(l) Permeability test with horizontal flow radially outwards to a pervious peripheral drainage layer (Fig. 22.5 (l)).
(m) As (l) but with horizontal flow radially inwards to a central drain (Fig. 22.5 (m)).

Permeability tests are usually carried out under 'equal strain' loading to maintain a uniform specimen thickness, but this does not preclude permeability measurements on a specimen undergoing a 'free strain' consolidation test.
   The calculation of permeability from vertical flow tests is similar to that for an ordinary permeameter test, but a different equation is necessary for radial flow tests (see Sections 22.2.7 and 22.7.3).

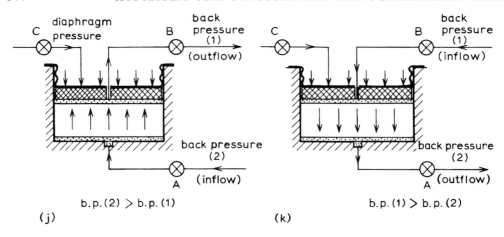

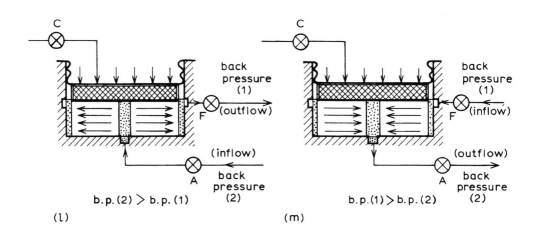

Fig. 22.5   *Flow conditions for permeability tests in Rowe cell; (j) and (k) vertical flow, (l) flow radially outwards, (m) flow radially inwards*

CONTINUOUS-LOADING CONSOLIDATION TESTS

The Rowe cell can be used or adapted for several types of 'rapid' consolidation test in which loading is applied continuously or in a single increment. These tests are beyond the scope of this book.

22.1.8 Applications

SOIL FABRIC

As stated in Section 22.1.3, the 150 and 250 mm diameter cells allow for the testing of undisturbed specimens that are large enough to include representative structural features referred to as the 'soil fabric'. Rowe (1968) found that tests on 250 mm diameter specimens gave much better agreement with observed in-situ behaviour than conventional oedometer tests.

The term 'soil fabric' relates to local variations in soil composition together with discontinuities such as joints and fissures which can dominate engineering behaviour. It is used here in the sense defined by Rowe (1972), and refers to the size, shape and arrangement of solid particles, organic inclusions and voids, especially when existing in layers having different particle sizes. Clays that are layered, laminated, varved or fissured, or contain inclusions such as organic matter which may provide seepage paths, are the more usual soil types in which fabric structure is significant.

In these soil types the effect of the fabric is the dominating influence on the drainage properties of the soil mass as a whole, and consequently on the consolidation (or swelling) characteristics. As an example, relatively permeable silt layers only 0.1 mm thick at intervals of 3 metres in a low permeability clay of 10 m or more thickness can drastically increase the drainage capacity and the rate of settlement of the deposit (Rowe, 1968). Meticulous inspection of the entire thickness is necessary to begin with (for example by using the Delft continuous sampling procedure in soft alluvial deposits (Delft Soil Mechanics Laboratory, 1966)) followed by very careful sampling at selected locations, in order to obtain representative samples of important features (Rowe, 1972). Tests on large specimens are then justified, because small specimens cannot adequately represent the fabric structure.

Large diameter undisturbed specimens of good quality, which are essential for these tests, can be obtained from soft clays by using special piston sampling equipment with sample tubes of about 160 or 260 mm internal diameter in boreholes; or by hand sampling from excavations, trial pits and trenches.

VALUE OF TESTS

Although sampling and testing of large diameter specimens are several times more expensive than procedures using smaller conventional specimens, the cost per unit volume of soil tested is many times less. In addition the results from tests on a few large specimens can be markedly more reliable than the sum total of those on a large number of conventional specimens.

USUAL APPLICATIONS

The facilities and degree of control provided by the large Rowe cells enable tests relating to numerous applications to be carried out, the most usual of which are summarised as follows. Items 1 to 4 have been covered in greater detail in Section 22.1.7.

(1) Consolidation with vertical drainage (single or double), or with horizontal drainage (radially outwards or radially inwards).

(2) Measurement of permeability in the vertical or horizontal directions.

(3) Choice of loading conditions, either 'free strain' or 'equal strain'.

(4) Simulation of drainage wells and estimation of the optimum spacing of vertical drains (Chalmers, 1982).

(5) Consolidation of soil initially deposited as slurry, to investigate the properties of fresh sediments, whether natural or man-made (e.g. 'tailings' deposits in a lagoon).

(6) Consolidation of overconsolidated clay that has been slurrified to obtain normally consolidated specimens for other tests.

(7) Imposition of cyclic loading and observation of consequent pore pressure response.

(8) Observation of instantaneous peak pore pressure readings in liquefaction tests.

(9) Consolidation under controlled continuously variable loading.

Items (7) to (9) are not covered in this volume.

The small (75 mm) diameter Rowe cell can be used for specimens prepared from traditional 100 mm diameter undisturbed tube samples for exploratory purposes. This size is also suitable for clays that are truly intact and homogeneous. The same pressure control and pore pressure measuring facilities as on the larger cells are used, but effective pressures lower than about 50 kPa should be avoided.

SELECTION OF TEST

Selection of the type of test depends on the particular conditions and requirements. The following comments give a general guide.

(1) For the determination of c_v, tests with drainage to the upper surface and with pore pressure measurement at the base are the most usual. 'Free strain' loading (Fig. 22.4 (a)) is normally preferred because the settlement measured at the centre of the specimen is not restrained by wall friction.

(2) Vertical drainage with 'equal strain' (Fig. 22.4) (b)) is the test most closely related to traditional oedometer tests, and settlement readings can be directly related to volume change measurements in the drainage line. Both measurements represent 'average' consolidation.

(3) For the determination of c_h, horizontal drainage to the periphery with 'free strain' loading is usually preferred (Fig. 22.4 (e)). Pore pressure is measured at the centre of the base.

(4) Tests with drainage to the centre, to simulate drainage wells, are usually carried out under 'equal strain' loading (Fig. 22.4 (h)), with measurement of pore pressure at $0.55R$ from the centre of the base.

(5) Consolidation tests with horizontal drainage theoretically drain nine times more rapidly outwards to the periphery than inwards towards a central drain.

(6) However, radial drainage under a rigid loading surface can result in non-uniformity across the diameter of the specimen after consolidation, especially when consolidating specimens from slurry (Sills, 1983). The pore pressure at the circumference falls rapidly when drainage starts, increasing the effective stress and giving an outer annulus of soil of greater stiffness than the remainder. Under subsequent increments of load this annulus supports an increasing proportion of the total stress and becomes progressively more stiff, leaving the middle part much softer. Radial drainage under 'equal strain' therefore cannot produce the same final specimen as is obtained by using verticle drainage.

(7) Permeability tests with either vertical or horizontal drainage are usually carried out under 'equal strain' loading in order to maintain a uniform thickness of specimen.

22.2 THEORY

22.2.1 Scope

The Terzaghi theory of consolidation and its application to the standard oedometer consolidation test (referred to here as the 'oedometer test') were outlined in Volume 2, Chapter 14. Some extensions to the theory relevant to the various conditions of test in the Rowe consolidation cell are discussed in this section. The emphasis is on the application of curve-fitting procedures for deriving the coefficient of consolidation from laboratory graphical plots. Some other topics are referred to briefly, together with references for further reading.

Most of the general calculations are similar to those described in Volume 2 for the oedometer test, and are illustrated in Section 22.6.6.

22.2.2 Coefficient of Consolidation

VERTICAL DRAINAGE

In a vertical double-drainage test in the Rowe cell (tests type (c) and (d)) the drainage conditions are the same as in the conventional oedometer test and the coefficient of consolidation, denoted by c_v, is derived in the same way if settlement or volume change data are used. With single drainage (tests type (a) and (b)) the principle is the same except that the drainage path h is equal to the full height of the specimen ($h = H$) instead of half the height.

HORIZONTAL DRAINAGE

For drainage with laminar flow in the horizontal direction the coefficient of consolidation, denoted by c_h, is often many times greater than for vertical drainage. In Rowe cell tests of types (e) to (h), drainage is not laminar (flow lines parallel) but radial (axially symmetrical). For the condition of drainage radially outwards to the periphery (Fig 22.4 (e) and (f)) the coefficient of consolidation is denoted by c_{ro}; for drainage radially inwards to a central well (Fig 22.4 (g) and (h)) it is denoted by c_{ri}.

The relationship between the coefficient of consolidation c_h and c_v in stratified clays was derived mathematically by Rowe (1964). This analysis showed that the value of c_h is dependent on the geometry and internal structure of the deposit, and confirmed earlier research (Rowe, 1959) on the consolidation of lacustrine clays. The relevance of this relationship to the design of sand drains was explained. The radial coefficients c_{ro}, c_{ri} approximate to c_h if the spacing of the more permeable layers is less than about 1/10th of the length of the radial drainage path.

GRAPHICAL ANALYSIS

From each stage of a Rowe cell consolidation test, graphical plots are obtained of settlement, volume change and (in most cases) pore pressure, against a function of time, which may be logarithmic or a power function. These graphs are used, as in the oedometer test, to derive the time corresponding to 50% or 90% of theoretical primary consolidation (t_{50} or t_{90}), from which the coefficient of consolidation can be calculated by using an equation with the appropriate multiplying factor.

Wherever possible it is better to use the pore pressure dissipation graph rather than a settlement or volume change curve because the end points (0 and 100% dissipation) are both clearly defined and t_{50} or t_{90} can be read directly from the graph. The t_{50} point is preferable because the middle portion of the curve has the best fit to the theoretical curve.

FACTORS FOR CALCULATION

The factors used for calculating the coefficient of consolidation depend upon:

Boundary conditions ('free strain' or 'equal strain')

Type of drainage (vertical, or radially inwards or outwards)

Location of relevant measurements.

The first two are self-evident (see Fig. 22.4), but the effect of the third on 'curve fitting' requires some comment.

Settlement and volume-change measurements are governed by the deformation of the specimen as a whole, and analysis is dependent on an overall 'average' behaviour. Some method of 'curve fitting' is necessary for graphs based on these measurements, the details depending upon the test conditions. On the other hand pore pressure measurements relate to the conditions at a particular point, usually the centre of the base for vertical and radial-outward drainage, or offset from centre at a specified radius for radial-inward drainage. 'Curve fitting' is not needed for pore pressure dissipation curves. However it is important not only to distinguish between the two types of measurement, but also to select the multiplying factor appropriate to the test conditions when using the equation for calculating c_v or c_{ro} or c_{ri}. A different multiplying factor is needed in every case.

The relevant data are summarised in Table 22.3, in which the consolidation location is either described as 'average' for settlement or volume-change measurements, or is stated as the point at which pore pressure is measured (e.g 'centre of base'). From these theoretical values, multiplying factors for applying to test data are obtained as explained below. Some applications are illustrated in Section 22.6.6, and the multiplying factors are summarised in Table 22.5.

22.2.3   Curve Fitting Procedures

SCOPE

Curve fitting procedures for use with graphical plots of settlement or volume change against a function of time are described below, for each of the eight test conditions shown in Fig. 22.4. The power function (usually square-root) method is generally used, and provides a value of t_{50} or t_{90}.

Use of pore pressure dissipation plots is also included. Pore pressure dissipation (percent) is usually plotted against log time, and the value of t_{50} is obtained directly from the graph.

VERTICAL DRAINAGE, ONE-WAY (Fig. 22.4 (a) and (b))

The equations governing the consolidation process in the Rowe cell are the same as in the standard oedometer test. Settlement or volume-change graphs are used in the same way, whether 'free strain' or 'equal strain' loading is applied. The relationship between the theoretical time factor, T_v (plotted to a logarithmic scale) and degree of consolidation is

Table 22.3 ROWE CELL CONSOLIDATION TESTS – DATA FOR CURVE FITTING

Test ref.	Drainage direction	Boundary strain	Consolidation location	Theoretical time factor T_{50}	T_{90}	Time function	Power curve slope factor	Measurements used	Coefficient of consolidation $m^2/year$
(a) and (b)	Vertical, one way	Free and equal	Average	0.197 (T_v)	0.848	$t^{0.5}$		ΔV or ΔH^*	$c_v = 0.526\dfrac{T_v H^2}{t}$
			Centre of base	0.379	1.031		1.15	p.w.p.	
(c) and (d)	Vertical, two way	Free and equal	Average	0.197 (T_v)	0.848	$t^{0.5}$	1.15	ΔV or ΔH^*	$c_v = 0.131\dfrac{T_v H^2}{t}$
								p.w.p.	
(e)	Radial, outward	Free	Average	0.0632 (T_{ro})	0.335	$t^{0.465}$		ΔV	$c_{ro} = 0.131\dfrac{T_{ro} D^2}{t}$
			Central	0.200	0.479		1.22	p.w.p.	
(f)		Equal	Average	0.0866 (T_{ro})	0.288	$t^{0.5}$		ΔV or ΔH	$c_{ro} = 0.131\dfrac{T_{ro} D^2}{t}$
			Central	0.173	0.374		1.17		
(g)	Radial, inward†	Free	Average	0.771 (T_{ri})	2.631	$t^{0.5}$		ΔV	$c_{ri} = 0.131\dfrac{T_{ri} D^2}{t}$
			$r = 0.55R$	0.765	2.625		1.17	p.w.p.	
(h)		Equal	Average	0.781 (T_{ri})	2.595	$t^{0.5}$		ΔV or ΔH	$c_{ri} = 0.131\dfrac{T_{ri} D^2}{t}$
			$r = 0.55R$	0.778	2.592		1.17	p.w.p.	

(see also Table 22.5)

†Drain ratio 1/20
* ΔH with equal strain only
T_v, T_{ro}, T_{ri} = theoretical time factors
t = time (minutes)
H = specimen height (mm)
D = specimen diameter (mm)

shown by curve A in Fig. 22.6. (See also Volume 2, Fig. 14.7; and see Fig. 14.8 for the corresponding square-root plot.) Values of theoretical time factors of T_{50} and T_{90} at 50% and 90% consolidation are 0.197 and 0.884 respectively (Table 14.2). When using a graph of settlement or volume change against square-root time, the 'slope factor' for obtaining t_{90} is 1.15, as shown in Volume 2, Figs. 14.10 and 14.11.

The theoretical time factor curve for pore pressure dissipation at the centre of the base of the specimen is shown by curve B in Fig. 22.6. This gives theoretical time factors T_{50} and T_{90} of 0.379 and 1.031 respectively. These factors are used in conjunction with the pore pressure dissipation curve for vertical drainage.

All the above factors apply to both 'free-strain' and 'equal strain' loading, and appear in Table 22.3.

Whichever type of plot is used, for one-way drainage the maximum length of drainage path h is equal to the mean height of the specimen H at any instant. The value of c_v may be calculated from either of the equations

$$c_v = \frac{T_{50} \cdot H^2}{t_{50}}$$ (22.1)

or

$$c_v = \frac{T_{90} \cdot H^2}{t_{90}}$$ (22.2)

depending on whether the 50% or the 90% consolidation point is used. In these equations, H is expressed in mm, t_{50} and t_{90} in minutes; therefore c_v is a mm^2/minute. Converting to m^2/year, Equation (22.1) becomes

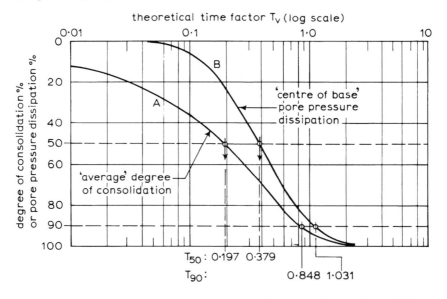

Fig. 22.6   *Theoretical relationships between time factor and degree of consolidation for vertical drainage for two methods of measurement*

$$c_v = \cfrac{T_{50} \times \left(\cfrac{H}{1000}\right)^2}{t_{50} \times 60/(31.56 \times 10^6)}$$

i.e.
$$c_v = 0.526 \frac{T_{50} \cdot H^2}{t_{50}} \quad \text{m}^2/\text{year} \tag{22.3}$$

As an example, if pore pressure measurements are made at the centre of the base, the second line of Table 22.3 applies and

$$c_v = 0.526 \times 0.379 \times \frac{H^2}{t_{50}} = \frac{0.2H^2}{t_{50}} \quad \text{m}^2/\text{year} \tag{22.4}$$

Similarly, Equation (22.2) becomes

$$c_v = \frac{0.54H^2}{t_{90}} \quad \text{m}^2/\text{year}$$

VERTICAL DRAINAGE, TWO-WAY (Fig. 22.4 (c) and (d))

The same principles apply as given above, except that the maximum length of drainage path h is now equal to $\frac{1}{2} H$. Equation (22.3) is therefore amended to

$$c_v = 0.131 \frac{T_{50} \cdot H^2}{t_{50}} \quad \text{m}^2/\text{year} \tag{22.5}$$

In this type of test, pore pressure is not measured so only the 'average' location factors are applicable. Theoretical values of T_{50} and T_{90} are the same as for one-way drainage provided that Equation (22.5) is used, and they are given in Table 22.3.

The value of T_{50} for 'average' measurements with two-way drainage (from the third line of Table 22.3) is 0.197. Substituting in Equation (22.5) gives

$$c_v = \frac{0.131 \times 0.197H^2}{t_{50}} = \frac{0.026H^2}{t_{50}} \quad \text{m}^2/\text{year} \tag{22.6}$$

This is the same as Equation (14.15) in Volume 2, Section 14.3.8, which applies to identical drainage conditions in the standard oedometer test.

RADIAL DRAINAGE TO PERIPHERY, 'EQUAL STRAIN' LOADING (Fig. 22.4 (f))

The relationship between the square-root time factor $\sqrt{T_{ro}}$ and degree of consolidation is shown in Fig. 22.7. The position of the T_{90} point in relation to the extended linear portion of the curve gives a 'slope factor' of 1.17. A construction similar to that used for the oedometer test, but with a slope factor of 1.17, is followed for locating the $\sqrt{t_{90}}$ point on a plot of settlement or volume change against square-root time obtained from a test. The $\sqrt{t_{50}}$ point is found by interpolation.

Values of the theoretical time factors, from Fig. 22.7, are as follows.

$$\sqrt{T_{50}} = 0.294 \quad \therefore \quad T_{50} = 0.0866$$

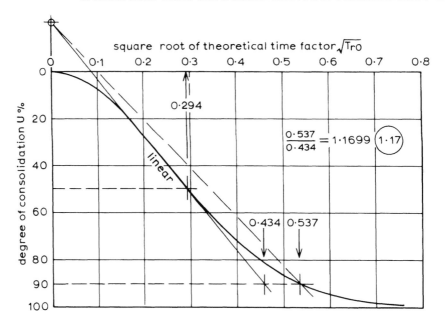

Fig. 22.7 *Theoretical curve relating square-root time factor to degree of consolidation for drainage radially outwards to periphery with 'equal strain' loading*

$$\sqrt{T_{90}} = 0.537 \quad \therefore \quad T_{90} = 0.288$$

The coefficient of consolidation, c_{ro}, is calculated from an equation similar to Equation (22.3):

$$c_{ro} = 0.526 \times \frac{T_{50} \cdot R^2}{t_{50}} \quad m^2/\text{year}$$

where R is the radius of the specimen (mm). In terms of the diameter D (mm):

$$c_{ro} = 0.131 \times \frac{T_{50} \cdot D^2}{t_{50}} \quad m^2/\text{year} \qquad (22.7)$$

Factors for the corresponding pore pressure dissipation readings taken at the centre of the base of the specimen are:

$$T_{50} = 0.173$$

$$T_{90} = 0.374$$

These values are included in Table 22.3.

RADIAL DRAINAGE TO PERIPHERY, 'FREE STRAIN' LOADING (Fig. 22.4 (e))

For this type of drainage a plot of settlement against square-root time ($t^{0.5}$) does not give a satisfactory initial linear relationship. However it was shown by McKinlay (1961) that a plot of settlement against time raised to the power of 0.465 (i.e. $t^{0.465}$) gave a close approximation to linearity. Results of his tests using a porous confining ring in an ordinary oedometer showed good agreement with the theoretical curve.

The theoretical time factor relationship, plotted as degree of consolidation against $T^{0.465}$, is shown in Fig. 22.8. The position of the 90% consolidation point related to the extension of the linear portion of the graph gives a 'slope factor' of 1.22, which can be used on a laboratory test plot of settlement or volume change against $t^{0.465}$ for obtaining t_{90}.

A slope factor of 1.25, using the above plot, was proposed by Tyrell (1969). From a study of the data the author favours McKinlay's factor but the difference is of little practical significance.

Values of the theoretical time factors derived from McKinlay's settlement relationship (Fig. 22.8) are:

$$T_{50} = (0.277)^{1/0.465} = 0.0632$$

$$T_{90} = (0.601)^{1/0.465} = 0.335$$

Corresponding values from pore pressure measurements taken at the centre of the base are:

$$T_{50} = 0.200$$

$$T_{90} = 0.479$$

Values are summarised in Table 22.3. Equation (22.7) is used for calculating c_{ro}, as shown by the example in Section 22.6.6.

RADIAL DRAINAGE TO CENTRE (Fig. 22.4 (g) and (h))

The theory of consolidation involving horizontal drainage to drainage wells, based on the Terzaghi assumptions, was presented by Barron (1947). His equations were applied to a laboratory consolidation test with drainage to a central well by Shields (1963), who showed that under 'equal strain' loading, at a certain radius r from the centre the excess

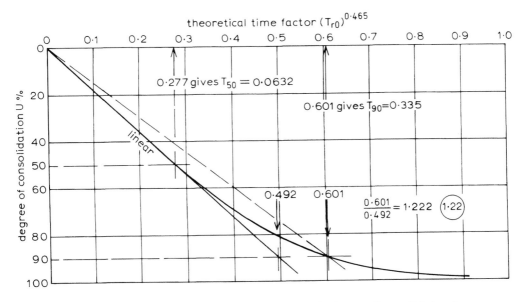

Fig. 22.8   Theoretical time-factor relationship with degree of consolidation for drainage radially outwards with 'free-strain' loading

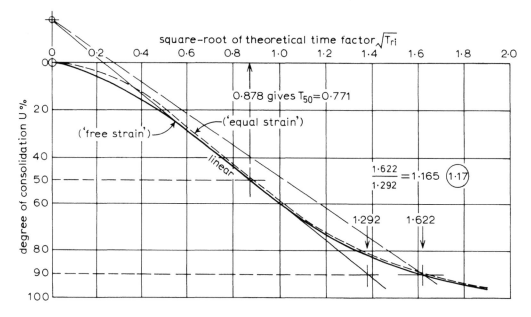

Fig. 22.9   Theoretical relationships between square-root factor and degree of consolidation for radial drainage to a central well, for 'free strain' and 'equal strain' loading (D/d = 20)

pore pressure is always equal to the average value. This radius depends on the ratio of specimen diameter D to drain diameter d; if the ratio D/d is 20, the radius r is equal to $0.55R$, where R is the specimen radius ($= D/2$). Thus in a 10 in (254 mm) diameter cell the pore pressure measuring point at 5.5 in (140 mm) from the centre will indicate the average pore pressure if the central drainage well is 0.5 in (12.7 mm) diameter.

The theoretical square-root time factor relationship with degree of consolidation for 'free strain' loading, for a diameter ratio of 20, is shown by the full line curve in Fig. 22.9. The slope factor here is again 1.17.

The dashed curve in Fig. 22.9 represents the 'equal strain' loading condition, but the two theoretical curves are so close together that for practical purposes the 'free strain' curve can be applied to both conditions. Theoretical time factors are approximately as follows.

$$T_{50} = (0.878)^2 = 0.771$$

$$T_{90} = (1.622)^2 = 2.631$$

These values, and the closely similar values for 'equal strain', are included in Table 22.3.

Values of T_{50} and T_{90} for drain diameter ratios from 5 to 100 are shown by the two graphs in Fig. 22.10.

The coefficient of consolidation is calculated from the equation

$$c_{ri} = \frac{0.131 D^2 T_{50}}{t_{50}} \quad \text{m}^2/\text{year} \tag{22.8}$$

22.2.4 Load Increments

In the conventional oedometer test, normal practice is to apply additional increments of pressure (Δp) equal to the pressure (p) already applied, giving a pressure increment ratio of unity, i.e. $\Delta p/p = 1$ (Volume 2, Section 14.8.2).

Another way of stating this is that the load ratio $p_1'/p_0' = 2$, where p_0' and p_1' are the effective stresses immediately before and after the application of a load increment.

For normally consolidated clays, Simons and Beng (1969) showed from tests in their hydraulic consolidation cells that the rate of dissipation of excess pore water pressure (and consequently the c_v value) was dependent on the load ratio. They suggested that in a laboratory test the load ratio should approximate to the changes in stress likely to occur in the field, rather than maintaining a standard value.

On the other hand, for laminated clays in which horizontal drainage is the dominating factor, the relevant coefficient of consolidation c_h is affected by structural viscosity (a 'creep' effect which contributes to secondary compression) in laboratory samples but not in the field (Berry and Wilkinson, 1969). This effect is emphasised with small pressure increment ratios; therefore they recommend the application of pressure increment ratios of not less than unity.

For many consolidation tests in the Rowe cell it seems to be advantageous to maintain consistency by using an increment ratio of unity, i.e. $\Delta p/p = 1$, unless there is a good reason for doing otherwise. If a different sequence of loading is used to reproduce field conditions, a constant pressure increment ratio should be maintained to enable compatible values of the coefficient of consolidation to be obtained.

22.2.5 Smear Effects

The effect of smear of clay soil adjacent to a porous plastic drainage layer was investigated by Berry and Wilkinson (1969). Smear occurs when the compressible soil moves relative to the fixed porous material, and is particularly significant in laminated soils where the smear zone can impede drainage from the layers of greater permeability. This leads to values of c_h lower than those of the undisturbed soil. Curves provided by Berry and Wilkinson enable corrections to be made to measured values.

Smearing effects on a central sand drain can be quite small if the sand is placed in a loose state so that it settles with the specimen. Correction curves for sand drains were also given by Berry and Wilkinson (1969).

22.2.6 Secondary Compression

In a Rowe cell consolidation test with pore pressure measurement the time at which the pore pressure dissipation U reaches 100% clearly defines the end of primary consolidation. The effect of secondary compression then becomes evident, and the coefficient of secondary compression, C_α, can be derived (see Volume 2, Section 14.3.13).

Rowe cell consolidation tests on remoulded amorphous peat were carried out by Berry and Poskitt (1972), and included an investigation of the coefficient of secondary compression (Volume 2, Section 14.7.1). Secondary compression of peat is discussed in detail by Hobbs (1985).

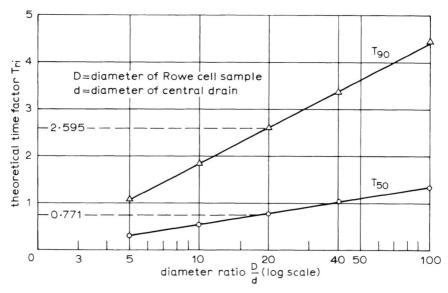

Fig. 22.10   *Values of theoretical time factors with radially-inward drainage for a range of values of D/d*

22.2.7   Radial Permeability

A cylindrical specimen of soil subjected to an inward radial-flow permeability test is represented in Fig. 22.11 (a). Symbols used in the following theoretical analysis are as follows:

Specimen diameter	$= D = 2r_2$
Central well diameter	$= d = 2r_1$
Height of specimen	$= H$
Rate of flow of water from radial boundary to central well	$= q = Q/t$
Head of water at radius r	$= h$
Velocity of flow at any radius r	$= v$
Soil permeability (radial)	$= k_r$ (to be determined)$_r$

The radial flow condition, which is axially symmetrical, is not the same as the laminar flow condition referred to in Volume 2, Chapter 10, because the velocity of flow increases towards the centre. However, Darcy's law may be applied to a thin annular element such as that shown in Fig. 22.11 (b). The rate of flow of water is equal to q and is independent of the radius.

Darcy's law states that

$$q = Aki$$

(from Equation (10.4)).

For the ring considered, of radius r,

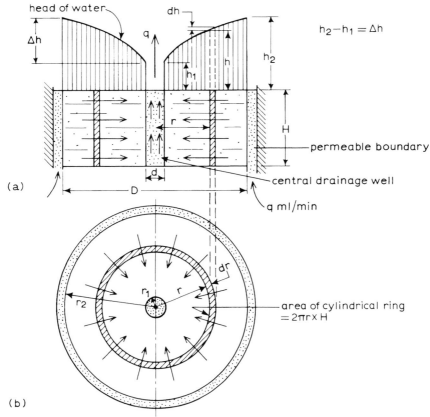

Fig. 22.11   *Radial permeability test: (a) representation of specimen and pressure distribution, (b) plan view showing annular element*

$$A = 2\pi r H$$

$$i = \frac{dh}{dr}$$

$$\therefore\ q = 2\pi r H \frac{dh}{dr} k_r \tag{22.9}$$

Rearranging,

$$k_r dh = \frac{q}{2\pi H} \cdot \frac{dr}{r}$$

Integrating between the limits $r = r_1$ and $r = r_2$,

$$k_r(h_2 - h_1) = \frac{q}{2\pi H} \log_e \left(\frac{r_2}{r_1} \right) \tag{22.10}$$

The term $(h_2 - h_1)$ is the drop in head Δh across the specimen, and may be expressed as a pressure difference Δp by using the relationship

$$\Delta p = \Delta h \,.\, \rho \,.\, g = (h_2 - h_1)\rho g$$

Substituting in Equation (22.10),

$$k_r = \left(\frac{q \cdot \rho \cdot g}{2\pi H . \Delta p} \right) \log_e \left(\frac{r_2}{r_1} \right)$$

The practical units normally used are:

q	ml/minute
H and Δh	mm
k_r	m/s
ρ	Mg/m^3 (for water $\rho = 1 \, Mg/m^3$)
Δp	kN/m^2
g	$9.81 \, m/s^2$

Radii are replaced by diameters, i.e. $r_1 = d/2$, $r_2 = D/2$ (mm).

Hence

$$k_r(\text{m/s}) = \frac{\dfrac{q}{60 \times 10^6} \times 1000 \times 9.81}{2\pi \dfrac{H}{1000} \times \Delta p \times 1000} \log_e \left(\frac{D}{d} \right)$$

i.e.

$$k_r = 0.26 \frac{q}{H . \Delta p} \log_e \left(\frac{D}{d} \right) \times 10^{-4} \quad \text{m/s} \qquad (22.11)$$

This equation is also applicable when the flow is radially outwards. It is used for calculating permeability from the test described in Section 22.7.3.

22.2.8   Vertical Permeability

Calculations for the determination of the coefficient of vertical permeability, k_v, are based on the equation in Section 20.3.1 derived for a triaxial specimen under similar conditions:

$$k_v = \frac{qL}{60A \times 102 \, \Delta p} \quad \text{m/s} \qquad (20.22)$$

in which q is the mean rate of flow of water (ml/minute), L the length of specimen (mm), A the area of cross-section of specimen (mm^2) and Δp is pressure difference across the specimen (kPa). In this case,

$$\Delta p = (p_1 - p_2) - p_c$$

where p_c is the pressure difference due to the pipeline head losses corresponding to the rate of flow q, and is derived from the flow calibration graph. The length L is the specimen height H. Hence

$$k_v = \frac{qH}{60A[(p_1 - p_2) - p_c] \times 102}$$

i.e.

$$k_v = \frac{1.63qH}{A[(p_1 - p_2) - p_c]} \times 10^{-4} \quad \text{m/s} \tag{22.12}$$

which is similar to the equation given in Clause 4.9.1.4 of BS 1377: Part 6: 1990.

22.3 PREPARATION OF EQUIPMENT
(BS 1377: Part 6: 1990: 3.2.6)

22.3.1 General

Before using a Rowe cell for the first time the measurements and calibrations described in Chapter 17, Sections 17.3.3 to 17.3.7, should be made. Pressure tests should be carried out to reveal any leakage from valves, seals, diaphragms and spindle bush.

When preparing the cell for a test, the first requirement is to check that all the necessary equipment, components and tools are readily available. These items are listed in Sections 22.1.4 to 22.1.6. A plentiful supply of de-aired water is also necessary.

Procedures for preparing and checking the Rowe cell immediately before a sample is set up for a test are outlined in Sections 22.3.2 and 22.3.4. These procedures relate to a type (a) consolidation test (Fig. 22.4(a)), for which the services to be connected to the cell are shown in Fig. 22.3. However the same principles apply to other types of test with various arrangements of apparatus and services indicated in Fig. 22.4 and 22.5.

Before connecting and de-airing the pore pressure and drainage lines within the cell (Section 22.3.4) the pressure systems themselves, and their ancillary equipment, should have been checked and made ready as described in Chapter 16, Sections 16.8.2 to 16.8.5. As in triaxial tests it is essential that the pore pressure measuring system is completely air-free and leak-free.

22.3.2 Cell components

The cell body, base and top should be clean and dry. Ends or flanges should be undamaged so as to make a good seal on the diaphragm or O-ring. The O-ring seal in the base should be in good condition and should fit properly into its groove. The cell body and O-ring seal may be lightly coated with silicone grease but must not be allowed to pick up dust or dirt. All clamping bolts, nuts and washers should be to hand, together with two spanners to fit them.

Check that there is no obstruction in any of the valve connecting ports, and that they are free from corrosion or soil particles.

Ensure that the diaphragm is free from leaks or kinks or areas of weakness, and that it is securely attached to the drainage spindle. Check that the spindle moves freely in its bush over its range of travel.

Ensure that the porous inserts in the base are securely fitted into their recesses, with no air gaps.

Connect the pore pressure transducer housing block to valve A on the outlet from the pore pressure measuring point in the cell base (Fig. 22.12). If necessary, use PTFE tape on the screw threads to ensure a watertight joint. The transducer must also be fitted into the block with a watertight joint, with its diaphragm uppermost.

Connecting ports to any ceramic inserts that are not being used should be completely filled with de-aired water, and sealed off with watertight blanking plugs.

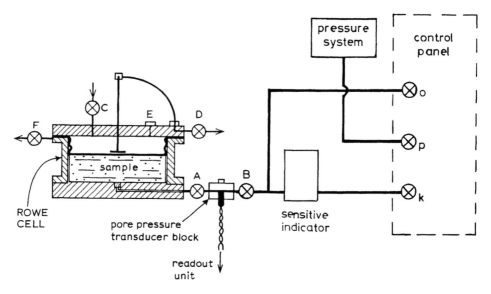

Fig. 22.12   *Arrangement for flushing and checking the Rowe cell pore pressure system*

22.3.3   Drainage Media

The sintered bronze porous disc should be boiled in distilled water for at least 10 minutes before use in order to remove air, then kept submerged in freshly de-aired water until it is required. Immediately after use it should be washed with running water and brushed with natural bristles or nylon (*not* a wire brush) to remove adhering soil. When a disc becomes clogged and cannot be cleared by boiling, it should be replaced. If water dripped on to the horizontal surface of a saturated disc does not readily pass through, the disc should be discarded.

Ceramic inserts should also be boiled before use, as described above, and replaced when clogged.

Porous plastic material, whether used as a peripheral drain or for a top or bottom drainage disc, should be boiled for at least 30 minutes. This material should be used once only, and then discarded. The smooth side should be placed in contact with the soil, and neither side should be greased. For tests requiring radial drainage to the periphery (test type (e), (f), Fig. 22.4; or (l), (m), Fig. 22.5) the porous liner is fitted to the cell wall as described in Section 22.5.3.

Sand intended for a central drainage well should be de-aired by boiling in distilled water, cooled in an airtight container, and kept submerged.

22.3.4   Pore Pressure System

NEED FOR CHECKING

Pore pressure is always measured at the base of the specimen, and drainage (whether from the top surface, the periphery or a central well) is usually taken to a back pressure system. It is essential at the outset that the pore pressure measuring system is completely free of air, and the back pressure system must be de-aired. The systematic checking and de-airing procedures described below are similar to those used for the triaxial test system (Section 18.3), and are facilitated by using a manual pressure control cylinder. A mercury null indicator device can be useful as an aid for checking whether a system is free from air and leaks. Checking procedures for the Rowe cell pressure systems are specified in BS 1377: Part 6:1990 in Clause 3.2.6. Checks fall into two categories, as follows:

(1) 'Complete' checks—to be carried out:

  (a) when an item of new equipment is introduced into a system;

  (b) if part of a system has been removed, stripped down, overhauled or repaired;

  (c) at intervals of not more than 3 months.

(2) 'Routine' checks—to be carried out before starting each test.

The valve designations quoted below relate to Figs. 22.1, 22.3 and 22.12. All valves initially are closed.

A plentiful supply of freshly de-aired water should be readily available for flushing, either direct from the de-airing apparatus or from an elevated reservoir.

One means of detecting movement of water due to leaks that is more positive than visual inspection is to connect a sensitive volume-change indicator between the source of flushing water and valve B. Another is to use a mercury null indicator, and this would provide even greater sensitivity. If either of these devices is used, it is connected to valve k on the control panel (Fig. 16.11). A separate connection to valve o enables the device to be by-passed when flushing the system. The suggested arrangement is shown in Fig. 22.12.

COMPLETE CHECK

(1) Connect the flushing system to valve B on the transducer housing block (Fig. 22.12) ensuring a tight joint without entrapping air. This connection should remain unbroken throughout a test series.

(2) Open valves A and B and using the control cyclinder, pass freshly de-aired water through the transducer mounting block and cell base and out through the base port. Continue until no entrapped air or bubbles are seen to emerge. This is to ensure that the system is filled with de-aired water. Close valve B.

(3) Close valve A and remove the bleed plug in the pore pressure transducer mounting block.

(4) Inject a solution of soft soap into the bleed plug hole. If a hypodermic needle is used, take care to keep it well clear of the transducer diaphragm.

(5) Open valve B to allow water from the de-aired supply to flow out of the bleed hole.

(6) Replace the bleed plug in the transducer mounting block while water continues to emerge, to avoid trapping air, and make it watertight.

(7) Open valve A briefly to allow about 500 ml of de-aired water to pass out of the pore pressure measurement port to waste, to ensure that any further air, or water containing air, in the mounting block is removed.

(8) While water is emerging from the pore pressure port, seal it without entrapping air by covering with a piece of latex rubber and a flat metal disc clamped down into place. A suitable arrangement is shown in Fig. 22.13.

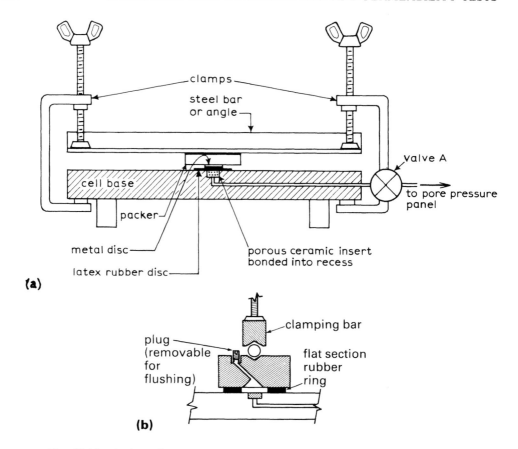

Fig. 22.13 Sealing the pore pressure measuring point for pressurising. (a) Clamping arrangement, (b) cap with provision for flushing (used at Manchester University)

(9) Raise the pressure in the system to 700 kPa, and again allow about 500 ml of water to pass out through the pore pressure port.

(10) Leave the system pressurised overnight, or for at least 12 hours. Record the reading of the volume-change indicator, or the level of the mercury thread in the null indicator, if fitted in the system, when the reading is steady, i.e. after allowing for expansion of the connecting lines.

(11) After this period, record the reading of the indicator, if fitted, and check the system carefully for leaks, especially at joints and connections. A change in the reading of the indicator will provide immediate confirmation of leakage. Any leaks that are detected should be dealt with immediately and rectified, and the above check repeated.

(12) When it is confirmed that the system is leak-free, open valves A and B, and apply the maximum available pressure (consistent with the limitations of the pressure system and transducer) to the cell base.

(13) Close valve B and record the reading of the pore pressure transducer. Leave for a minimum of 6 hours.

(14) If after this period the pressure reading remains constant the pore pressure connections can be assumed to be free of air and free of leaks.

(15) A decrease in the pressure reading indicates that there is a defect in the system— either a leak, or air was present and has been forced into solution. The defect must be rectified, and steps (2) to (13) must be repeated until it is confirmed that the system is free of entrapped air and leaks.

ROUTINE CHECK

Immediately before setting up each test specimen, follow steps (2) to (12) above to ensure that the cell base connection is flushed clear of any remaining soil particles and that the system is still free of leaks and entrapped air. Close valve B.

Keep the base covered with de-aired water until ready for setting up the test specimen. This is to ensure that no air enters the base ports.

22.3.5   Checking the Back Pressure System

COMPLETE CHECK

(1) Ensure that the back pressure system is fully charged with a supply of freshly de-aired water.

(2) Open valve D and flush freshly de-aired water from the back pressure system, through the volume-change indicator and the specimen drainage line and out through the drainage stem. Apply enough pressure to maintain a reasonable rate of flow. During this operation the volume-change indicator should be worked to the limits of its travel at least twice. Replenish the water in the system with freshly de-aired water as necessary.

(3) Continue until the absence of emerging bubbles indicates that the system is substantially free of air. If bubbles persist, the volume-change indicator might need de-airing (see below).

(4) Insert a suitable watertight plug securely into the end of the drainage stem.

(5) Increase the pressure in the back pressure system to 750 kPa with valve D open. Observe the volume-change indicator and record the reading when steady, i.e. after allowing for initial expansion of the connecting lines.

(6) Leave the system under pressure overnight, or for at least 12 hours, then record the reading of the volume-change indicator again.

(7) Take the difference between the two readings and deduct the volume change due to further expansion of the tubing. If this corrected difference does not exceed 0.1 ml the system can be considered to be leak-free and air-free, and ready for a test.

(8) If the corrected difference exceeds 0.1 ml, the system needs to be investigated and the leaks rectified until the above requirement is achieved.

ROUTINE CHECK

Immediately before setting up each test specimen, follow steps (1) and (3) above to ensure that the drainage line is free of air and obstructions. Close valve D.

(2) Increase the pressure in the back pressure line to 750 kPa, and after 5 minutes record the reading of the volume-change indicator.

(3) Leave the system under pressure, and determine the change in indicated volume, as in steps (6) to (8) above. Carry out remedial measures if found necessary, then recheck.

The above checking procedures should also be used for flushing and checking the rim drainage line when horizontal drainage is required. In this case valve F is used instead of valve D.

If it is necessary to remove air from the volume-change indicator, follow the procedure given in Section 18.3.3. The manufacturer's instructions should also be followed carefully.

Finally, adjust the volume-change indicator on the back pressure line (Fig. 22.3) so that the liquid interface is near one end of its range of travel, and ascertain in which direction the reading changes as water enters the specimen. This represents a negative volume change. The cell is then ready for receiving the test specimen, as described in Section 22.4.

22.4 SPECIMEN PREPARATION

22.4.1 Preparation of Undisturbed Specimen

The preparation of undisturbed specimens for the large diameter Rowe consolidation cells, i.e. 150 mm and 250 mm diameter, is outlined below. Specimens for 75 mm diameter cells can be prepared by more conventional means such as are described in Chapter 9, but handling of the larger specimens and cells requires special procedures and some adaptations of equipment. Methods of preparation appropriate to two types of undisturbed specimen are suggested, namely:

(a) Extrusion from piston samples;

(b) Hand trimming from block samples.

It may be necessary to devise other methods for the transfer of an undisturbed sample into the cell body, and details will depend upon local conditions and available equipment.

The method given in Section 22.4.2 for assembly of the cell on its base applies to any type of undisturbed specimen once it has been fitted in the cell body.

The procedures described relate to specimens for a vertical single-drainage consolidation test. Modifications for specimens requiring peripheral or base drainage for other types of test are covered in Section 22.5.2 to 22.5.5.

(a) *Piston sample*

An undisturbed sample taken on site in a piston sampler should if possible be extruded, trimmed to the correct diameter and pushed into the cell body in one operation. In the absence of a purpose-designed sample extruder for 250 mm diameter samples the author devised a means of using a large compression machine as an extrusion jack, as outlined below. The general arrangement is shown in Fig. 22.14.

The upper face of the sample in the tube is trimmed flat and level. A cutting ring about 50 mm long and the same internal diameter as the cell is bolted to one flange of the cell, and the inside faces (which must be in precise alignment) are lightly coated with silicone grease. The other flange is bolted to an extension piece which has an internal diameter slightly larger than that of the cell, and is provided with large observation holes in the side. The extension is attached to a rigid plate supported from the crosshead of the machine.

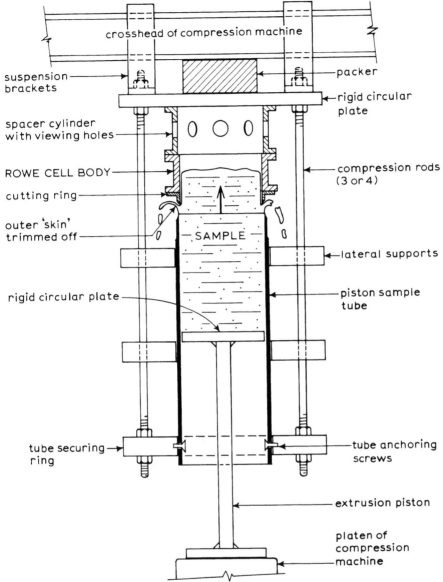

Fig. 22.14   General arrangement of an extrusion device for 250 mm diameter piston specimens (devised by the author)

Three substantial rods hanging from the plate support collars from which the sampling tube is suspended. The tube is mounted securely with its cutting edge uppermost and about 25 mm below the cutting ring. The lower end of the sample is supported by a rigid circular steel plate resting on a spacer carried by the machine platen.

As the platen of the machine is moved steadily upwards the sample is pushed out of the tube and the outer 2 or 3 mm is skimmed off by the cutting ring, enabling the sample to enter the cell body without restraint. The observation holes in the top spacer allow escape of air and enable the surface of the sample to be seen when it has completely filled the cell.

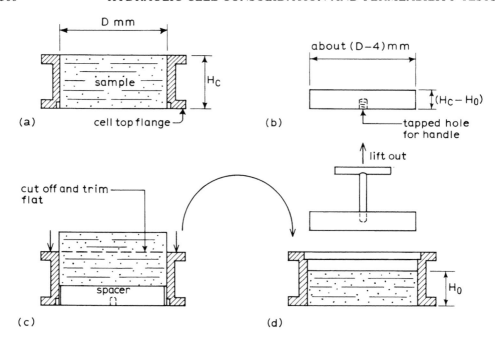

Fig. 22.15   Preparing sample for Rowe cell test: (a) soil surfaces trimmed flush with both faces of cell, (b) requirements for spacer disc, (c) sample displaced by pushing down on cell body, (d) spacer removed after trimming sample and inverting the cell

When the sample has been extruded far enough it is severed close to the cutting ring. Two spatulas, or a sheet of metal or plywood, are inserted to hold the specimen in place and prevent distortion by sagging. The assembly can then be dismantled and the cell body, cutting ring and spacer ring removed as one unit.

Both ends of the specimen are trimmed so that they are flush with the cell body flange. The specimen height is reduced to that required for the test by using a cylindrical spacer of appropriate thickness to push out a length of soil for cutting away, leaving the desired height of specimen in the cell (Fig. 22.15). The effect of surface smear on the cut surfaces can be reduced by lightly scarifying with a brass brush. If the soil contains particles of gravel size, any protruding particles should be removed carefully when trimming. Resulting voids should be filled with fine material from the trimmings, remoulded to the same density as the undisturbed soil. Trimmings are used for the determination of initial moisture content and particle density.

With very organic clays, oxidation can take place rapidly in contact with air, releasing bubbles of gas which can affect the permeability and therefore the results of consolidation tests. In addition air can be drawn in during trimming, reducing the degree of saturation. These effects can be minimised by extruding samples into the cell body under water, using a special horizontal extruding arrangement as devised at Manchester University and shown in principle in Fig. 22.16. Final trimming of the specimen and assembly of the cell can also be carried out under water. Saturated soils of this kind are not likely to swell much, but any swelling will draw in water, not air, thus ensuring that saturation is maintained.

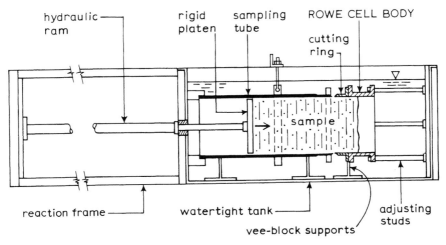

Fig. 22.16 Arrangement for extruding organic soils into Rowe cell under water (Manchester University method)

(b) *Block sample*

A sample in the form of an undisturbed block can be hand trimmed and jacked into the cell (fitted with a cutting ring) using the procedure described in Chapter 11 for preparing an undisturbed CBR test specimen. For soft soils the weight of the cell may provide enough force to advance it downwards as the soil is trimmed ahead of the cutting edge, but the axis of the cell must be kept vertical; a spirit level placed on the uppermost flange provides a convenient guide. Final trimming to the required height is as described above.

A similar procedure can be used for taking a sample from a trimmed exposure on site. The sample should completely fill the cell, and the end faces should be trimmed flush, protected from the atmosphere and clamped with rigid end plates, before transporting to the laboratory.

22.4.2 Fitting the Cell Base

Determine the height of the trimmed specimen (H_0) by measuring down from the top of the cell body to the soil surface at several points, reading to 0.5 mm. Weigh the cell body with the specimen to an accuracy of within 0.1%. The specimen volume and initial bulk density can then be calculated. If the initial moisture content and particle density have been determined from sample trimmings, the dry density, initial voids ratio and degree of saturation can also be calculated (see Volume 2, Chapter 14).

The cell body is transferred and secured to its base as follows:

(1) Cover the de-aired cell base with a thin film of de-aired water.

(2) Open valve A and valve B connecting the pore pressure transducer to the de-aired water supply line, or to a column of water, to avoid damage by sudden over-pressurising (Fig. 22.12).

(3) Place two thin steel spatulas under the bottom flange of the cell body to retain the specimen flush with the flange while it is lifted. Slide the specimen on to the flooded cell base without entrapping any air, and remove the spatulas.

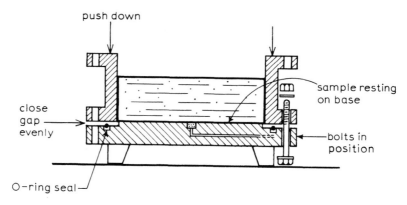

Fig. 22.17   Assembly of cell body on to base

(4) When the cell body is correctly registered with the base, press the body down so that the gap between flanges is reduced evenly all round until the body rests on the O-ring seal (Fig. 22.17). The specimen remains stationary as the cell body moves down by a few millimetres.

(5) Fit the flange securing bolts and nuts, with washers, and tighten them systematically as follows.

(a) Make all nuts finger-tight, and check that the gap between flanges is equal all round.

(b) Using two spanners of the correct size, make two diametrically opposite bolts fairly tight.

(c) Tighten the remaining pairs in the same way.

(d) Continue by progressively tightening opposite pairs, until the flanges are tightly bolted together. For a cell with 9 bolts instead of 8, bolts should be tightened progressively in groups of 3 forming equilateral triangles (tighten one, miss two), instead of in pairs.

(e) Ensure that the gap between the metal flanges is closed uniformly all round the perimeter.

(6) Close valve B. Flood the top of the cell above the specimen with de-aired water, if appropriate (see Section 22.5.1, step (1)).

Assembly of the cell top, and the remaining preparations for various types of test, are described in Section 22.5.

22.4.3 Preparation of Specimen from Slurry

Fully saturated specimens, often required for experimental purposes, can be formed from slurry in the cell in which they are to be tested. A clay soil is first broken down to a fine powder, by grinding if appropriate, and mixed to a slurry with distilled or de-ionised water. Use of a motorised rotary mixer for a period of 2–3 hours, followed by sealing and storage for about a week, should ensure that the mixture is homogeneous. The slurry should be well shaken just before use. A water content of up to twice the liquid limit is generally suitable, and gives a viscosity low enough to allow removal of air by the application of vacuum. However, if segregation of certain particle sizes is evident a lower moisture content might be better.

The volume change of the slurry due to initial consolidation is very large and it might be necessary to bolt two cell bodies together to provide enough volume at the start. The initial thickness of slurry may need to be about twice the thickness of specimen required after initial consolidation.

The lower cell body is bolted to its base, and the inside of the body is coated with silicone grease. The pore pressure connections should already have been de-aired as described in Section 22.3.4, and the valves on the cell base are closed.

The slurry is poured into the cell to a depth of about 25–40 mm. A Perspex lid with vacuum and water inlet connections is greased and bolted to the flange of the cell body to make an airtight seal. The lid must be rigid enough to sustain full atmospheric pressure. A vacuum desiccator lid, if of the right size, could be used. Vacuum is applied through the lid, and the surface of the slurry will be seen to 'boil' as air is removed.

When the action has virtually stopped, water is introduced through the other inlet in the Perspex lid, to just cover the specimen. The vacuum can then be released to enable the lid to be removed and a further layer of slurry poured in. The above process is repeated layer by layer until the required thickness of slurry has been deposited and de-aired.

After de-airing the final layer, the Perspex cover is removed, the surface of the specimen is levelled, the saturated sintered bronze disc is placed on top, and the cell is topped up with de-aired water. The diaphragm and cell top are fitted and tightened down, as described in Section 22.5.1. Valve A is opened and the initial pore water pressure is recorded.

The diaphragm pressure is raised to provide a small consolidation pressure, e.g. 5 kPa. The pore pressure transducer should indicate a pressure change equal to this increase almost at once. The drainage outlet at D is led into a measuring cylinder. When valve D is opened the specimen will start to consolidate and the consolidation process can be monitored from three sets of observations:

Volume of water collected;

Settlement (as measured by the dial gauge);

Pore water pressure change

Readings are taken at time intervals used for a consolidation test. When the pore pressure has fallen to the initial value, that consolidation stage is complete and further stages can be applied as required. An alternative procedure is to increase the consolidation pressure gradually to the desired value, once some initial consolidation has taken place under a small pressure. The application of too high a pressure all at once could cause displacement of slurry around the edge of the drainage disc.

Settlement readings should be observed continuously as a check on the movement of the diaphragm. If it is apparent that the diaphragm extension is nearing the limit of its working range, consolidation should be interrupted for removal of the cell top so that a spacer can be inserted between the diaphragm and the drainage disc on the specimen. The spacer should allow free passage of water from the drainage disc to the drainage stem in the diaphragm. Similarly it may be necessary at some stage to remove the extension cell if two bodies were used initially.

If radial drainage to the periphery is to be carried out the cell walls should first be lined with porous plastic material, as described in Section 22.5.3. The inside face must not be greased. If two cell bodies are bolted together, each should be fitted with its own liner to facilitate separation when the extension is removed. (See Section 22.1.8, item 6 under 'Selection of test', regarding non-uniformity resulting from consolidation with radial drainage.)

22.4.4 Preparation of Compacted Specimen

When a Rowe cell consolidation test is to be carried out on compacted soil, the specimen can be prepared in the cell by applying either static compression or dynamic compaction, in a manner similar to that used for preparing specimens in a CBR mould, as described in Volume 2, Chapter 11.

The cell base should be prepared as described in Section 22.3.2, and bolted to the body. While the specimen is being compacted, valve A should be kept closed and the pore pressure transducer connected to a column of water open to atmosphere. Very high instantaneous pore water pressures can be induced by compaction, and this precaution is to protect the transducer from sudden overload. Vibration could also damage the transducer, and compaction by vibration should be avoided unless a very light vibrator is used.

Before compacting the soil, any particles larger than one-sixth of the specimen height should be removed by passing the soil through the appropriate sieve, if necessary. Otherwise preparation of the soil is as described in Volume 2, Section 11.6.2, including determination of the moisture content. The prepared specimen is weighed to an accuracy of 0.1%, covered with an airtight seal, and stored for at least 24 hours before starting a test.

Compaction to a specified density, whether static or dynamic, is straightforward because a known mass has to fill a specified volume (Volume 2, Sections 11.6.3 to 11.6.6). If dynamic compaction with a specified effort is to be used (e.g. BS 'light' compactive effort, Volume 1, Chapter 6), the number of blows to be applied depends upon the volume of soil compacted (Volume 2, Section 11.6.7). Calculations relating to the usual cell sizes are given below. In all cells the pattern of blows should be similar to that described in Volume 1, Section 6.5.5, compacting first the edge, then the middle, then giving systematic overall coverage.

After compaction, trim the top surface flat and level. A levelling template of the kind shown in Volume 2, Fig. 12.41, and rigid enough to act as a trimming tool, is useful for this purpose. Protruding coarse particles should be removed carefully by hand and replaced by fine material from the trimmings, well pressed in.

Return all trimmings and unused material to the remains of the prepared sample, which is then weighed so that the mass of soil in the cell can be determined by difference. Use representative portions of the remains to determine the moisture content.

Determine the height of the test specimen by measuring down to the trimmed surface from the cell body flange, as described in Section 22.4.2.

Cover the specimen with an airtight seal and allow to stand for at least 24 hours before starting a test, to allow any excess pore pressure to dissipate.

250 mm DIAMETER CELL

Assuming a cell of 252.3 mm diameter, the circular area $= (\pi/4) \times 252.3^2 = 50{,}000 \text{ mm}^2$.

Height to give 1 litre (10^6 mm^3), i.e. the volume of a BS compaction mould $= 10^6/50{,}000 = 20 \text{ mm}$.

For BS 'light' compaction this volume of soil requires a total of $27 \times 3 = 81$ blows of the 2.5 kg rammer. For 'heavy' compaction the number of blows is $25 \times 5 = 135$ using the 4.5 kg rammer.

If the compacted height of specimen in the cell is h mm, the volume of soil is $h/20$ litres, and the total number of blows required are:

for BS 'light' compaction: $\dfrac{h}{20} \times 81 = 4.05\,h$

for BS 'heavy' compaction: $\dfrac{h}{20} \times 135 = 6.75\,h$

For a specimen 90 mm high, the total number of blows required are 365 and 608 respectively.

The height/diameter ratio of a specimen of this size is only $90/252.3 = 0.357$ compared with

$$\frac{115.5}{105} = 1.10 \quad \text{for the BS compaction mould,}$$

and

$$\frac{127}{152} = 0.836 \quad \text{for the CBR mould.}$$

It therefore seems reasonable to compact the soil into a cell of this size in fewer layers, and it is suggested that for 'light' compaction 2 layers should be used instead of 3, and for 'heavy' compaction, 3 layers instead of 5.

The number of blows to apply to each layer of a specimen of 90 mm final height are:

'Light' compaction — in 3 layers: 122 per layer

in 2 layers: 182 per layer

'Heavy' compaction — in 5 layers: 122 per layer

in 3 layers: 203 per layer

The above data are summarised in Table 22.4, together with corresponding data for cells of 254 mm (10 in) diameter.

150 mm DIAMETER CELL

Assuming a cell of 151.4 mm diameter, the circular area $= \dfrac{\pi}{4} \times 151.4^2 = 18,000\,\text{mm}^2$.

Height to give 1 litre $= 10^6/18,000 = 55.56$ mm.

Thus the volume of soil in a specimen of 50 mm height is almost equal to that of the BS compaction mould.

The volume of soil for a specimen height of h mm is $h/55.56$ litres, and the total number of blows required are:

for BS 'light' compaction $\dfrac{h}{55.56} \times 81 = 1.458\,h$

for BS 'heavy' compaction $\dfrac{h}{55.56} \times 135 = 2.43\,h$

For a specimen 50 mm high, the total number of blows required are 73 and 122 respectively.

Table 22.4. COMPACTION DATA FOR ROWE CONSOLIDATION CELLS

Cell diameter	mm	254	**252.3**	152.4	**151.4**	76.2	**75.7**
area	mm²	50,671	**50,000**	18,241	**18,000**	4560	**4500**
Sample height	mm	90	**90**	50	**50**	30	**30**
volume	cm³	4560	**4500**	912	**900**	137	**135**
Degree of compaction		'light' (2.5 kg rammer)	'heavy' (4.5 kg rammer)	'light' (2.5 kg rammer)	'heavy' (4.5 kg rammer)		
Total blows per mm height		4.10 **4.05**	6.84 **6.75**	1.48 **1.46**	2.46 **2.43**	(not appropriate)	
Blow per layer for recommended specimen height:							
in 2 layers		185* **182**	—	37* **37***	—		
in 3 layers		123 **122**	205* **203***	25 **24**	41* **41***		
in 5 layers		—	123 **122**	—	25 **24**		

*Recommended procedure
Bold figures related to cells with rational areas

The height/diameter ratio of a specimen of this size is $50/151.4 = 0.33$, which is similar to that for the 250 mm cell specimen. Compaction in 2 or 3 layers, as suggested above, would be reasonable.

The number of blows to apply to each layer of a specimen of 50 mm final height are:

BS 'light' compaction —   in 3 layers: 24 per layer

in 2 layers: 37 per layer

BS 'heavy' compaction —   in 5 layers: 24 per layer

'in 3 layers: 41 per layer

The above data are summarised in Table 22.4, together with corresponding data for 152 mm (6 in) diameter cells.

75 mm DIAMETER CELL

Assuming a cell of 75.5 mm diameter, the circular area $= (\pi/4) \times 75.7^2 = 4500 \text{ mm}^2$.

Typical specimen height $= 30$ mm, and volume $= (30 \times 4500)/1000 = 135 \text{ cm}^3 = 0.135$ litre.

This size of specimen is too small for compaction by a standard compaction rammer. The soil should be first compacted into a BS compaction mould, then extruded and trimmed into the cell as for an undisturbed specimen. Alternatively a small scale compactor, such as the Harvard compactor, could be used for compaction directly into the cell. The appropriate number of blows should be obtained from calibration check tests against standard compaction procedures.

22.5 CELL ASSEMBLY AND CONNECTIONS

Assembly of the Rowe cell, and preparations for five different types of test, are described in the following sections. These procedures follow on from the end of the last stage of the preparation procedure described in Section 22.4, relevant to the type of sample. Ancillary equipment should have been made ready beforehand as described in Section 22.3. Assembly operations should be carried out over a sink or large tray to confine the inevitable splashing of water.

22.5.1 Vertical Drainage (one way)

The following procedure (stages 1 to 13) is for the consolidation test in which vertical drainage takes place from the top surface of the specimen, and pore pressure is measured at the base (test type (a), Fig. 22.4). Variations in certain details to suit other types of test are given in the ensuing sections.

GENERAL PROCEDURE

(1) Flood the space at the top of the cell above the specimen with de-aired water, if not already done. This should be omitted if the soil is susceptible to swelling, or is to be tested in an unsaturated condition, or is sensitive to moisture content change at zero stress. (The procedure for these soils is outlined separately below.)

(2) Place a saturated drainage disc through the water onto the specimen, without entrapping air. For a 'free' strain test, use a disc of porous plastic material of 3 mm thickness. In a 250 mm diameter cell the sintered bronze disc allows some degree of flexibility. For an 'equal' strain test, use the sintered bronze disc, or a porous plastic disc, covered by the circular steel plate which is lowered into position using the lifting handle. Avoid trapping air under the plate. The central drainage hole must be left open, and aligned with the settlement stem drainage outlet.

Ensure that there is a uniform clearance all round between the disc or discs and the cell wall. Filter paper should not be inserted between the specimen and drainage disc.

(3) Connect a length of tubing to valve F (Fig. 22.1) and immerse the other end in a beaker or measuring cylinder containing de-aired water. The tube should first be completely filled with de-aired water making sure that there are no entrapped air bubbles.

(4) Support the cell top at three points so that it is level, and with more than enough clearance underneath for the settlement spindle attached to the diaphragm to be fully extended downwards (Fig. 22.18). The cell top should be supported near its edge so that the flange of the diaphragm is not restrained. About one-third fill the diaphragm with water using rubber tubing connected to a header tank or water tap. Open valve C.

(5) Place three or four spacer blocks, about 30 mm high, on the periphery of the cell body flange. Lift the cell top, keeping it level, and lower it onto the spacers, allowing the diaphragm to enter the cell body and in doing so to displace water over the flange (Fig. 22.19). Bring the bolt holes in the cell top into alignment with those in the body flange.

(6) Add more water via the tubing to fill the inside of the diaphragm so that the weight of water brings the diaphragm down and its periphery is supported by the cell body. By inserting a spatula blade as shown in Fig. 22.19, check that the cell body is completely filled with water. The whole of the extending portion of the diaphragm should be inside the

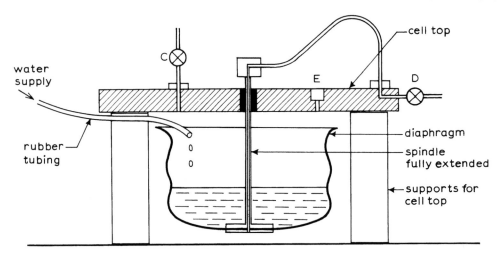

Fig. 22.18   Preparation of diaphragm before assembly of cell top

cell body, and the diaphragm flange should lie perfectly flat on the cell body flange, concentric with it and not obscuring any of the bolt holes.

(7) Hold the cell top while the supporting blocks are removed, then carefully lower it to seat onto the diaphragm flange without entrapping air (Fig. 22.20 (a)) or causing ruckling or pinching (Fig. 22.20 (b)). Align the bolt holes. When correctly seated the gap between top and body should be uniform all round and equal to a diaphragm thickness. (Fig. 22.20 (c)). Open valve F (Fig. 22.19) to permit escape of excess water from under the diaphragm.

(8) Fit the flange securing bolts and nuts with washers (Fig. 22.20 (c)). Tighten the bolts systematically, as described in Section 22.4.2 stage 5 for bolting the body to the base.

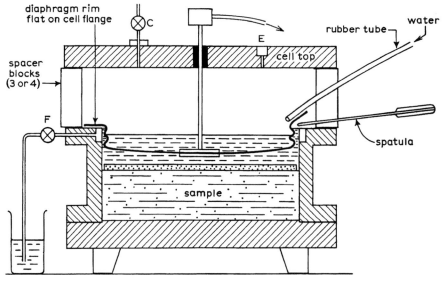

Fig. 22.19   Diaphragm inserted into cell body

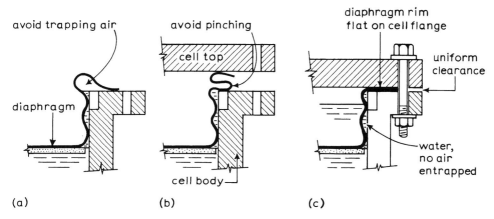

Fig. 22.20   *Seating the diaphragm:* (*a*) *avoid trapping air under flange,* (*b*) *avoid ruckling and pinching,* (*c*) *diaphragm correctly seated*

Ensure that the diaphragm remains properly seated, and that the gap between the metal flanges remains constant all round the perimeter.

As a safeguard against trapping air under the diaphragm, fitting the cell top (stages 5 to 8) could be carried out with the cell completely immersed in a tank of water.

(9) Open valve D (Fig. 22.1), and press the settlement stem steadily downwards until the diaphragm is firmly bedded on top of the plate covering the specimen. Close valve D when no more water emerges.

(10) Connect valve C to a header tank of water having a free surface about 1.5 m above the specimen. Alternatively a constant-pressure system set at 10 to 15 kPa could be used.

(11) Completely fill the space above the diaphragm with water through valve C with bleed screw E removed. Tilt the cell so that the last pocket of air can be displaced through E. To ensure complete removal of air, connect E to a moderate vacuum. Maintain the supply of water at C when subsequently replacing the bleed screw.

(12) Maintain the head of water at C, and as the diaphragm expands allow the remaining surplus water from above the specimen to emerge through valve F. Open valve D for a moment to allow the escape of any further water from immediately beneath the diaphragm.

Escape of water from F due to diaphragm expansion may take some considerable time because of the barrier formed by the folds of the diaphragm pressing against the cell wall. The barrier effect can be overcome by fitting a layer of porous plastic material above the level of the specimen, or by inserting a short 'wick' of the same material, as shown in Fig. 17.10 (b). A clay specimen of low permeability is not likely to consolidate within the few seconds during which the excess water drains out. However, special care is needed with peat, in which rapid initial consolidation can occur within a few seconds.

(13) Close valve F when it is evident that the diaphragm has fully extended. Observe the pore pressure at the base of the specimen, and when it has reached a constant value record it as the initial pore pressure, u_0. This corresponds to the initial or seating pressure p_{d0} under the head of water connected to C.

If the height from the top of the specimen to the level of water in the header tank is h mm, then:

$$p_{d0} = \frac{h \times 9.81}{1000} = \frac{h}{102} \quad \text{kPa}$$

(14) Connect the lead from the diaphragm constant pressure system to valve C, without entrapping any air (Fig. 22.21). Set the pressure system to the same pressure as that of the head of water previously connected.

(15) Connect the lead from the back pressure system to valve D without entrapping any air.

The arrangement of the apparatus is now as shown in Fig. 22.21. The test procedure is described in Section 22.6.2.

SOILS SUSCEPTIBLE TO SWELLING

When setting up a soil with a potential for swelling, which may include compacted material, the cell top may be fitted without first flooding the specimen. Water can be admitted to the base of the specimen via one of the ceramic inserts while an appropriate diaphragm pressure is maintained to prevent swelling. Much of the air from behind the diaphragm can be displaced although valve F, and removed by carefully applying suction after bolting down the top cover. Any remaining air can be forced into solution under the subsequent application of back pressure.

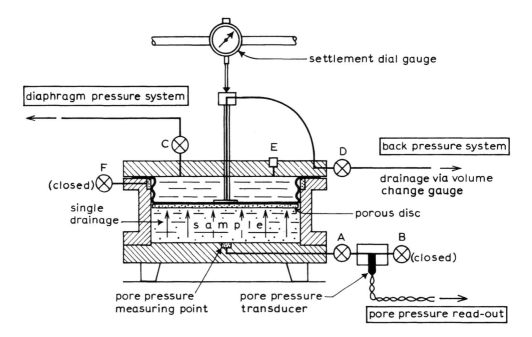

Fig. 22.21   Arrangement of Rowe cell for consolidation test with drainage to top face, and measurement of pore pressure at base ('basic' test)

Swelling pressure can be measured by adjusting the diaphragm pressure to maintain the height of the specimen constant while water percolates through the soil, until the equilibrium condition is reached. This is likely to take much longer than a swelling pressure test in an oedometer (Volume 2, Section 14.6.1).

22.5.2 Vertical Drainage (two way)

In this type of test drainage takes place from both top and bottom faces of the specimen, and pore pressure is not measured during consolidation. A porous drainage disc is placed under the specimen, and is connected to the same back pressure system as the top drainage line for the consolidation stages.

Setting up and assembly are very similar to the procedure given in Section 22.4.2, with the following variations:

(Step 1) After covering the base with a film of water, place a saturated porous disc of sintered bronze or porous plastic on the cell base without entrapping any air.

(Step 2) Keep valve A closed (Fig. 22.22). Connect a length of tubing, completely filled with water and immersed in a beaker of water at the other end, to the outlet from valve A.

(Step 3) Ensure that the porous disc remains centrally on the cell base while the specimen and cell body are placed in position. The disc will displace the sample upwards relative to the cell body by its own thickness as the body flange is bolted down on to the base. Alternatively a firm or stiff specimen can be displaced by that amount beforehand. When the cell body is secured to the base, open valve A momentarily to release any excess pressure of water.

Determination of specimen thickness must take into account the thickness of the bottom porous disc.

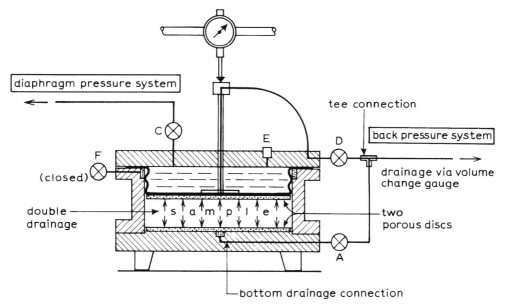

Fig. 22.22 Arrangement of Rowe cell for consolidation test with two-way vertical drainage

The remainder of the assembly operation is the same as that described in Section 22.5.1, steps 1 to 14.

Outlets from valves D and A are both connected to the same back pressure system, via a volume-change indicator, which measures the total volume of water draining out of the specimen from both the top and bottom faces combined, as shown in Fig. 22.22. Valve F remains closed throughout the test.

The test procedure is given in Section 22.6.3.

22.5.3 Radial Drainage to Periphery

GENERAL PREPARATION

The procedure for fitting a porous plastic peripheral drain to the Rowe cell is described below.

Extrusion and preparation of an undisturbed specimen for peripheral drainage is then identical to that given in Section 22.4.1, except that the smaller trimming ring must be used to allow for the thickness of the porous plastic lining. The ring should be distinguished by a conspicuous mark such as bright red paint on the flange. It should be clean and dry and *not* greased.

Preparation of remoulded specimens consolidated from slurry, and compacted samples, are identical to the procedures given in Sections 22.4.3 and 22.4.4 respectively.

FITTING PERIPHERAL DRAIN

(1) Cut a strip of the plastic material of width equal to the depth of the cell body, and about 20 mm longer than its internal circumference. Cut the ends square using a sharp blade and metal straight-edge.

(2) Fit the plastic tightly against the wall of the cell body. Mark the end of the overlap with a sharp pencil (Fig. 22.23).

(3) Lay the plastic material on a flat surface and mark another line exactly parallel to the first (i.e. square to the edges) at the following distance outside it (denoted by x in Fig. 22.23):

  For the 75 mm cell: 1.5 mm

  For the 150 mm cell: 3 mm

  For the 250 mm cell: 5 mm

Make a clean square cut on this line.

(4) Fit the plastic in the cell body again, smooth face inwards and trimmed ends butting. Allow the additional length to be taken up in the form of a loop opposite the butt joint (Fig. 22.23).

(5) Push the loop outwards and the plastic material will spring against the wall of the cell. Check that it fits tightly, with no gaps.

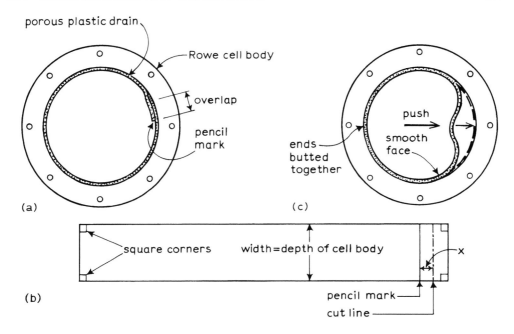

Fig. 22.23 *Fitting porous plastic liner in Rowe cell: (a) initial fitting and marking,
(b) locating line of cut, (c) final fitting*

(6) Immediately before inserting the soil, remove the porous plastic for saturating and de-airing in boiling water, then replace it in the cell. The inside face of porous plastic must *not* be greased, because grease will prevent drainage.

ASSEMBLY

(7) The specimen is prepared and set up in the cell by one of the procedures described in Section 22.4, and is covered with water if appropriate.

(8) Place an impervious membrane such as a disc of latex rubber of the kind used for triaxial sample membranes, or of a plastic material, through the water on to the specimen, without entrapping air. For a 'free strain' test, if plastic material is used it should be flexible.

For an 'equal strain' test, cover the membrane with the circular steel plate, lowered into position using the lifting handle and without entrapping air. The central hole is then plugged.

(9) Fit and assemble the cell top to the body as described in Section 22.5.1, stages 1 to 14. The tubing between valve D and the end of the hollow stem should be filled with de-aired water. Valve D is not used and remains closed.

(10) Connect the back pressure system to valve F, without trapping any air. The cell connections are shown in Fig. 22.24.

The test procedure is given in Section 22.6.4.

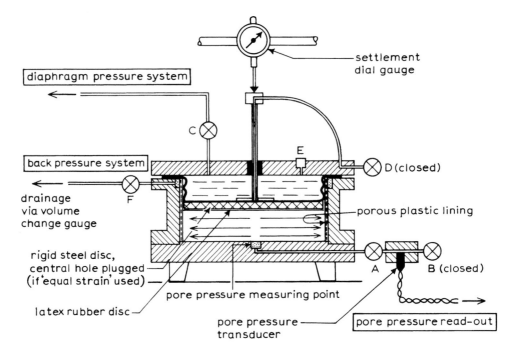

Fig. 22.24   Arrangement of Rowe cell for consolidation test with radial drainage to periphery

22.5.4.   Radial Drainage to Centre

GENERAL PREPARATION

The cell base is made ready as described in Section 22.3.4, except that the ceramic insert situated at $0.55R$ from the centre is prepared for measuring pore pressure. The transducer block, with valve B and the connection to the pore pressure panel, is fitted to valve G (Fig. 22.25). The back pressure system (drainage line) is connected to valve A. The ports connecting to ceramic inserts at the centre and at $0.55R$ should both be de-aired as described in Section 22.3.4. The connection to valve D is not used but should be filled with de-aired water, and valves D and F remain closed.

The specimen (undisturbed, remoulded or compacted) is prepared and set up in the cell as described in Section 22.4.

DRAINAGE WELL

The central drainage well is prepared as outlined below. The hole diameter should normally be about 5% of the specimen diameter, i.e. 12.5 mm for the 250 mm cell or 7.5 mm for the 150 mm cell. (This procedure would not normally be used in a 75 mm cell.)

(1) A vertical hole is formed in the centre of the specimen by using a template to guide a suitable mandrel, as indicated in Fig. 22.26. Detailed procedures for forming the hole depend upon the type of mandrel used. Some examples were described by Singh and

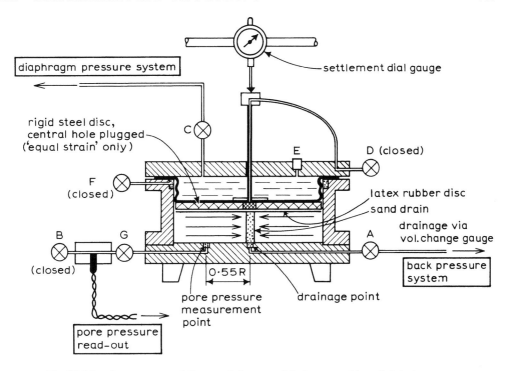

Fig. 22.25   *Arrangement of Rowe cell for consolidation test with radial drainage to central well; pore pressure measurement offset from centre*

Hattab (1979). Seven types of mandrel are shown in Fig. 22.27, and their use is outlined below.

(a) *Solid* A solid mandrel forms a hole by displacing the soil as it is advanced. The mandrel should be fitted with an air vent to release the suction which could otherwise cause partial collapse of the hole on withdrawal, leaving a hole of non-uniform diameter.

This method should be used only when disturbance is not critical, for instance in soil reconstituted from slurry.

(b) *Hollow* A hollow mandrel should be advanced by a distance equal to half the diameter at a time, then rotated through 90° and withdrawn for removal of the core of soil. In more permeable soils it may be advantageous to pass water up from the pore pressure point at the base of the hole as the mandrel is advanced. This type of mandrel is suitable for most soil types (except very soft soils) not containing stony material.

(c) *Cruciform* A cruciform section mandrel should be pushed straight into the full depth, then rotated through 90° before withdrawal.

(d) *Star* As for (c) except that the mandrel is rotated through an angle equal to that between successive blades (i.e. 60° for the type shown). Cruciform and star mandrels are not suitable if fibrous or stony material is present.

(e) *Auger* An auger mandrel should be advanced by a distance less than the pitch of the helix for each revolution, and withdrawn several times for clearing the soil removed. This type is appropriate for laminated soil.

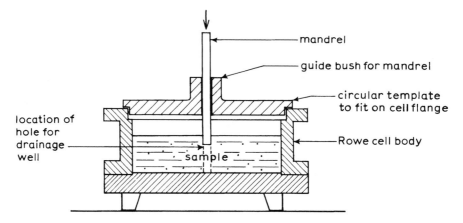

Fig. 22.26   *Template used for forming central drainage well*

(f) *Jetting (double wall type)* and (g) *Jetting (Dutch type)* A mandrel using the jetting process should normally be allowed to advance under its own weight, with no additional applied force. The rate of flow of water used should be gradually increased until this occurs. Jetting is suitable for fine cohesionless soils but should be undertaken with great care.

(2) Flush water through the porous insert at the bottom of the hole to wash off any smeared material and to ensure that there is no obstruction. Remove water from the hole with a vacuum line if necessary.

(3) About two-thirds fill the hole with de-aired water. Place the de-aired saturated sand steadily in the hole under water by using a pipette. Allow the sand a free fall of about 10 mm onto the sand already deposited, to obtain a loose state of packing. Avoid disturbance to the sand, and jolting or vibrating the cell after placing, which could densify the loose sand.

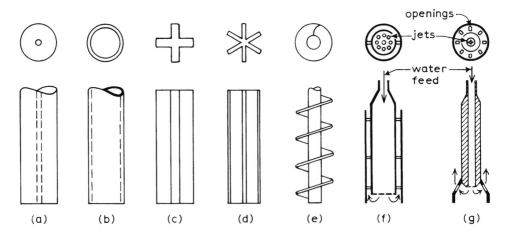

Fig. 22.27   *Types of mandrel for forming drainage well: (a) solid, (b) hollow tube, (c) cruciform, (d) star, (e) auger, (f) double-wall jetting, (g) Dutch type jetting*

(4) Check that the drain is operating satisfactorily by allowing water to drain down through the sand and out through valve A. Level the surface of the sand drain flush with the top of the specimen.

ASSEMBLY

(5) Place a disc of latex rubber or similar impermeable material to cover the whole surface of the specimen.

(6) For an 'equal strain' test, place the rigid circular steel plate centrally over the impermeable disc, and plug the drainage hole.

(7) Fill the cell above the steel disc with water.

The specimen is then ready for assembly of the top, as described in Section 22.5.1, stage 3 onwards. Connections to the cell are as shown in Fig. 22.25.

The test procedure is given in Section 22.6.5.

22.5.5 Permeability Tests

Permeability measurements can be carried out on a specimen in a Rowe cell either with laminar flow of water in the vertical direction (upwards or downwards) or with radial flow horizontally (inwards or outwards). The procedures for preparing specimens for each type of test are outlined below, with reference to earlier sections for details. A permeability test may be carried out during a consolidation test sequence at the end of a loading stage, provided that the appropriate drainage facilities have been incorporated.

When the difference between inlet and outlet pressures is small, it can be measured more accurately from a differential pressure gauge or transducer, or a pressurised manometer, than by using individual pressure gauge readings. The differential gauge or manometer is connected as shown in Fig. 22.28.

If the rate of flow is relatively large, allowances must be made for the head losses in connecting pipework and porous plates, as described in Chapter 17, Section 17.3.7.

(a) *Vertical permeability*

The arrangement for a permeability test with vertical flow is shown in Fig. 22.29. The preparation of the specimen and assembly of the cell are summarised as follows:

(1) Fit a bottom drainage disc on the cell base (Section 22.5.2).

(2) Set up the specimen in the cell by the appropriate method (Section 22.4).

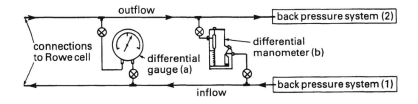

Fig. 22.28 Method of connecting differential pressure gauge or pressurised differential manometer for permeability tests

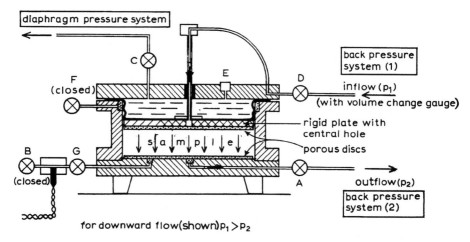

Fig. 22.29   *Arrangement of connections to Rowe cell for permeability test with downward vertical flow (for upward flow, $p_2 > p_1$)*

(3) Fit a drainage disc and rigid steel plate on top of the specimen (Section 22.5.1, stage 2). The hole in the plate must coincide with the settlement stem drainage outlet.

(4) Assemble the cell top (Section 22.5.1, stages 3 to 13).

Three independently controlled constant pressure systems, each with a calibrated pressure gauge, are required for the permeability test. One system is connected to valve C (Fig. 22.29) to provide pressure on the diaphragm. Two separate back pressure systems are used, one connected to valve D, the other to valve A. Each system should incorporate a volume-change indicator, as specified in BS 1377. However, if only one indicator is available it should be connected into the inlet line, and a satisfactory test can be carried out provided that it has been confirmed that the specimen is fully saturated.

Pore pressure readings are not required, except as a check on the B value if incremental saturation is applied before starting the test. Valve F remains closed.

The difference between the inlet and outlet pressures should be appropriate to the vertical permeability of the soil, and should be determined by trial until a reasonable rate of flow is obtained. Normally the direction of flow is downwards, but the pressures can be reversed to provide upward flow if required. If the specimen is fully saturated initially and maintenance of an elevated back pressure is not necessary, the outflow back pressure system can be replaced by an elevated water reservoir, fitted with an overflow to maintain a constant water level.

The test procedure is given in Section 22.7.2.

(b) *Horizontal permeability*

The arrangement for horizontal (radial) permeability is shown in Fig. 22.30. Specimen preparation and assembly of the cell are summarised as follows.

(1) Fit a peripheral drain in the cell as described in Section 22.5.3.

(2) Set up the specimen in the cell (Section 22.4).

(3) Form a central drain (Section 22.5.4).

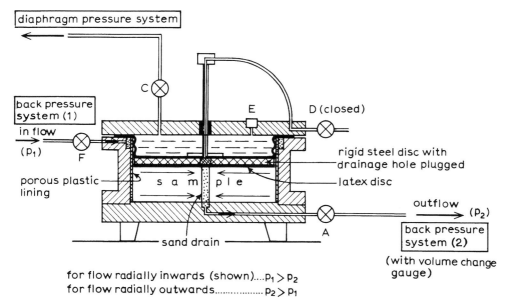

C

back pressure
system (1)

in flow
(P₁)

E

D (closed)

F

porous plastic
lining

s a m p l e

rigid steel disc with
drainage hole plugged

latex disc

outflow
(P₂)

sand drain

A

back pressure
system (2)

(with volume change
gauge)

for flow radially inwards (shown)....$P_1 > P_2$
for flow radially outwards.................$P_2 > P_1$

*Fig. 22.30   Arrangement of Rowe cell for permeability tests with horizontal (radial)
flow (inwards or outwards)*

(4) Place a disc of impervious material such as latex rubber to cover the specimen. Then place the rigid steel disc, with its central hole plugged.

(5) Assemble the cell top (Section 22.5.1, stages 3 to 13).

Three independently controlled pressure systems are required, each with a calibrated pressure gauge. One system is for applying the diaphragm pressure, and is connected to valve C (Fig. 22.30). The second pressure system is connected to valve F, and the third to valve A. The system on the inlet should incorporate a volume-change indicator. Pore pressure readings are not necessary. Valve D remains closed.

Permeability measurements may be made with the flow either radially inwards or radially outwards, according to whether the pressure at B is less than or greater than at F. The difference between these two pressures should be appropriate to the horizontal permeability of the soil, and should be determined by trial until a reasonable rate of flow is obtained. If the specimen is fully saturated initially, and maintenance of an elevated back pressure is not necessary, the lower of the two pressure lines can be an elevated reservoir as described in (a) above.

22.6 CONSOLIDATION TESTS

22.6.1 Scope

Consolidation test procedures using the four types of drainage indicated in Fig. 22.4 are described in Sections 22.6.2 to 22.6.5. The single vertical drainage test, Fig. 22.4 (a) and (b), is described in detail in Section 22.6.2, and is used as the 'basic' test to which reference is made in the descriptions of the other tests with different drainage conditions. Analysis of data from all four types of test is described in Section 22.6.6.

The following test conditions and requirements should be specified by the engineer at the time the tests are requested:

Size of test specimen

Loading conditions ('free strain' or 'equal strain')

Drainage conditions

Location of pore pressure measurement point (when required)

Whether saturation is to be applied, and if so the method to be used

Whether voids ratios are to be calculated and plotted

Sequence of effective pressure increments and decrements

Criterion for terminating each primary consolidation and swelling stage

Whether secondary compression characteristics are required.

22.6.2   Consolidation with Vertical One Way Drainage (BS 1377:Part 6:1990:3.5)

PRINCIPLES

This is the most common type of test carried out in the Rowe cell. Drainage takes place vertically upwards and from the top face, while pore water pressure is measured at the centre of the base. The principle is shown in Fig. 22.4 (a) and (b), for the two types of applied loading. The general arrangement of the Rowe cell is shown in Fig. 22.21, and the ancillary systems are connected as shown in Fig. 22.3. A photograph of a typical test set-up is shown in Fig. 22.31.

The specimen is first prepared by one of the methods given in Section 22.4, and the cell is assembled and connected as described in Section 22.5.1.

The test is described under the following stages, which follow on from step (15) of Section 22.5.1.

(1) Preliminaries

(2) Saturation

(3) Undrained loading

(4) Consolidation

(5) Further load increments

(6) Unloading

(7) Conclusion

(8) Dismantling

(9) Graphical plots

(10) Calculations

(1) PRELIMINARIES

(a) The pore pressure transducer must be isolated from the flushing system throughout the test by keeping valve B closed (Fig. 22.21).

(b) Set the vertical movement dial gauge at a convenient initial reading near the upper limit of its travel, but allow for some upward movement if saturation (see below) is to be applied. Record the reading as the zero (datum) value under the seating pressure p_{d0}.

Fig. 22.31   *Rowe cells and ancillary equipment set up for tests (courtesy LTG Laboratories)*

(c) Set the back pressure to the required initial value, with valve D closed. This will normally be not less than the initial pore pressure u_0, or if saturation is to be applied the back pressure should be 10 kPa less than the first increment of cell pressure.

(d) Record the initial reading of the volume-change indicator when steady.

(e) Record the initial pore water pressure, u_0, corresponding to p_{d0}, when steady.

(f) Ascertain the actual stress p_0 applied to the specimen corresponding to the diaphragm pressure p_{d0}, from the diaphragm calibration data (see Section 17.3.4).

(g) If measurement of the swelling pressure is required, allow access of de-aired water to one or both ends of the specimen and increase the diaphragm pressure to maintain the reading of the settlement dial gauge constant at the initial value. The principle is similar to that given in Volume 2, Section 14.6.1, for measurement of swelling pressure in the oedometer test. A back pressure may be applied to facilitate saturation. When equilibrium is established the difference between the applied pressure and the back pressure (if used) is the effective stress required to prevent swelling, and is reported as the swelling pressure.

(2) SATURATION

Principles

The reason for and objectives of saturation were given in Section 15.6. Saturation is usually achieved by the application of increments of back pressure, alternating with increments of diaphragm pressure, so that the pore water pressure in the specimen is raised sufficiently to enable the water to absorb into solution all the air originally in the voids. The procedure is similar to that described for the saturation of triaxial test specimens (Section 18.6.1, method (1)), but with diaphragm pressure substituted for cell pressure. Saturation might not always be appropriate but it is usually desirable for undisturbed samples taken from above the water table, and for compacted specimens.

The degree of saturation is related to the pore pressure ratio $\delta u/\delta \sigma$, where δu is the pore pressure response to an increment of total vertical stress of $\delta \sigma$ when no drainage is allowed. The ratio is not referred to as the pore pressure parameter B, which by definition relates to increments of isotropic stress. Saturation is usually accepted as being complete when the ratio reaches about 0.95. However, in some soils it might not be possible to achieve this value, and a lower value would then be appropriate (see Section 15.6.5).

Diaphragm pressure increments of 50 or 100 kPa are usually applied. The back pressure for each 'water in' stage should normally be 10 kPa less than the vertical stress. During saturation of a large specimen it might be necessary to reverse the volume-change indicator several times.

Saturation by increments of back pressure

(a) Ensure that valve D is closed, and increase the diaphragm pressure from the initial seating pressure p_{d0} to a value which gives the required first-stage pressure p_1 (typically an increase of 50 kPa) on the specimen. The appropriate diaphragm pressure is obtained from the diaphragm calibration data.

(b) Record the pore pressure when it reaches a steady value (u_1), and calculate the ratio $\dfrac{\delta u}{\delta \sigma}$ from the equation

$$\frac{\delta u}{\delta \sigma} = \frac{u_1 - u_0}{p_1 - p_0} \tag{22.13}$$

(c) With valve D remaining closed, increase the pressure in the back pressure line to a value equal to $(p - 10)$ kPa (assuming that 10 kPa is the desired differential pressure). Record the reading of the volume-change indicator (v_1) when it reaches a steady value, to allow for expansion of the connecting lines.

(d) Open valve D to admit the back pressure into the specimen. Observe readings of pore pressure and the volume-change indicator, and if necessary plot these values against time to ascertain when equilibrium conditions are reached. This might take some considerable time.

(e) When the pore pressure becomes virtually equal to the back pressure, and the volume-change indicator shows that movement of water into the specimen has virtually ceased, record the readings of pore pressure (u_2) and the volume-change indicator (v_2). The difference between volume-change indicator readings v_1 and v_2 gives the volume of water taken in by the specimen during this increment. Close valve D.

(f) Increase the diaphragm pressure to give a further increment of pressure (e.g. 50 kPa) on the specimen, as in step (a). Observe the change in pore pressure, and when equilbrium is achieved calculate the new value of the ratio $\delta u/\delta \sigma$ as in step (b).

(g) Repeat steps (c) to (f) until the pore pressure ratio $\delta u/\delta \sigma$ reaches a value of 0.95, or such other value which indicates that saturation has been achieved according to the type of soil (see Section 15.6.5). Increments of diaphragm pressure can be inceased to 100 kPa after the first two increments.

(h) Calculate the total volume of water taken up by the specimen by totalling the differences of volume-change indicator readings obtained in step (e). This change in volume can be compared with the volumetric swell calculated from the vertical movement measured by the dial gauge. The former will generally exceed the latter if air was present in the voids initially.

Prevention of swell

An alternative method for achieving saturation, if necessary, is by applying increments of back pressure while adjusting the diaphragm pressure to prevent swell. To check the degree of saturation, a small back pressure increment is applied while the diaphragm pressure is increased to maintain a constant effective stress in the specimen. Immediate and equal response of the pore pressure transducer indicates that saturation is essentially complete.

(3) UNDRAINED LOADING

(a) With valves A and C open, and all other valves closed, record the readings of pore pressure, diaphragm pressure and the compression gauge.

(b) Close valve C and set the pressure in the diaphragm pressure system to the value needed to apply the desired vertical stress on the specimen for the first stage of consolidation, taking into account the diaphragm pressure calibration data.

If the required total vertical stress is denoted by σ, and the corresponding diaphragm pressure pressure correction by δp (see Section 17.3.4), the diaphragm pressure to be applied (p_d) is obtained by rearranging equation (17.3):

$$p_d = \sigma + \delta p$$

For a desired effective stress σ' when a known back pressure (u_b) is applied to the specimen, this equation becomes

$$p_d = \sigma' + u_b + \delta p \tag{22.14}$$

(c) Open valve C to admit the pressure to the diaphragm, and at the same instant start the timer. The additional applied stress is carried by the pore water pressure.

(d) Open the rim drain valve F briefly (see Section 17.3.6) to allow excess water to escape from behind the diaphragm into a measuring cylinder. Release of this water should take no more than 2 or 3 seconds, but care is needed with soils of high permeability and with peats.

(e) Observe and record readings of pore pressure at time intervals suitable for plotting a curve of pore pressure against time during this build-up stage. If the soil is saturated the increase in pore pressure should eventually become almost equal to the increment of vertical pressure applied to the specimen.

(f) When the pore pressure becomes steady, record it, and the reading of the compression gauge, as the final readings for the undrained loading stage.

(4) CONSOLIDATION

Set the timer to zero and record readings of the diaphragm pressure, back pressure, compression gauge and volume-change indicator corresponding to zero time.

Start the consolidation stage by opening the drainage outlet (valve D in Fig. 22.21) and at the same instant start the clock. Water then drains from the specimen as the applied stress is transferred from the pore water to the soil 'skeleton', increasing the effective stress, while the total applied vertical stress is held constant. Read the following data at time intervals similar to those used for a conventional oedometer test (Volume 2, Section 14.5.5, Table 14.11).

Vertical settlement gauge

Pore water pressure

Volume-change indicator on back pressure line

Diaphragm pressure (check)

If consolidation continues beyond 24 hours, record further readings at about 28 hours and 32 hours from the start (as suggested in Table 14.11), and on subsequent days at least twice a day, morning and evening.

When the limit of the volume-change indicator is approached, reverse the direction of flow and record the reading at the instant of reversal.

The primary consolidation phase is theoretically completed when the pore pressure has fallen to the value of the back pressure, i.e. when 100% dissipation of the excess pore pressure is achieved. For most practical purposes, 95% dissipation of the excess pore pressure is sufficient. The percentage pore pressure dissipation, denoted by $U\%$, is given by

$$U = \frac{u_0 - u}{u_0 - u_b} \times 100\% \tag{15.28}$$

where u is the pore water pressure at the time considered, u_b is the back pressure against which drainage takes place, and u_0 is the pore water pressure at start of consolidation stage.

For low permeability clays draining vertically, even 95% dissipation may take several days to achieve. Nevertheless if undrained pore pressure ratios are to be derived it is necessary to extend the primary consolidation to reach as close to 100% as possible. Graphical plots as described in stage 9 below should be made while consolidation is in progress.

If the secondary compression coefficient C_{sec} is required, the consolidation stage should be continued after 100% dissipation is achieved. This will enable further readings of settlement and volume change to be taken until the plot of settlement against log time is seen to be linear and the slope can be derived (see Volume 2, Section 14.3.13).

To terminate the consolidation stage, close the drainage line valve D. Record the final reading of pore pressure and of the compression gauge and volume-change indicator.

(5) FURTHER LOAD INCREMENTS

Increase the diaphragm pressure to give the next value of effective stress, as described in stage (3) above. Allow excess water to drain from behind the diaphragm if necessary. The pore pressure should then be allowed to reach equilibrium before proceeding to the next consolidation stage, especially if less than 100% dissipation was achieved on the previous stage. The effective stress is normally doubled at each stage; i.e. the pressure increment is made equal to the effective stress already applied, as in the conventional oedometer test (see Section 22.2.4).

Consolidation is as described in stage (4). Repeat the undrained loading and consolidation stages for each subsequent increment of pressure.

At least four loading stages should be applied. The range of loadings should be such that the voids ratio/log pressure curve derived from the test more than covers the range of in-situ and post-construction effective stresses.

(6) UNLOADING

At the end of the consolidation stage under the maximum required stress, and after recording the final readings, the specimen is unloaded in a series of decrements. The procedure follows the same sequence as that described above. In each unloading stage the diaphragm pressure is reduced with valve D closed (i.e. with undrained conditions), and as a result the pore pressure decreases until a steady value is reached. This is followed by a swelling stage (the counterpart of stage (4)) with valve D open, during which upward movement, volume increase and increasing pore pressure readings are taken in the same way as for consolidation. The pore pressure should be allowed to reach equilibrium at the end of each stage before proceeding to the next.

The number of unloading stages (stress decrements) should normally be at least half the number of applied increments, and should follow a constant unloading stress ratio.

(7) CONCLUSION OF TEST

For the final unloading stage the diaphragm pressure is reduced to the initial seating pressure. When equilibrium has been achieved record the final settlement, volume change and pore pressure readings.

Close valve A (Fig. 22.21) and open valves C, D and F to atmosphere, allowing surplus water to escape.

(8) DISMANTLING

Unbolt and remove the cell top and place it on the bench supports (Fig. 22.18).

Remove the porous or rigid plates and remove any free water, to expose the specimen surface. Measure down to the surface along two or more diameters from a straight edge placed across the cell flange (Fig. 22.32), using a steel ruler or depth gauge reading to 0.5 mm. From these measurements plot the surface profile of the specimen, which can be used for calculating the final specimen volume.

Remove the cell body from the base and weigh the specimen in the cell body. Remove the specimen intact from the cell, using an extrusion device if necessary. Split the specimen in two along a diameter by cutting to about one-third of its depth, then breaking open. Take representative portions from two or more points for moisture content measurements. Record a description of the soil, including details of the soil fabric with illustrative sketches, and colour photographs if required. It is often advantageous to allow one half of the split specimen to air-dry to reveal the fabric and any preferential drainage paths which may have affected the test behaviour. These features usually show up best for photographs after being exposed to atmosphere for up to 24 hours, during which time silty material

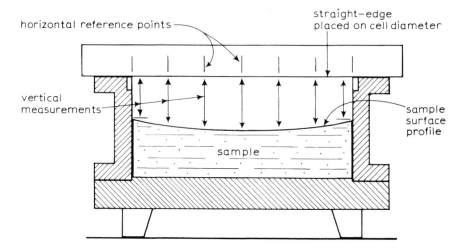

Fig. 22.32   Determination of surface profile of Rowe cell sample after test

dries more quickly than clay and acquires a lighter colour. An example is shown in Fig. 22.33. The other half of the specimen may be weighed and oven-dried for determination of the overall moisture content.

The cell components should be cleaned and dried before putting away, giving careful attention to the sealing ring in the base. Porous bronze and ceramic discs and inserts should be boiled and brushed; used porous plastic should be discarded. Connecting ports and valves should be washed out to remove any soil particles. Any corrosion growth on exposed metal surfaces should be scraped off, and the surface made smooth and lightly oiled.

(9) GRAPHICAL PLOTS

During each undrained loading phase, plot pore pressure against log time.

As each consolidation stage proceeds, plot the following graphs from the observed data.

Settlement (ΔH mm) against log time

Fig. 22.33   Rowe cell sample split open after test to reveal soil fabric (courtesy of LTG Laboratories)

*Settlement against square-root time

Volume change (ΔV mm) against log time

*Volume change against square-root time

*Pore water pressure dissipation ($U\%$) against log time

These graphs should be kept up to date during each stage so that the approach to 100% primary consolidation can be monitored. The plots marked with an asterisk * are those generally preferred for analysis.

A typical set of data from one consolidation stage of a test is shown in Figs. 22.34, 22.35 and 22.36. It is usually convenient to plot ΔH and ΔV together on the same time base as shown. Settlement and volume-change graphs are plotted cumulatively. Pore pressure dissipation curves are plotted from 0 to 100%, for each increment stage.

Achievement of 100% pore pressure dissipation, if reached, represents the end of the primary consolidation phase. The slope of the line following the 100% point on the log time/settlement plot gives the coefficient of secondary compression, C_{sec} (Section 14.3.13). When the pore pressure is not measured, or 100% dissipation is not achieved, the 100% primary consolidation point can be estimated as described in Section 22.6.6.

For each unloading stage, graphs similar to the above should be plotted for both the undrained unloading and the drained pore pressure equalisation phases. The same axes as

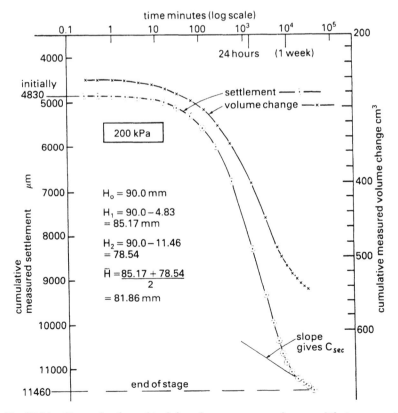

Fig. 22.34 Example of graphical data from one stage of a consolidation test with vertical drainage: settlement and volume change plotted against log time

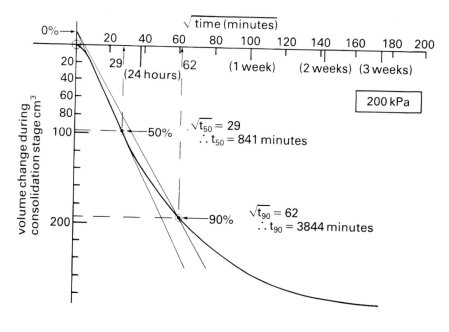

Fig. 22.35   *Volume change plotted against square-root time*

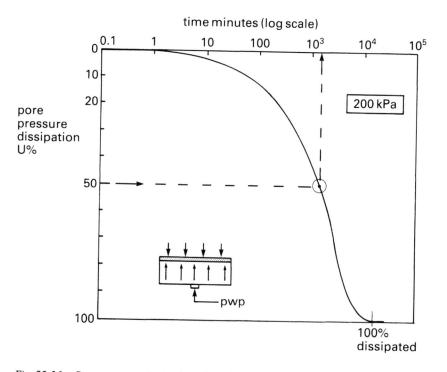

Fig. 22.36   *Pore pressure dissipation plotted against log time ($t_{50} = 1400$ minutes)*

used for consolidation may be used, but the values will change in the opposite sense (see Volume 2, Section 14.5.5, stage (20)).

(10) CALCULATIONS

At the end of each drained loading or unloading phase the voids ratio is calculated as described in Section 22.6.6. Voids ratio is plotted against effective pressure on a log scale to give the $e/\log p'$ curve.

Values of the coefficient of consolidation, c_v or c_{ro} or c_{ri}, are calculated as described in Section 22.6.6 and may be plotted against log effective pressure (using the same axes as the $e/\log p'$ curve).

22.6.3 Consolidation with Vertical Two-Way Drainage (BS 1377 : Part 6 : 1990 : 3.6)

For consolidation tests with double drainage in the vertical direction (tests type (c) and (d) Fig. 22.4) the procedure is similar to the foregoing but with the following modifications:

(1) A porous drainage disc is placed at the base, as well as on top, of the specimen, as shown in Fig. 22.22.

(2) If the specimen is to be saturated under increments of back pressure, the initial connection of ancillary systems is as shown in Figs. 22.21 and 22.3, enabling the saturation procedure described in stage (2) of Section 22.6.2 to be followed.

(3) After saturation the pore pressure transducer housing is disconnected from valve A, and the back pressure system is connected instead, without entrapping air.

(4) The top drainage line from valve D is connected to the same back pressure system by means of a tee-piece, as shown in Fig. 22.22.

(5) Valves A and D are opened simultaneously when starting a consolidation stage, and both are closed at the end of the stage.

(6) Pore water pressure is not measured during consolidation; therefore the end of primary consolidation and values of t_{50} or t_{90} can be obtained only from analysis of the settlement/time curves. Appropriate factors are given in Table 22.3, (c) and (d).

22.6.4 Consolidation with Radial Drainage to Periphery (BS 1377 : Part 6 : 1990 : 3.7)

The principle is illustrated diagrammatically in Fig. 22.4 (e) and (f), with 'free strain' and 'equal strain' loading respectively. The arrangement of the cell and ancillary equipment is shown in Fig. 22.24. Details differ from the arrangement for vertical drainage tests in the following ways.

(1) The specimen is surrounded by a drainage layer of porous plastic material.

(2) The top surface of the specimen is covered by an impermeable membrane.

(3) The back pressure is connected to the rim drain at the top of the cell, via valve F.

(4) The top drainage line is not used and valve D remains closed.

The diaphragm pressure line and pore water pressure connection to the centre of the base are the same as used for the single vertical drainage test described in Section 22.6.2.

Equipment is made ready as described in Section 22.3. The specimen is prepared as described in Section 22.4, bearing in mind that if an undisturbed sample is used it must be finally trimmed with the slightly smaller cutting ring to allow for the thickness of the peripheral drainage layer. Assembly in the cell is as described in Section 22.5.3.

Saturation of the specimen if appropriate, and the test itself, are carried out as described in Section 22.6.2. If the specimen contains approximately horizontal laminations, consolidation is likely to be very much more rapid than for similar material draining vertically. Enough time must be allowed during the saturation stages for saturation and pore pressure build-up to extend into the clay layers between laminations.

A special settlement/time relationship is plotted for this test when 'free strain' is used. Details are given in Section 22.6.6, together with appropriate curve fitting factors for both types of loading. Consolidation with radial drainage under 'equal strain' loading can produce non-uniformities within the specimen, for the reason given at item 6 under 'Selection of test' in Section 22.1.8.

22.6.5   Consolidation with Radial Drainage to Centre   (BS 1377 : Part 6 : 1990 : 3.8)

The principle is illustrated in Fig. 22.4 (g) and (h) with 'free strain' and 'equal strain' loading respectively. The arrangement of the cell and ancillary equipment is shown in Fig. 22.25.

Details differ from the vertical drainage test (Section 22.6.2) as follows:

(1) The top surface of the specimen is covered by an impermeable membrane.

(2) The back pressure system is connected to the centre of the base, via valve A (Fig. 22.25).

(3) Pore water pressure is measured at the porous insert at 0.55 R from the centre (i.e. at a radius of 70 mm in the 250 mm cell). The pore pressure transducer housing block is connected to valve G which replaces the blanking plug at that cell outlet (Fig. 22.25).

(4) The top drainage line and the rim drain are not used, and valves D and F remain closed.

(5) If the drainage capacity of the central drain and base outlet is considered to be inadequate, drainage can also be provided from the top end of the central drain. A central hole is made in the impervious membrane, and a filter paper should be placed over the sand drain to prevent movement of particles into the drainage system. The outlet from valve D is connected to the same back pressure system as the outlet from valve A, in a similar manner to that shown in Fig. 22.22.

(6) The central sand drain is prepared by one of the methods given in Section 22.5.4.

Equipment is made ready as described in Section 22.3. Saturation if appropriate, and the consolidation test, are as described in Section 22.6.2. If the specimen contains horizontal laminations, consolidation is likely to be more rapid than when draining vertically, although (theoretically) about nine times slower than when draining radially outwards (Berry and Wilkinson, 1969). Enough time must be allowed during saturation to allow saturation and pore pressure build-up to extend into the clay layers between laminations, and into the zone beyond the pore pressure measurement point.

22.6.6 Analysis and Presentation of Consolidation Test Data

GENERAL

The parameters derived from consolidation tests in the Rowe cell are basically the same as those obtained from conventional oedometer tests. These comprise the usual specimen details, the coefficients of consolidation and volume compressibility, and the relationship between effective pressure and voids ratio. However the methods by which some of these parameters are derived differ from those used in the oedometer test, especially with regard to the coefficient of consolidation, c_v, for which a different curve fitting procedure is needed for each type of drainage and loading condition.

SPECIMEN DETAILS

The initial and final conditions of the specimen are calculated in the same way as explained for an oedometer test sample in Volume 2, Section 14.5.7. The symbols used are as listed in Table 14.13, the main ones being as follows:

	initial	final	
Volume	V_0	V_f	cm^3
Density	ρ	ρ_f	Mg/m^3
Moisture content	w_0	w_f	%
Dry density	ρ_D	ρ_{Df}	Mg/m^3
Voids ratio	e_0	e_f	
Degree of saturation	S_0	S_f	%

VOIDS RATIO CHANGES

The voids ratio e at the end of each loading or unloading stage can be calculated from the initial voids ratio and the overall change in height of the specimen, using the equations from Chapter 14 (Section 14.3.9):

$$\Delta e = \frac{1 + e_0}{H_0} \Delta H \qquad (14.19)$$

and

$$e = e_0 - \Delta e \qquad (14.21)$$

in which the symbols Δe and ΔH represent *cumulative* changes of e and H, as defined in Volume 2, Section 14.39 and Table 14.13. This calculation is valid only for the 'equal strain' loading condition, in which specimen volume change is directly proportional to vertical settlement.

In a saturated soil under either type of loading, volume change can be obtained directly from measurements of the volume of water draining out of the specimen. Voids ratio changes are calculated from the measured volume changes by using an equation similar to the above, i.e.

$$\Delta e = \frac{1 + e_0}{V_0} \Delta V \qquad (22.15)$$

where ΔV is the cumulative volume change from the initial volume V_0.

From the definition of voids ratio (Volume 1, Equation (3.1) and Fig. 3.2) the volume of solid particles V_s in a volume of soil V_0 is given by

$$V_s = \frac{V_0}{1 + e} \qquad (22.16)$$

The term

$$\frac{1 + e}{V_0} = \frac{1}{V_s}$$

can be calculated from the initial specimen measurements and the voids ratio change equation can be written

$$\Delta e = \frac{\Delta V}{V_s} \qquad (22.17)$$

Incremental changes in voids ratio for each loading or unloading stage, denoted by δe, are calculated as explained in Section 14.5.7, i.e.

$$\delta e = e_1 - e_2$$

The coefficient of volume compressibility, m_v, for each stage is calculated by using Equation (14.24), Section 14.3.10:

$$m_v = -\frac{\delta e}{\delta p'} \times \frac{1000}{1 + e_1} \quad \mathrm{m^2/MN} \qquad (14.24)$$

Alternatively, the coefficient m_v can be calculated directly from the changes in height (for 'equal strain' loading) or changes in volume (for 'free strain' loading) using one of the following equations:

$$m_v = \frac{\Delta H_2 - \Delta H_1}{H_0 - \Delta H_1} \times \frac{1000}{p'_2 - p'_1} \qquad (22.18)$$

$$m_v = \frac{\Delta V_2 - \Delta V_1}{V_0 - \Delta V_1} \times \frac{1000}{p'_2 - p'_1} \qquad (22.19)$$

In these equations, subscript 1 denotes the values at the end of the previous loading increment, and subscript 2 the values at the end of the increment being considered.

GRAPHICAL ANALYSIS

Graphical plots relating settlement or volume change, and pore pressure dissipation, to some function of time are used for determining the coefficient of consolidation for each load increment. Procedures are similar in principle to those used for the oedometer consolidation test, but the theoretical time factors used may not be the same, and depend upon

Boundary conditions ('free strain' or 'equal strain')

Type of drainage (vertical or horizontal; and direction)

Location of relevant measurements.

These factors are discussed in Section 22.2.2, and the factors appropriate to each test condition and interpretation method are explained in Section 22.2.3. Three empirical methods of analysis are given in BS 1377: Part 6: 1990, Clause 3.5.8.5.1, for the derivation of the coefficient of consolidation. These are based on:

(a) the pore pressure dissipation curve;

(b) curve fitting from the plot of settlement or volume change against log time;

(c) curve fitting from the plot of settlement or volume change against square-root time.

These procedures enable the time (t_{50}) corresponding to 50% pore pressure dissipation to be evaluated.

Method (a) is based on pore pressure readings at a particular point (in most cases the centre of the base). Methods (b) and (c) depend on the 'average' behaviour of the whole specimen, and require empirical methods of analysis (curve fitting procedures) of the curves of settlement or volume change against a function of time (logarithmic or a power factor). The theoretical time factors in methods (b) and (c) depend upon the type of drainage and the boundary strain conditions, and are different for each type of test. Method (a) is preferable wherever possible because the value of t_{50} is obtained directly from the graph.

An example of method (a), and examples of methods (b) and (c) for two different drainage conditions, are given below.

Example (i)   Pore pressure dissipation plots

On a pore pressure dissipation plot ($U\%$ against log time) the final level at which 100% primary consolidation is reached is known, even if it is not achieved. The value of t_{50} can therefore be read directly off the graph, as shown by the example in Fig. 22.36. This procedure applies to any type of test in which pore pressure readings are recorded and plotted.

Example (ii)   Curve fitting: vertical drainage

An example of a graph of volume change against square-root time for one stage of a vertical drainage consolidation test is shown in Fig. 22.35. A similar curve would be obtained by plotting settlement against square-root time, as in the standard oedometer test.

From Table 22.3 (Section 22.2.3) the 'slope factor' for vertical drainage, either one-way or two-way, is 1.15. Derivation of the d_0 and d_{90} points is therefore similar to that given in Volume 2, Section 14.5.6 and Fig. 14.31. The d_0 point representing theoretical 0% consolidation should be about the same as the initial reading because undrained loading should eliminate initial bedding errors. The d_{100} point is obtained by extrapolation, and enables $\sqrt{t_{50}}$ to be read off the graph, and t_{50} and t_{90} (minutes) can be calculated. Either of these values is used for calculating c_v as described later.

Example (iii)   Curve fitting: radial drainage

The example given below illustrates the curve-fitting method for one stage of a radial-outward drainage test using 'free strain' loading. From Table 22.3, reference (e), volume change should be plotted against $t^{0.465}$ and the slop factor is 1.22. The graph is shown in Fig. 22.37 and the corresponding pore pressure dissipation plot (pore pressure measured at the centre of the base) is given in Fig. 22.38. The relevant theoretical curve of 'average' consolidation is shown in Fig. 22.8 (Section 22.2.3).

The linear portion of the graph in Fig. 22.37 is extended to form the line QA which intersects the vertical axis at Q, the theoretical 0% consolidation point. The line QB, which has abscissae 1.22 times greater than QA, intersects the laboratory curve at C. The ordinate of C represents 90% consolidation, and the abscissae is 35.8, i.e.

$$(t_{90})^{0.465} = 35.8$$

$$\therefore \quad t_{90} = (35.8)^{1/0.468} = (35)^{2.15}$$

$$= 2192 \text{ minutes}$$

Extrapolation beyond the 90% point to fix the 100% point enables the point of 50% consolidation to be obtained. From Fig. 22.37

$$(t_{50})^{0.465} = 16.7;$$

$$\therefore \quad t_{50} = 426 \text{ minutes.}$$

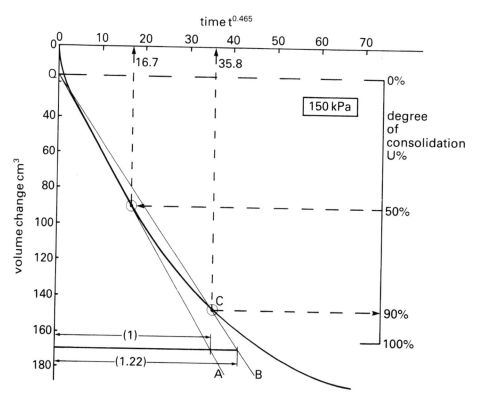

Fig. 22.37   Example of graphical data from one stage of a consolidation test with radial-outward drainage under 'free strain' loading; volume change plotted against time to a power factor of 0.465

Table 22.5. FACTORS FOR CALCULATING COEFFICIENT OF CONSOLIDATION

Test reference	Drainage direction	Boundary strain	Consolidation location	Multiplying factors for:		Coefficient
				$\bar{H}^2/t_{50}$	$\bar{H}^2/t_{90}$	
(a) and (b)	vertical, one way	free and equal	average central	0.10 0.20	0.45	c_v
(c) and (d)	vertical, two way	free and equal	average	0.26	0.11	c_v
				Multiplying factors for:		
				D^2/t_{50}	D^2/t_{90}	
(e)	radial outward	free	average centre	0.0083 0.026	0.044	c_{ro}
(f)	radial outward	equal	average centre	0.011 0.023	0.038	c_{ro}
(g) and (h)	radial inward	free and equal	average $r = 0.55R$	0.10 0.10	0.34	c_{ri}

$\bar{H}$ is the mean specimen height; D is the diameter (mm).
t_{50}, t_{90} are the times for 50% and 90% pore pressure dissipation (minutes).
The factors in (a) to (d) multiplied by $\bar{H}^2/t_{50}$ or $\bar{H}^2/t_{90}$ give c_v (m^2/year).
The factors in (e) to (h) multiplied by D^2/t_{50} or D^2/t_{90} give c_{ro} or c_{ri} (m^2/year).

The other approach is to use the value of t_{50} based on central pore pressure measurements. This can be read directly from the pore pressure dissipation graph as shown in Fig. 22.38 where $t_{50} = 1250$ minutes.

Values of t_{90} and t_{50} derived in this way are used for calculating c_{ro} as described below.

COEFFICIENT OF CONSOLIDATION

The coefficient of consolidation (c_v, c_{ro} or c_{ri}) is calculated by using the following general equation:

$$(c_v \text{ or } c_{ro} \text{ or } c_{ri}) = \frac{(\text{multiplying factor}) \times (H^2 \text{ or } D^2)}{(t_{50} \text{ or } t_{90})}$$

The relevant multiplying factors for each type of test are summarised in Table 22.5. These factors are derived from the data given in Table 22.3 (Section 22.2.2). They are calculated by multiplying the appropriate theoretical time factor (T_{50} or T_{90}), by 0.526 or 0.131 (shown in the right-hand column of Table 22.3), and are rounded to two significant figures.

For example, if using volume-change or height-curves ('average' measurements) for a test with vertical one-way drainage, the factor for t_{50} would be

$$0.197 \times 0.526 = 0.1036 \quad (\text{rounded to } 0.10)$$

For t_{90} it would be

$$0.848 \times 0.526 = 0.4460 \quad (0.45)$$

These factors are shown in the first line of Table 22.5. For vertical drainage the height $\bar{H}$ is the mean height of the specimen during the loading stage, i.e. $\bar{H} = \frac{1}{2}(H_1 + H_2)$ where H_1 and H_2 are the specimen heights at the beginning and end of the stage. In a 'free strain' test, H_1 and H_2 are the average heights taking into account the depression formed in the top surface.

For horizontal drainage the diameter D is the diameter of the specimen itself, making allowance for the thickness of the peripheral drain in a radial-outward test. Where a central drain is used the factors given in Table 22.5 are valid for a diameter ratio of 20. If the ratio differs significantly from 20, multiplying factors obtained from Fig. 22.10 should be used.

If the mean specimen height $\bar{H}$ or diameter D is expressed in millimetres, and the time t_{50} in minutes, the value of the coefficient of consolidation is in m^2/year.

In all cases a correction for temperature may be applied if appropriate (see Section 14.3.16 in Vol 2).

Two example calculations are given below, based on the curve-fitting examples already described.

(1) Vertical (one-way) drainage

The coefficient of consolidation for the load stage considered is calculated from the equation in Section 22.2.3:

$$c_v = 0.526 \, \frac{T_v \bar{H}^2}{t} \quad \text{m}^2/\text{year} \tag{22.3}$$

Using data from the pore pressure dissipation curve in Figs. 22.34 and 22.36 (example 1 above)

$$t_{50} = 1400 \text{ minutes}$$

$$\bar{H} = 81.86 \text{ mm}$$

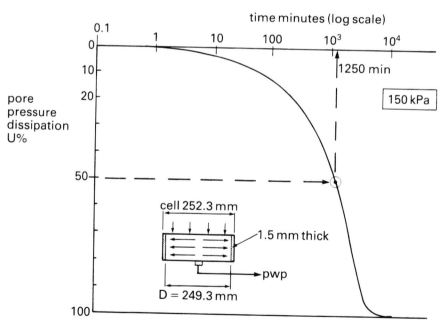

Fig. 22.38   Pore pressure dissipation plotted against log time

Multiplying factor from Table 22.25 is 0.20:

$$\therefore \quad c_v = \frac{0.20 \times (81.86)^2}{1400} \text{ m}^2/\text{year}$$

$$= 0.95 \text{ m}^2/\text{year}$$

Using data from the volume-change curve in Fig. 22.35 (example 2 above),

$$t_{50} = 841 \text{ minutes}$$

The relevant factor is 0.10

$$\therefore \quad c_v = \frac{0.10 \times (81.86)^2}{841}$$

$$= 0.83 \text{ m}^2/\text{year.}$$

Alternatively, using t_{90} ($= 3884$ minutes) the factor is 0.45:

$$c_v = \frac{0.45 \times (81.86)^2}{3884} = 0.77 \text{ m}^2/\text{year}$$

(2)   *Horizontal (radial outward) drainage, 'free strain'*

The coefficient of consolidation for the load stage considered is calculated from the equation in Section 22.2.3.

$$c_{ro} = \frac{0.131 \, T_{ro} D^2}{t} \tag{22.7}$$

Using data from the volume-change curve in Fig. 22.37 (example 3 above)

$$(t_{50})^{0.456} = 16.7$$

$$\therefore \quad t_{50} = 426 \text{ minutes}$$

Multiplying factor from Table 22.5 is 0.0083

$$D = (252.3 - 3) = 249.3 \text{ mm}$$

$$\therefore \quad c_{ro} = \frac{0.0083 \times (249.3)^2}{426}$$

$$= 1.21 \text{ m}^2/\text{year}$$

Alternatively, using $(t_{90})^{0.465} = 35.8$ gives

$$t_{90} = 2196 \text{ minutes}$$

The relevant factor is 0.044

$$\therefore \quad c_{ro} = \frac{0.044 \times (249.3)^2}{2196}$$

$$= 1.24 \text{ m}^2/\text{year}$$

Using data from the pore pressure dissipation curve in Fig. 22.38,

$$t_{50} = 1250 \text{ minutes}$$

The relevant factor is 0.026

$$\therefore \quad c_{ro} = \frac{0.026 \times (249.3)^2}{1250}$$

$$= 1.30 \text{ m}^2/\text{year}$$

REPORTING RESULTS (VERTICAL DRAINAGE TESTS)

A complete set of results from a Rowe cell consolidation test with vertical drainage consists of the following. The test report should confirm that the test was carried out in accordance with BS 1377: Part 6: 1990, Clause 3.5 or 3.6, as appropriate. Items marked * are additional to those listed in Clauses 3.5.9 or 3.6.9 of the BS.

Sample details

   Sample identification, reference number, location and depth
   Type of sample
   Soil description
   Comments on the condition and quality of the sample
   Method of preparation of test specimen
   Comments on difficulties experienced during preparation.

Test specimen

Initial dimensions
Initial moisture content, density, dry density
Particle density, and whether measured or assumed
Initial voids ratio and degree of saturation (if required)
Type of loading ('free strain' or 'equal strain')
Drainage conditions, direction and surface from which drainage took place
Location of pore water measuring point or points (if relevant)
*Classification test data (Atterberg limits; particle size curve).

Saturation

Swelling pressure, if determined
Method of saturation, with applied pressure increments and differential pressure, if applicable
Volume of water taken in by the specimen during saturation
Diaphragm pressure, pore pressure and value of the ratio $\delta u/\delta\sigma$ at the end of saturation.

Consolidation

For each loading stage, the following tabulated data:
    effective diaphragm pressure and back pressure
    effective stress at termination of loading stage
    settlement and pore pressure increase due to undrained loading
    voids ratio and percentage pore pressure dissipation at end of loading stage
    values of the coefficients m_v and c_v
Method used for determining the coefficient of consolidation
Graphical plots for each consolidation stage:
    pore pressure dissipation (%) against log time (if relevant)
    volume change and settlement against log time or square-root time
Graphical plot of voids ratio (or settlement) against effective stress (log scale) at the end of each consolidation or swelling stage
*Values of the coefficient of secondary compression, C_{sec}, if required.

Final details

Final density and overall moisture content
Moisture contents from specified zones or layers within the specimen (with sketch for identification)
Colour photographs illustrating features of the soil fabric (if required)
In-situ total and effective stresses in the ground at the depth from which the sample was taken, if known.

REPORTING RESULTS (RADIAL DRAINAGE TESTS)

For radial drainage tests the following details should be included, as appropriate, in addition to those listed above for vertical drainage tests. The report should confirm that the test was carried out in accordance with Clause 3.7 or 3.8 (as appropriate) of BS 1377: Part 6: 1990. Items marked * are additional to those listed in Clauses 3.7.9 or 3.8.9 of the BS.
    Material used for peripheral drain, and its thickness
    *Method used for forming the well
    *Description of material used for the drain, including grading curve
    *Method of placing drainage material.

Values of the coefficient of consolidation, c_{ro} or c_{ri}

Graphical plot of volume change and settlement against time raised to the power of 0.465 (radial outward drainage only).

22.7 PERMEABILITY TESTS

22.7.1 Scope

Tests for the determination of the coefficient of permeability of soil specimens can be carried out in Rowe consolidation cells under conditions of known effective stress and under the application of a back pressure. The direction of flow of water can be either vertical or horizontal (radially outwards or radially inwards). It is usual to carry out these tests under 'equal strain' loading, but permeability can also be measured at some stage of a consolidation test under either 'equal strain' or 'free strain' loading. The methods of test are similar in principle to those described in Chapter 20 for permeability tests in a triaxial cell, and are suitable for soils of low and intermediate permeability. Test specimens are usually either prepared from undisturbed samples, or formed by recompaction of disturbed soil (categories (1) and (3) of Section 22.1.7).

The apparatus, consisting of the Rowe cell and its accessories, pressure systems, and instrumentation, is the same as that described in Sections 22.1.4 to 22.1.6, and is prepared and checked as in Section 22.3. Calibration of the cell for head losses in the pipelines and connections for various rates of flow should have been carried out as described in Section 17.3.7.

Two or more 250 mm diameter cell bodies can be bolted together, if necessary, to form a large permeameter for granular soils with particle sizes up to about 40 mm (see Volume 2, Section 10.6.5).

The following test conditions and requirements should be specified by the engineer at the time the tests are requested:

Size of test specimen

Loading conditions ('free strain' or 'equal strain')

Drainage conditions and direction of flow of water

Effective stress at which each permeability measurement is to be carried out

Whether saturation is to be applied, and if so the method to be used

Whether voids ratios are to be calculated.

The tests described in Sections 22.7.2 and 22.7.3 below are as given in Clause 4 of BS 1377: Part 6: 1990.

22.7.2   Vertical Permeability (BS 1377: Part 6: 1990: 4.8.3)

APPARATUS

The arrangement of the cell and ancillary equipment is shown in Fig. 22.29 (Section 22.5.5(a)). The ceramic insert normally used for pore pressure measurement is replaced by a small disc of porous plastic material of much higher permeability. Three independent constant pressure systems are normally required, one for applying the vertical stress, the other two (each including a volume-change indicator) on the inlet and outlet flow lines. The specimen is prepared by one of the methods given in Section 22.4, and is set up as described in Section 22.5.5(a).

SATURATION

If saturation by incremental back pressure is to be carried out initially, the pore pressure transducer housing should be connected to valve G. The presence of the porous disc at the base means that the measured pore pressure relates to the mean value over the base area, whether the pore pressure ceramic is central or not. During the saturation stage valve A should remain closed and water admitted to the specimen through valve D as usual.

For a smaller cell with only the central pore pressure point, the transducer housing is connected to valve A during saturation, as described in Section 22.5.1, stage 2. When saturation is achieved the pore pressure housing is disconnected and the second back pressure system is connected to valve A, preferably under water to avoid trapping any air.

If only two pressure systems are available the outlet from the specimen can be connected to an elevated reservoir of water, or to an elevated open burette. The back pressure system is connected to the flow inlet at valve A. Much of the air contained in a silt or sand specimen can be displaced by allowing water under a small pressure to permeate slowly upwards from the base, preferably before fitting the cell cover. Air can then be completely removed by the application of vacuum as described in Section 22.4.3 However, this process should be applied with caution, because granular soils, especially if loosely placed, could be sensitive to disturbance or collapse if inundated before being loaded, or when air is removed by vacuum.

APPLIED PRESSURE

The arrangement shown in Fig. 22.29 allows water to flow vertically through the specimen under the application of a differential pressure between the base and top, while the specimen is subjected to a vertical stress from the diaphragm pressure as in a consolidation test. Two independent back pressure systems, one connected to each end, enable the pressure difference to be maintained at a constant value to give the required hydraulic gradient. Flow of water may be upwards or downwards, depending on which pressure is the greater. The mean of the two pressures gives the average pore water pressure within the specimen, and the difference between this and the vertical stress due to the diaphragm pressure is the mean vertical effective stress. If the flow is to an open burette, the outlet pressure is zero if the free water surface in the burette is maintained at the same level as the specimen face from which the water emerges.

When the difference between inlet and outlet pressure is small it can best be measured by using a differential pressure gauge or manometer, as described in Section 22.5.5 (Fig. 22.28).

In the test described below, the flow of water through the specimen is downwards, as specified in BS 1377.

CONSOLIDATION

The specimen is first consolidated to the required effective stress in one or more stages by the application of diaphragm loading, as described in Section 22.6.2. Consolidation should be virtually completed, i.e. the excess pore pressure should be at least 95% dissipated, before starting a permeability test.

CONSTANT HEAD PERMEABILITY TEST PROCEDURE

The test is carried out by adjusting the pressure difference across the specimen to provide a reasonable rate of flow through it. The hydraulic gradient required to induce flow should be ascertained by trial, starting with equal pressures on the inlet and outlet lines and progressively increasing the inlet pressure, which must never exceed the diaphragm pressure. Intact clay soils generally require higher gradients than silty and sandy soils, and a gradient of $i = 20$ might be required to achieve a measurable rate of flow. The inlet pressure should be increased carefully, keeping the rate of flow under observation so as to avoid piping and internal erosion which would cause soil disturbance. Suggested maximum hydraulic gradients are given in Section 20.4.1, step (6). Pressure differences required to give unit hydraulic gradient ($i = 1$) across specimens of recommended height in each cell size are given in Table 22.6, together with the value of i corresponding to unit (1 kPa) pressure difference.

Readings of the volume-change indicators on both inlet and outlet pressure lines are taken during the test. They are observed at suitable regular intervals of time, and graphs are plotted of the cumulative flow of water Q (ml) against elapsed time t (minutes) for each gauge. The test is continued until both graphs are linear and parallel. The temperature adjacent to the Rowe cell is recorded to the nearest $0.5\,°C$.

The test is stopped by closing valves A and D. An additional test at a lower effective stress can be carried out by increasing the pressures p_1 and p_2 as appropriate. An additional test at a higher effective stress can be carried out by applying a further consolidation stage (as described in Section 22.6.2) after either raising the diaphragm pressure, or reducing the back pressure, to the appropriate values.

If only one back pressure system is used, with the outlet draining to an open burette, the burette may be used for flow volume measurements. The elevation of the water surface in the burette should be maintained at a constant level for the 'constant head' condition to be valid. If the outlet is connected to a large elevated reservoir the constant head assumption is valid and flow measurement readings are taken from the volume-change indicator in the inlet pressure line.

If the rate of flow of water exceeds about 20 ml/minute it might be necessary to take into account the head losses which occur in the porous discs, valves and connecting pipework before calculating the permeability (see Section 17.5.3).

CALCULATIONS

The coefficient of vertical permeability as measured, k_v (m/s), is calculated from the following equation, derived in Section 22.2.8.

$$k_v = \frac{1.63qH}{A[(p_1 - p_2) - p_c]} \times 10^{-4}\,\text{m/s} \tag{22.12}$$

in which q is the mean rate of flow of water through the specimen (ml/minute), H the height of the specimen (mm), A the area of cross-section of the specimen (mm^2), p_1 and p_2 are the inlet and outlet pressures (kPa) and p_c is the pressure loss in the system for the rate

Table 22.6. PERMEABILITY TEST DATA (VERTICAL FLOW)

Nominal specimen diameter mm	Recommended specimen height mm	Pressure difference (k Pa) for i=1	Gradient i for 1 k Pa pressure difference
75	30	0.29	3.4
150	50	0.49	2.04
250	90	0.88	1.13
D	H	$\dfrac{H}{102}$	$\dfrac{102}{H}$

of flow q (obtained from the calibration graph). The measured permeability may be corrected to the equivalent value at 20 °C by using a correction factor as described in Volume 2, Section 10.3.4.

REPORTING RESULTS

The test report should confirm that the test was carried out in accordance with Clause 4 of BS 1377 : Part 6 : 1990. The permeability value is reported to two significant figures, together with the following data. (Items additional to those listed in Clause 4.10 of the BS are indicated by *.)

Sample details

Sample identification, reference number, location and depth
Type of sample
Soil description
Comments on the condition and quality of the sample
Method of preparation of test specimen
Comments on difficulties experienced during preparation.

Test specimen

Initial dimensions
Initial moisture content, density, dry density
*Particle density, and whether measured or assumed
*Initial voids ratio and degree of saturation (if required)
*Type of loading ('free strain' or 'equal strain')
*Classification test data (Atterberg limits; particle size curve).

Saturation

*Swelling pressure, if determined
Method of saturation, with applied pressure increments and differential pressure, if applicable
Diaphragm pressure, pore pressure and value of the ratio $\delta u/\delta \sigma$ at the end of saturation.

Consolidation

Relevant data from the consolidation stage or stages, as appropriate, including direction of drainage.
*Percentage pore pressure dissipation achieved.

Permeability

Direction of flow of water (whether upwards or downwards)
Value of the coefficient of vertical permeability, k_v, to two significant figures, corrected to 20 °C if necessary
Vertical stress applied to the specimen for the test, and the mean pore water pressure

during the test
Difference between the inlet and outlet pressures, or the hydraulic gradient, during the test.

FALLING HEAD PERMEABILITY TEST

When the rate of flow through the specimen is very small, the falling head principle can be used. Procedures are similar to those described in Chapter 20 for falling head tests in a triaxial cell (Sections 20.3.2 and 20.3.3). This procedure is not included in BS 1377.

If a falling head method is used the permeability is calculated either from Equation (20.27) of Section 20.4.2, or from Equation 20.28 of Section 20.4.3. Results are reported as above.

22.7.3 Horizontal Permeability (BS 1377 : Part 6 : 1990 : 4.8.4)

Horizontal permeability can be measured with the flow of water either radially outwards from a central well to the peripheral drain, as indicated in Fig 22.5(l); or radially inwards, as in Fig. 22.5(m). The 'equal strain' loading condition is usually applied.

The arrangement of the cell and ancillary equipment for both kinds of test is shown in Fig. 22.30 (Section 22.5.5(b)). The ceramic insert normally used for pore pressure measurement is replaced by a small disc of porous plastic material of much higher permeability. The specimen is prepared by one of the methods given in Section 22.4, and is set up as described in Section 22.5.5(b). The top surface of the specimen is sealed with an impermeable membrane. Three independent constant pressure systems are required, as for a vertical permeability test. One back pressure system is connected to the rim drain valve at F, and one to the central base outlet at A. Valve D remains closed.

SATURATION

If saturation by incremental back pressure is to be carried out in order to assess the degree of saturation, a pore pressure transducer housing should be connected to valve A initially. During saturation, water is admitted to the periphery of the specimen from the back pressure line through valve F.

When saturation is achieved the pore pressure housing is disconnected and the second back pressure system is connected to valve A, preferably under water to avoid trapping any air.

APPLIED PRESSURE

The arrangement shown in Fig. 22.30 allows water to flow horizontally (in elevation) and radially (in plan) under the application of a differential pressure between the centre and periphery, while the specimen is subjected to a vertical stress from the diaphragm pressure. Flow of water may be from the centre outwards, or from the periphery inwards, depending on which of the two back pressure systems is set at the greater pressure. The mean of the two pressures gives the average pore water pressure within the sample, and the difference between this and the vertical stress due to the diaphragm pressure gives the mean vertical effective stress.

As with vertical permeability measurements (Section 22.7.2) the difference between the two back pressures is best obtained by using a differential pressure gauge or pressurised manometer (see Section 22.5.5).

CONSOLIDATION

The specimen is first consolidated to the required effective stress with drainage to the periphery, as described in Section 22.6.4. At least 95% pore pressure dissipation should be achieved.

TEST PROCEDURE

The pressure difference across the specimen is adjusted to give a reasonable rate of flow by progressively increasing the inlet pressure without allowing it to equal or exceed the diaphragm pressure. Unless the rate of flow is small, a correction for head losses in connections and porous media should be applied to the measured pressure difference, as outlined in Section 17.5.3. Determine the rate of flow from graphs of cumulative flow against time, as in Section 22.7.2.

CALCULATIONS AND REPORTING

Calculate the horizontal (radial) permeability from the equation derived in Section 22.2.7:

$$k_h = 0.26 \frac{q}{H . \Delta p} \log_e \left(\frac{D}{d} \right) \times 10^{-4} \, \mathrm{m/s} \qquad (22.11)$$

where q = measured rate of flow = Q/t (ml/minute)
   t = time (minutes)
   Δp = pressure difference (kPa) = $(p_1 - p_2) - p_c$
   D = diameter of specimen (mm)
   d = diameter of central drain well (mm)
   H = height of specimen (mm)

The result and test data are reported in the same way as for the vertical permeability test (Section 22.7.2), except for the following:
The value of the horizontal permeability, k_h, is reported
The direction of flow is reported as horizontal and radially inwards or radially outwards as appropriate
*Details of the central drainage well should include its diameter; the method by which it was formed; and the material used, including the grading curve.

REFERENCES

Barron, R. A. (1947). 'Consolidation of fine-grained soils by drain wells'. *Proc. Am. Soc. Civ. Eng.*, Vol. 73:6:811.
Berry, P. L. and Poskitt, T. J. (1972). 'The consolidation of Peat'. *Géotechnique*, 22:1:27.
Berry, P. L. ad Wilkinson, W. B. (1969). 'The radial consolidation of clay soils'. *Géotechnique*, 19:2:253.
Delft Soil Mechanics Laboratory (1966), 'A new approach for taking a continuous soil sample'. Paper No. 4, Laboratorium voor Grondmechanica, Delft, Holland.

Escario, V. and Uriel, S. (1961). 'Determining the coefficient of consolidation and horizontal permeability by radial drainage'. *Proc. 5th Int. Conference of Soil Mechanics & Foundation Eng.*, Paris. Vol. 1, pp 83–87.

Gibson, R. E. and Shefford, G. C. (1968). 'The efficiency of horizontal drainage layers for accelerating consolidation of clay embankments'. *Géotechnique*, 18:3:327.

Hobbs, N. B. (1986). 'Mire morphology and the properties and behaviour of some British and foreign peats'. *Quarterly Journal of Eng. Geology*, Vol. 19, No. 1.

Leonards, G. A. and Girault, P. (1961). 'A study of the one-dimensional consolidation test'. *Proc. 5th Int. Conference of Soil Mechanism & Foundation Eng.*, Paris. Vol. 1, Paper 1/36, pp 213–218.

Lo, K. Y., Bozozuk, M. and Law, K. T. (1976). 'Settlement analysis of the Gloucester test fill'. *Canadian Geotechnical Journal*, Vol. 13: 339, Nov. 1976.

Lowe, J., Zaccheo, P. F. and Feldman, H. S. (1964). 'Consolidation testing with back pressure'. *J. Soil Mech. Fdn. Div. ASCE*, Vol. 90, SM5: 69.

McGown, A., Barden, L., Lee, S. H. and Wilby, P. (1974). 'Sample disturbance in soft alluvial Clyde Estuary clay'. *Canadian Geotechnical Journal*, Vol. 11: 651, Nov. 1974.

McKinlay, D. G. (1961). 'A laboratory study of rates of consolidation in clays with particular reference to conditions of radial porewater drainage'. *Proc. 5th Int. Conference on Soil Mechanics & Foundation Eng.*, Vol. 1, Paper 1/38, Dunod, Paris.

Rowe, P. W. (1954). 'A stress–strain theory for cohesionless soil, with applications to earth pressure at rest and moving walls'. *Géotechnique*, 4:2:70.

Rowe, P. W. (1959). 'Measurement of the coefficient of consolidation of lacustrine clay'. *Géotechnique*, 9:3:107.

Rowe, P. W. (1964). 'The calculation of the consolidation rates of laminated, varved or layered clays, with particular reference to sand drains'. *Géotechnique*, 14:4:321.

Rowe, P. W. (1968), 'The influence of geological features of clay deposits on the design and performance of sand drains'. *Proc. I.C.E. Supplementary Paper* No. 7058S.

Rowe, P. W. (1972). 'The relevance of soil fabric to site investigation practice'. Twelfth Rankine Lecture. *Géotechnique*, 22:2:195.

Rowe, P. W. and Barden, L. (1966). 'A new consolidation cell'. *Géotechnique*, 16:2:162.

Rowe, P. W. and Shields, D. H. (1965). 'The measured horizontal coefficient of consolidation of laminated, layered or varved clays'. *Proc. 6th Int. Conference of Soil Mechanics & Foundation Eng.*, Vol. 1, Paper 2/44. University of Toronto Press.

Shields, D. H. (1963). 'The influence of vertical sand drains and natural stratification on consolidation'. PhD thesis, University of Manchester.

Shields, D. H. (1976). 'Consolidation tests'. Technical Note, *Géotechnique*, 26:1:209.

Shields, D. H. and Rowe, P. W. (1965). 'A radial drainage oedometer for laminated clays'. *J. Soil Mech. Fdns. Div. ASCE*, Vol. 91, SM1: 15.

Sills, G. C. (1983). Private communication to author.

Simons, N. E. and Beng, T. S. (1969). 'A note on the one dimensional consolidation of saturated clays'. Technical Note, *Géotechnique*, 19:1:140.

Singh, G. and Hattab, T. N. (1979). 'A laboratory study of efficiency of sand drains in relation to methods of installation and spacing'. *Géotechnique*, 29:4:395.

Tyrrell, A. P. (1969). 'Consolidation properties of composite soil deposits'. PhD thesis, University of Manchester.

Whitman, R. V., Richardson, A. M. and Healy, K. A. (1961). 'Time lags in pore pressure measurements'. *Proc. 5th Int. Conference of Soil Mechanics and Foundation Eng.*, Paris, Vol. 1, Paper 1/69, pp 407–411.

Appendix

Units, symbols, reference data

C.1. METRIC (SI) UNITS

The SI units and prefixes used in this volume are the same as those used in Volumes 1 and 2, as summarised in Tables B.1 and B.2 of the Appendix to Volume 2. Conversion factors for converting other units to SI, and vice versa, are given in Table B.3 of that Appendix.

Throughout this volume the unit of stress or pressure most often used is kPa (kilopascal), which is identical to kN/m^2 (kilonewtons per square metre). For high stresses MPa (megapascal), which is identical to MN/m^2, is used. This usage is in accordance with current international recommendations.

C.2. SYMBOLS

Symbols used in this volume are summarised in Table C.1 (English) and Table C.2 (Greek). These lists do not include some symbols that are used only for a specific application, where they are defined in the text.

C.3. TEST SPECIMEN DATA

Data relating to specimens required for the tests described in this volume are given in Table C.3, which lists the areas, volumes and approximate mass of specimens of the sizes most often used.

C.4. MISCELLANEOUS DATA

Some useful miscellaneous data for quick reference are given in Table C.4.

Table C.1. ENGLISH SYMBOLS

Symbol	Measured quantity or item	Usual unit of measurement
A	Pore pressure coefficient	—
A_f	Pore pressure coefficient A at failure	—
$\bar{A}$	Pore pressure coefficient in partially saturated soil	—
$\bar{A}_f$	Pore pressure coefficient $\bar{A}$ at failure	—
A	Area of cross-section of test specimen	mm^2
A_0	Area of cross-section of specimen initially	mm^2
A_c	Area of cross-section of specimen after consolidation	mm^2
A	Area of consolidation cell	mm^2
A_s	Area of contact	mm^2
a	Area of cross-section of piston	mm^2
B	Pore pressure coefficient	—
$\bar{B}$	'Overall' pore pressure coefficient	—
C_s	Volume compressibility of soil skeleton	m^2/MN
C_w	Volume compressibility of water	m^2/MN
CD	Consolidated-drained triaxial test	—
CU	Consolidated-undrained triaxial test	—
CCV	Consolidated constant-volume triaxial test	—
C-QU	Consolidated quick-undrained triaxial test	—
C_R	Load ring calibration (average)	N/div
CSL	Critical state line	—
C_{sec}	Coefficient of secondary compression	—
c'	Cohesion intercept based on effective stresses	kPa
c_d	Apparent cohesion derived from drained tests	kPa
c'_r	Cohesion intercept for residual condition	kPa
c_u	Undrained shear strength of saturated soil	kPa
c_v	Coefficient of consolidation (vertical drainage)	$m^2/year$
c_h	Coefficient of consolidation (horizontal drainage)	$m^2/year$
c_{ri}	Coefficient of consolidation (drainage radially inwards)	$m^2/year$
c_{ro}	Coefficient of consolidation (drainage radially outwards)	$m^2/year$
c_{vi}	Coefficient of consolidation (isotropic)	$m^2/year$
D	Diameter of specimen	mm
D_0	Initial diameter of specimen	mm
D_{50}	Particle diameter at which 50% of soil is finer	mm
d	Diameter of drainage well	mm
ESP	Effective stress path	—
e	Voids ratio	—
e_c	Critical voids ratio; Voids ratio after consolidation	—
e_f	Final voids ratio	—
e_s	Voids ratio after saturation	—
e_0	Initial voids ratio	—
e_1	Voids ratio at start of load increment	—
e_2	Voids ratio at end of load increment	—
e	Base of natural logarithms	—
F	Applied force	N
F_0	Initial force	N
F	Factor of safety	—
f (subscript)	Failure condition	—

Table C.1. — *continued*

Symbol	Measured quantity or item	Usual unit of measurement
f	Frictional force in cell bushing	N
f_{cv}	Factor relating c_v to c_{vi}	—
f_s	Factor for single-plane slip in triaxial compression	—
g	Acceleration due to gravity	m/s^2
H	Height of specimen	mm
H_0	Initial height of specimen	mm
H_s	Height of specimen after saturation	mm
$\bar{H}$	Mean height of specimen during load increment	mm
H	Henry's coefficient of solubility of gases in water	—
h	Depth of static water table below ground surface	m
h	Compacted height of specimen;	mm
	Length of drainage path	mm
h_0	Initial height of water level in burette or standpipe	mm
h_f	Final height of water level in burette or standpipe	mm
h, h_1, h_2	Head of water	mm
i	Hydraulic gradient	—
I_L	Liquidity index	—
I_P	Plasticity index	—
K	Lateral effective stress ratio (σ'_h/σ'_v)	—
K_f	Lateral effective stress ratio at failure	—
K_0	Coefficient of earth pressure at rest	—
k	Coefficient of permeability of soil	m/s
k_v	Coefficient of vertical permeability	m/s
k_h	Coefficient of horizontal permeability	m/s
k_D	Coefficient of permeability of porous disc	m/s
L	Length of specimen	mm
L_0	Initial length of specimen	mm
L_c	Length of specimen after consolidation	mm
LL	Liquid limit	%
m	Mass	g
m_0	Initial mass of specimen	g
m_c	Mass of specimen after consolidation	g
m_D	Dry mass of specimen	g
m_f	Final mass of specimen	g
m_s	Mass of specimen after saturation	g
m_h	Mass of load hanger	g
m'	Additional mass applied to hanger	g
m_p	Mass of top cap and piston	g
m_v	Coefficient of volume compressibility (one-dimensional)	m^2/MN
m_{vi}	Coefficient of volume compressibility (isotropic)	m^2/MN
m_w	Volume displaced by top cap and submerged length of piston	g
max (subscript)	Maximum value	—
min (subscript)	Minimum value	—
NC	Normally consolidated	—
n	Porosity	—

Table C.1. — *continued*

Symbol	Measured quantity or item	Usual unit of measurement
OC	Overconsolidated	—
OCR	Overconsolidation ratio	—
P	Applied force	N
P_0	Initial applied force;	N
	Axial force to counteract cell pressure	N
PL	Plastic limit	%
PI	Plasticity index	%
p	Mean total principal stress;	kPa
	Net applied pressure	kPa
p_a	Atmospheric pressure	kPa
p_b	Back pressure	kPa
p_c	Cell confining pressure	kPa
p'_c	Greatest previous effective consolidation stress;	kPa
	Isotropic consolidation stress	kPa
p'_s	Isotropic swelling stress	kPa
p_t	Total axial stress	kPa
p'_0	Present mean effective stress	kPa
p_1, p_2, etc.	Applied pressures or total stresses;	kPa
	Inlet and outlet water pressures	kPa
p'_1, p'_2, etc.	Applied effective stresses	kPa
$[p_0]$	Initial absolute pressure	kPa
p	Stress path parameter ($\frac{1}{3}(\sigma_1 + \sigma_2 + \sigma_3)$)	kPa
p'	Stress path parameter ($\frac{1}{3}(\sigma'_1 + \sigma'_2 + \sigma'_3)$)	kPa
p_d	Pressure applied to diaphragm	kPa
Q	Applied force;	N
	Force due to effective mass of piston and top cap	N
Q	Cumulative volume of flow of water	ml
QU	Quick-undrained triaxial compression test	—
q	Rate of flow of water	ml/min
q_m	Measured rate of flow	ml/min
q_f	Maximum deviator stress at failure	kPa
q_m	Maximum allowable working deviator stress	kPa
q, q'	Stress path parameter ($\sigma_1 - \sigma_3$)	kPa
q_0	Intercept of failure envelope on (p', q) stress path plot	kPa
R	Specimen radius	mm
R	Load ring reading	divs.
r	Ratio of specimen length to diameter	—
r	Rate of strain	%/min
r	Radius	mm
r_1	Radius of drainage well	mm
r_2	Radius of specimen	mm
S	Degree of saturation	%
S_0	Initial degree of saturation	%
S_f	Final degree of saturation	%
s	Stress path parameter $\frac{1}{2}(\sigma_1 + \sigma_3)$	kPa
s'	Stress path parameter $\frac{1}{2}(\sigma'_1 + \sigma'_3)$	kPa
T	Temperature	°C
T_v	Theoretical time factor for vertical drainage	—

Table C.1. — *continued*

Symbol	Measured quantity or item	Usual unit of measurement
T_{ri}	Theoretical time factor for drainage radially inwards	—
T_{ro}	Theoretical time factor for drainage radially outwards	—
T_{50}	Theoretical time factor at 50% consolidation	—
T_{90}	Theoretical time factor at 90% consolidation	—
TSP	Total stress path	—
t_f, t_f'	Maximum shear stress	kPa
t, t'	Stress path parameter $\frac{1}{2}(\sigma_1 - \sigma_3)$	kPa
t_0	Intercept of failure envelope on (s', t) stress path plot	kPa
t	Time	min
t_{50}	Time for 50% primary consolidation	min
t_{90}	Time for 90% primary consolidation	min
t_{100}	Time for theoretical 100% consolidation	min
t_f	Theoretical time to failure	min
t	Thickness	mm
U	Pore pressure dissipation	%
U_f	Pore pressure dissipation at failure	%
UU	Unconsolidated undrained triaxial test	kPa
u	Pore pressure	—
u_1, u_2	Values of pore water pressure	kPa
u_b	Back pressure applied to specimen	kPa
u_c	Component of pore pressure due to confining pressure	kPa
u_d	Component of pore pressure due to deviator stress	kPa
u_f	Pore water pressure at failure	kPa
u_i	Initial pore water pressure	kPa
u_s	Pore water pressure after saturation	kPa
$\bar{u}$	Mean pore water pressure	kPa
u_d	Pore water pressure at drained face	kPa
u_u	Pore water pressure at undrained face	kPa
V	Volume	cm^3
V_0	Volume of specimen initially	cm^3
V_a	Volume of specimen at start of consolidation	cm^3
V_c	Volume of specimen after consolidation	cm^3
V_f	Volume of specimen finally	cm^3
V_s	Volume of specimen after saturation	cm^3
V_w	Volume of water	cm^3 or ml
v	Velocity of flow of water	mm/s
w_0	Initial moisture content	%
w_c	Moisture content after consolidation	%
w_f	Final moisture content	%
w_L	Liquid limit	%
w_p	Plastic limit	%
x	Horizontal displacement; Axial compression or extension	mm
x'	Stress path parameter (σ_3')	kPa
x_0	Intercept of failure envelope on x' axis of (x', y) stress path plot	kPa
y	Axial deformation	mm
y	Equivalent height of air voids	mm
y	Stress path parameter $(\sigma_1' - \sigma_3')$	kPa

Table C.1. — *continued*

Symbol	Measured quantity or item	Usual unit of measurement
y_0	Intercept of failure envelope on (x', y) stress path plot	kPa
y	Height of earth embankment	m
z	Depth below ground surface	m
z	Movement of mercury thread in null indicator	mm
z	Equivalent height of total voids	mm

Table C.2. GREEK SYMBOLS

Symbol	Quantity	Uusual unit of measurement
α	Stiffness of pressure transducer	mm^3/kPa
α	Inclination of slip surface to horizontal	deg
α	Slope of failure envelope on (x', y) stress path plot	deg
β	Ratio of vertical and horizontal total stresses (σ_v/σ_h)	—
β	Subtended angle in single-plane slip	deg
β	Slope of envelope on (x', y) stress path plot representing ϕ'_m	deg
γ	Shear strain	radians
Δ	Change (e.g. $\Delta\sigma$); Cumulative change	—
Δp	Pressure difference	kPa
ΔH	Axial deformation	mm
ΔV_p	Volume displaced by piston movement	cm^3
δ	Incremental change	—
δp_c	Pressure loss in pipeline connections	kPa
$\delta V_1, \delta V_2$, etc.	Cell volume corrections	cm^3
δp	Diaphragm pressure correction	kPa
δu	Excess pore pressure	kPa
δ	Load ring dial gauge deflection from zero load	kPa
δ	Slope of line on (s, t) stress path plot representing the K_0 condition	—
ε	Strain	%
ε_f	Axial strain at failure	%
ε_{lim}	Limiting axial strain	%
ε_s	Axial strain from start of slip in triaxial test	%
ε_v	Volumetric strain	—
ε_{vs}	Volumetric strain due to shear	—
$\varepsilon_1, \varepsilon_2, \varepsilon_3$	Principal strains	—
ε_h	Lateral strain	—
η	Drainage coefficient during compression related to boundary conditions	—
η	Slope of failure envelope on (p', q) stress path plot	deg
θ	Inclination of slip surface relative to specimen axis	deg
θ	Slope of failure envelope on (s', t) stress path plot	deg
θ_K	Slope of line on (s', t) stress path plot representing a lateral stress ratio of K	deg

Table C.2. — *continued*

Symbol	Quantity	Usual unit of measurement
λ	Drainage coefficient related to specimen boundary conditions	—
ρ	Bulk density	Mg/m^3
ρ_0	Initial specimen density	Mg/m^3
ρ_D	Dry density	Mg/m^3
ρ_{D0}	Initial dry density	Mg/m^3
ρ_{Dc}	Consolidated dry density	Mg/m^3
ρ_s	Particle density	Mg/m^3
ρ_{sat}	Saturated density	Mg/m^3
ρ_w	Density of water	Mg/m^3
σ	Normal stress	kPa
σ'	Normal effective stress	kPa
$\sigma_1, \sigma_2, \sigma_3$	Principal stresses	kPa
$\sigma_1, \sigma_2, \sigma_3$	Principal effective stresses	kPa
σ_v, σ_h	Vertical and horizontal stresses	kPa
σ_v', σ_h'	Vertical and horizontal effective stresses	kPa
σ_n	Stress normal to surface of failure	kPa
σ_1	Axial stress	kPa
σ_3, σ_c	Confining pressure	kPa
σ_{3max}	Maximum working pressure of triaxial cell	kPa
$(\sigma_1 - \sigma_3)$	Deviator stress	kPa
$(\sigma_1 - \sigma_3)_f$	Deviator stress at failure	kPa
σ_{dr}	Correction to deviator stress for side drains	kPa
σ_{ds}	Correction to deviator stress for side drains with single plane slip failure	kPa
σ_{mb}	Barrelling correction for membrane	kPa
σ_{ms}	Slip correction for membrane	kPa
τ	Shear stress	kPa
τ_f, τ_f'	Shear stress on failure surface at failure	kPa
ω	Slope of failure envelope on (σ_3', σ_1') stress path plot	deg

Table C.3.   SPECIMEN DIMENSIONS, AREA, VOLUME, MASS

Type of test	Diameter in	Diameter mm	Height mm	Height in	Area mm^2	Volume cm^3	Approximate mass
Triaxial shear		35	70		962.1	67.35	140 g
	1.4			2.8	993.1	70.63	150 g
		38	76		1134	86.19	180 g
	1.5			3	1140	86.87	180 g
		50	100		1963	196.3	410 g
	2			4	2027	205.9	430 g
		70	140		3848	538.8	1.1 kg
	2.8			5.6	3973	565.1	1.2 kg
		100	200		7854	1571	3.3 kg
	4			8	8107	1647	3.5 kg
		105	210		8659	1818	3.8 kg
		150	300		17 671	5301	11 kg
	6			12	18 241	5560	12 kg

Table C.3. — *continued*

Type of test	Diameter in	Diameter mm	Height mm	Height in	Area mm²	Volume cm³	Approximate mass
Triaxial consolidation							
		70	70		3848	269.4	560 g
		100	100		7854	785.4	1.6 kg
	4	101.6	101.6	4	8107	823.7	1.8 kg
		105	105		8659	909.2	1.9 kg
		150	150		17 671	2651	5.6 kg
	6	152.4	152.4	6	18 241	2780	5.8 kg
Rowe cell consolidation							
		75.7	30		4500	135.0	280 g
	3	76.2	30		4560	136.8	290 g
		151.4	50		18 000	900	1.9 kg
	6	152.4	50		18 241	912	1.9 kg
		252.3	90		50 000	4500	9.4 kg
	10	254	90		50 671	4560	9.6 kg

Table C.4.   USEFUL DATA

Time	1 day	= 1440 minutes
	1 week	= 10 080 minutes
	1 month (average)	= 43 920 minutes
	1 year	= 525 960 minutes
		= 31.56 × 10⁶ seconds

Fluid pressure	Density of mercury (20° C)	= 13.546 g/cm³
	1 kPa = 1 kN/m²	= 102 mm of water
		= 7.53 mm of mercury
	1 m of water	= 9.807 kPa
	Mercury manometer with limb open to atmosphere:	
	1 mm difference	= 0.128 kPa
	1 kPa	= 7.82 mm difference
	Mercury manometer with limb connected to water reservoir or to constant pressure system:	
	1 mm difference	= 0.123 kPa
	1 kPa	= 8.13 mm difference
	Standard atmosphere at 0° C (1 atm)	= 101.325 kPa
		= 760 mm of mercury

General	Circumference/diameter of a circle	
	π	= 3.142
	Base of natural logarithms	
	e	= 2.718
	Standard acceleration due to terrestrial gravity	
	g	= 9.807 m/s²

Index

Errata and Amendments for Volume 1, Second Edition and Volume 2, Second Edition

(including those arising from BSI Amendments to BS 1377: 1990 issued in 1996)

VOLUME 1, SECOND EDITION

Page	Section	Amendment
47	* 1.5.4	Sentence under Fig. 1.33, for 'us' write 'is'.
171	Fig. 4.7	For 'S.G.' substitute 'Q_s'.
179	* Table 4.7	Third line, for 'm_s2' write 'm_{s2}'.
179	* 4.6.1 (B)	First sentence, for 'substracted' write 'subtracted'.
226	4.8.4	Second paragraph, line 5, for '1_1', '1_2' substitute 'd_1', 'd_2'.
236	Table 5.1	Opposite 5.6, for 'BRE Digest 250' substitute 'BRE Digest 363'.
263	5.6.2	First line, for 'v/v', substitute 'm/v'.
263	5.6.2. (B)	Delete paragraphs (2) to (6) and substitute the following.

(2) Cover the beaker with a cover-glass, bring to the boil and simmer gently for 15 minutes in a fume cupboard. Rinse the underside of the cover-glass with distilled water back into the beaker.

(3) Filter the suspension through a Whatman no. 42 filter paper into a 500 ml conical beaker. Wash the first beaker and the residue with distilled water until the washings are free from chloride, as indicated by absence of turbidity when a drop is added to a small volume of silver nitrate solution. Collect all the washings. The filtrate, together with the washings, is then normally ready for the gravimetric analysis described in Section 5.6.5.

If the soil contains sesquioxides in appreciable quantity (e.g. as in some tropical residual soils) these should be precipitated before proceeding with the analysis, as follows.

(4) Add a few drops of nitric acid while the suspension continues to boil.

Page	Section	*Amendment*

(5) Add ammonia solution slowly (preferably from a burette), with constant stirring, to the boiling suspension until the sesquioxides are precipitated and red litmus is turned to blue by the liquid. Filter and proceed as described in (3) above.

(6) If a voluminous precipitate of sesquioxides forms when ammonia is added in step (5), some sulphate might be entrapped which will not be removed by washing and could lead to low results. In this case a second precipitation is recommended. This is done by carefully removing the filter paper with the precipitate and replacing it in the original beaker. Add 10% solution of hydrochloric acid and stir the contents until the sesquioxides have gone into solution (20 ml of 10% hydrochloric acid should be sufficient). Bring the contents to the boil and repeat step (5).

265   5.6.3 (B)   Delete existing paragraph (7) and substitute the following.

(7) Transfer exactly 50 ml of the water extract to a clean dry 250 ml conical beaker, using the 50 ml pipette.

268   5.6.5 (D)   The equation in step (18) should read

$$SO_3 \ (\%) = 1.372 \, m_4$$

The next step but one should be number (20), not (2).

269   5.6.5 (F)   First sentence, after '0.01%' add 'of the fraction passing a 2 mm sieve.'

271   5.6.6   In step (8), second sentence, for '50 ml' substitute '100 ml'.

272   5.6.7   Step (3), for '25 ml' substitute '50 ml'.

Step (7), the equation should read

$$SO_3 = 0.8 \ B \ V \ g/litre$$

283   * 5.8.3   Under 'Calculations', first word should be 'Calculate'. After the equation, 'increased' should be 'increase'.

315   6.5.1   First sentence, for 'BS 1377: 1990: Part 3' substitute 'BS 1377: 1990: Part 4'.

317   6.5.2   Table 6.6, heading to right-hand column, for '(kf)' substitute '(kg)'.

357   7.1.2   Section heading should read 'Origin of Soils'.

388   * Index   Vacuum — page numbers should be 19, 53.

VOLUME 2, SECOND EDITION

Page	Section	Amendment
Page	*Section*	*Amendment*
69	9.4.2	Section should be numbered 9.4.2, not 9.4.5.
161	* 11.6.2 (2)	Line 6, for 'rouch' substitute 'rough'.
437	* Index	For 'Humidifier' substitute 'Humidified'.